NEUROMETHODS

For further volumes:
http://www.springer.com/series/7657

Discovering Hidden Temporal Patterns in Behavior and Interaction

T-Pattern Detection and Analysis with THEME™

Edited by

Magnus S. Magnusson

University of Iceland, Reykjavik, IS 101, Iceland

Judee K. Burgoon

University of Arizona, Tucson, AZ, USA

Maurizio Casarrubea

Department of Bio.Ne.C., University of Palermo, Palermo, Italy

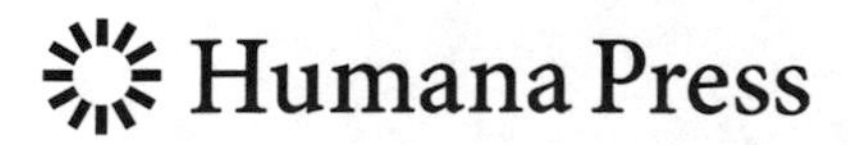

Editors
Magnus S. Magnusson
University of Iceland
Reykjavik, Iceland

Judee K. Burgoon
University of Arizona
Tucson, AZ, USA

Maurizio Casarrubea
Department of Bio.Ne.C.
University of Palermo
Palermo, Italy

ISSN 0893-2336 ISSN 1940-6045 (electronic)
Neuromethods
ISBN 978-1-4939-8005-5 ISBN 978-1-4939-3249-8 (eBook)
DOI 10.1007/978-1-4939-3249-8

Springer New York Heidelberg Dordrecht London

Softcover reprint of the hardcover 1st edition 2016

Printed on acid-free paper

Humana Press is a brand of Springer
Springer Science+Business Media LLC New York is part of Springer Science+Business Media (www.springer.com)

Series Preface

Experimental life sciences have two basic foundations: concepts and tools. The *Neuromethods* series focuses on the tools and techniques unique to the investigation of the nervous system and excitable cells. It will not, however, shortchange the concept side of things as care has been taken to integrate these tools within the context of the concepts and questions under investigation. In this way, the series is unique in that it not only collects protocols but also includes theoretical background information and critiques which led to the methods and their development. Thus it gives the reader a better understanding of the origin of the techniques and their potential future development. The *Neuromethods* publishing program strikes a balance between recent and exciting developments like those concerning new animal models of disease, imaging, *in vivo* methods, and more established techniques, including, for example, immunocytochemistry and electrophysiological technologies. New trainees in neurosciences still need a sound footing in these older methods in order to apply a critical approach to their results.

Under the guidance of its founders, Alan Boulton and Glen Baker, the *Neuromethods* series has been a success since its first volume published through Humana Press in 1985. The series continues to flourish through many changes over the years. It is now published under the umbrella of Springer Protocols. While methods involving brain research have changed a lot since the series started, the publishing environment and technology have changed even more radically. Neuromethods has the distinct layout and style of the Springer Protocols program, designed specifically for readability and ease of reference in a laboratory setting.

The careful application of methods is potentially the most important step in the process of scientific inquiry. In the past, new methodologies led the way in developing new disciplines in the biological and medical sciences. For example, Physiology emerged out of Anatomy in the nineteenth century by harnessing new methods based on the newly discovered phenomenon of electricity. Nowadays, the relationships between disciplines and methods are more complex. Methods are now widely shared between disciplines and research areas. New developments in electronic publishing make it possible for scientists that encounter new methods to quickly find sources of information electronically. The design of individual volumes and chapters in this series takes this new access technology into account. Springer Protocols makes it possible to download single protocols separately. In addition, Springer makes its print-on-demand technology available globally. A print copy can therefore be acquired quickly and for a competitive price anywhere in the world.

Wolfgang Walz

Preface

I am delighted to write the preface to this publication, which marks another step in the long and diligent scientific career of a group of investigators who have spent decades studying the behavioral structures underlying social interaction.

The publication of Anolli, Duncan, Magnusson, and Riva's reading path a decade ago marked an era. Now, with the heartfelt losses of Luigi Anolli and Starkey Duncan, we are set to take another step, united in our interest to further explore the vast opportunities offered by T-pattern detection and to take stock of the already extensive areas of knowledge in which T-pattern analysis has proven to be an exceptional analytical tool, while enabling a permanent exchange with the respective conceptual framework.

The detection of structures in behavior patterns forms a nexus between studies carried out in very diverse fields and contexts, involving humans (with highly diversified characteristics, and by analyzing the relation with hormonal levels, personality, culture, and so forth), animals (dogs, cats, primates, shoals of fish, wolves, rodents, chickens, insects), interaction studies between hormones and behavior, or neurons. Most studies are observational, with a perfectly established methodology. However, T-pattern detection is also used in laboratory-based experimental studies, where observation is simply a technique. A unifying factor is the observation of visually or even acoustically perceptible events or behaviors, nearly always arranged in clusters, which often correspond to interactive situations. The scale is extraordinarily rich and varied, ranging from micro-movements of individual facial expressions to broad migratory movements in the marine environment.

As the following chapters confirm, there are many forms of communication, and none is resistant to T-pattern detection. These forms range from the basic dyadic interaction between two individuals who regularly communicate and take decisions (be they two people, or a cat and its owner) to gregarious interaction (a shoal of fish) in a real or virtual situation, group conduct, or "self-interaction" (if that is a valid term), in the study of personal style.

Communication flows offer enormous research potential thanks to their multiple dimensions or levels of response (Anguera and Izquierdo 2006) and their extraordinary dynamic nature. Their study, however, poses methodological challenges, beginning with the establishment of dimensions or response levels (known as variables in THEME) and the criteria used to segment episodes into behavioral units, which give rise to event types and their arrangement into separate blocks. For good reason, the complexity of interactive behavior results in an episode or chain of episodes being expressed in a code matrix, adapted to the syntactic rules of THEME, so that the invisible structure to which it adapts and by which it is regulated can be extracted and studied quantitatively. An important aid to recording is the recently created freeware program LINCE (Gabín et al. 2012), which facilitates researchers' work by enabling the direct export of data to THEME.

We want to foreground the ample possibilities of the multivariate approach of T-pattern detection in data recorded using *ad hoc*, highly flexible observation instruments (Anguera et al. 2007), comprising field format or field formats combined with category systems. The invisible nature of T-patterns increases the potential for discovery, as researchers are interested in extracting the internal structure that unveils the key to the target behaviors. The following chapters show how T-pattern detection can be used alone or in combination with other techniques, such as variance analysis, covariance analysis, and time series analysis.

We believe it is important to stress the multifaceted interest of the chapters on T-pattern detection, which are heterogeneous in terms of substantive scope and are mostly applied, although several contain proposals to develop the knowledge of structure revealed through T-patterns and the logistics of THEME program management.

A particularly salient novel feature is synfire patterns. Though reminiscent of certain aspects of T-patterns, synfire patterns have a different algorithm and a structure that positions them between T-patterns and random distribution. The critical intervals of synfire patterns are constant and preestablished, and very recent works, such as that by Nicol et al. (2015), have postulated that they may be artifacts linked to the nature of the situation being examined. This topic is currently at the center of much scrutiny.

Another question for which there is empirical evidence in complex reality is the existence of different overlapping structures (Casarrubea et al. 2015), each of which gives rise to T-pattern detection; this reality cannot be analyzed by other multivariate techniques. Consequently, rather than one structure generating noise to detect another, each one can be clearly identified and differentiated, thereby marking a significant development in the meticulous analysis of reality.

The scientific dialogue concerning differences between T-pattern detection and lag sequential analysis, already recorded in other studies (Kemp et al. 2008; Lapresa et al. 2013), is also highly significant and will doubtless be followed up in subsequent works that may focus on the type and sensitivity of the information obtained, as part of a common goal to achieve a more complete picture of the organization of behavior, from both a synchronic and diachronic perspective. The significant sequences detected though time lag sequential analysis should be first level special cases of T-Patterns, and detectable by Theme.

I feel extremely honored to have had the privilege of writing this preface and to bear witness to the quality of the chapters in this book. Moreover, I am fully aware of the fresh impetus it brings to a field of knowledge, whose long history dates back to 1979 and is a result of the initial efforts and endeavors of Magnus S. Magnusson. His work was nearly always in direct collaboration with Starkey Duncan and for years also with Hubert Montagner, Karl Grammer, Rodolf Ghiglione, and a number of others. All his work at Musée de l'Homme and with U. Paris V and Paris XIII occurred well before MASI created, but he was still collaborating with Paris V and Baudichon; also in a second stage it has been relevant the collaboration of Gudberg K. Jonsson. This field is now reinforced by the developments of a wide range of research projects, not only in this book but also in several publications and works in progress, spanning a variety of areas and subareas (psychology, biology, medicine, sport, ethology, sociology, education, robotics, anthropology, finance, seismology, pharmacology, entertainment, agriculture, ergonomics, police, music, cuisine, tourism, and mass media). All reveal the high intrinsic value that T-pattern detection contributes to research, and we are convinced that the enhancements included in THEME Version 6 will lead to a new stage of successful qualitative and quantitative research.

The most of essential development of the T-Pattern model and T-Pattern Analysis occurred before MASI and the principal additions, since then such as the T-Associates and T-Packets have, unfortunately, not yet been used in research. Even the much older T-Markers have, regrettably, hardly been used either. We hope that this book is an invite to readers to work on these.

In terms of qualitative research, the new options of THEME Version 6 permit high-level scrutiny (see Casarrubea et al. 2015; Nicol et al. 2015) and have given rise to several components of the T-pattern model: T-Bursts, T-Markers, T-Associates, Satellites and Taboos, T-Packet structure (gravity and repulsion zones), Drifters and T-Kappa.

In terms of quantitative research, the presentation in November 2012—coinciding with the MASI meeting held in Guadalajara, Mexico—of the open access, educational version of THEME Version 6 for noncommercial purposes has enabled researchers and professionals the world over to use the program in their everyday work. Evidently, this straightforward program should ideally be used as an integral part of conceptual and methodological training, since it is vital to have knowledge of the basic structure involved in detecting patterns and of certain fundamental features of parameter assignment (such as the level of significance, the minimum number of occurrences, lumping factor, how to reduce redundancy, and deciding between modalities by shuffling or rotation in order to randomize data).

We firmly believe that the "great THEME family" will continue to strive to further develop the potential of T-pattern detection and further explore the hidden structures that emerge from data.

Barcelona, Spain
January 1st 2015

M. Teresa Anguera

References

Allister N, Segonds-Pichon A, Magnusson MS (2015) Complex spike patterns in olfactory bulb neuronal networks. J Neurosci Methods 239:11–17

Anguera MT, Magnusson MS, Jonsson GK (2007) Instrumentos no estándar. Avances en Medición 5(1):63–82

Anguera MT, Izquierdo C (2006) Methodological approaches in human communication. From complexity of situation to data analysis. In: Riva G, Anguera MT, Wiederhold BK, Mantovani F (eds) From communication to presence. Cognition, emotions and culture towards the ultimate communicative experience. IOS Press, Amsterdam, pp 203–222

Anolli L, Duncan S, Magnusson MS, Riva G (eds) (2005) The hidden structure of interaction. From neurons to culture patterns. IOS Press, Amsterdam

Casarrubea M, Jonsson GK, Faulisi F, Sorbera F, Di Giovanni G, Benigno A, Crescimanno G, Magnusson MS (2015) T-pattern analysis for the study of temporal structure of animal and human behavior: A comprehensive review. J Neurosci Methods 239:44–46

Gabín B, Camerino O, Anguera MT, Castañer M (2012) Lince: multiplatform sport analysis software. Procedia Social Behav Sci 46:4692–4694

Kemp AS, Fillmoreb PT, Lenjavia MR, Lyond M, Chicz-DeMeta A, Touchettec PE, Sandmana CA (2008) Temporal patterns of self-injurious behavior correlate with stress hormone levels in the developmentally disabled. Psychiatry Res 157(1-3):181–189

Lapresa D, Arana J, Anguera MT, Garzón B (2013) Comparative analysis of the sequentiality using SDIS-GSEQ and THEME: a concrete example in soccer. J Sports Sci 31(15):1687–1695

Introduction

Discovering hidden recurring patterns in observable behavioral processes is an important issue frequently faced by numerous advanced students and researchers across many research areas, such as, for example, psychology, biology, sports, robotics, media, finance, and medicine. As generally, the many powerful methods included in statistical software packages were not developed for this kind of analysis, discovering such patterns has proven a particularly difficult task, due to a lack of (a) adequate formalized models of the kinds of patterns to look for, (b) corresponding detection algorithms, and (c) their implementation in available software.

The research described in this book is based on the application of such pattern types, algorithms, and software developed over decades or since the late 1970s and until this day in the context of research in collaboration with human and animal behavioral research teams at internationally leading universities in the USA and Europe, thus testing the usefulness and validity of the pattern types, algorithms, and software in numerous research areas.

With the (scale-independent statistical hierarchical and fractal-like) T-Pattern at its heart, a set of proposed pattern types, called the T-System, forms the basis for the search algorithms implemented as the software THEME™ (v 6), which is easily available in free educational and full commercial versions (copyright www.patternvision.com). Recent original additions to the T-System and Theme are the T-Burst, the T-Packet (with its gravity and repulsion zone), T-Associates, T-Satellites, and T-Taboos.

As each chapter of this book describes a different research application of T-Pattern Detection and Analysis with THEME™, it can be seen as a sequel to Anolli et al. eds. book *The Hidden Structure of Interactions: From Neurons to Culture Patterns.* Both books can also be seen as products of an international research network, called "Methodology for the Analysis of Social Interaction" (MASI), based on a formal international interuniversity collaboration convention between leading European universities, with "Magnusson's analytical model" as the common reference, initiated in 1995 by the University of Paris V, René Descartes and first signed by the rectors and presidents of seven universities, but now involves 24 universities in Europe and the Americas. Both books include a number of contributions from collaborators outside the MASI network, for example, in the area of human interaction at the University of Chicago in continuation of collaboration since the beginning of this R&D effort in the 1970s; moreover, the University of Arizona (deception in interactions, Burgoon et al.); the University of California, Irvine (psychiatry, Sandman et al.); the University of Palermo, Italy (behavior and brain research, Casarrubea et al.); and the University of Cambridge, UK (neuroscience; multi-cell interaction patterns in living brains, Nicol et al.).

The chapters of this book provide advanced students and researchers highly varied models for their own research with easily available software tools (including free educational version) and should be a natural addition to university and research libraries.

Reykjavik, Iceland
Tucson, AZ, USA
Palermo, Italy

M.S. Magnusson
J.K. Burgoon
M. Casarrubea

Contents

Contributors

CÉSAR ADES • *Institute of Psychology, University of Sao Paulo, Sao Paulo, Brazil*
LUIGI ANOLLI • *Department of Human Sciences for Education "Riccardo Massa", CESCOM (Centre for Studies in Communication Sciences), University of Milano-Bicocca, Milan, Italy*
M. TERESA ANGUERA • *Methodology of the Behavioral Sciences, Faculty of Psychology, University of Barcelona, Barcelona, Spain*
ARNAUD ARABO • *UFR des Sciences et Techniques, Université de Rouen, Mont-Saint-Aignan, France*
ISABELLE BARAUD • *UMR 6552 CNRS, Biological Station of Paimpont, University of Rennes 1, Paimpont, France*
CLAUDE BAUDOIN • *Laboratory of Experimental and Comparative Ethology, University of Paris 13 Sorbonne Paris Cité, Villetaneuse, France*
CATHERINE BLOIS-HEULIN • *UMR 6552 CNRS, Biological Station of Paimpont, University of Rennes 1, Paimpont, France*
MICHAEL BRILL • *Wuerstburg University, Wuerstburg, Germany*
JUDEE K. BURGOON • *Center for the Management of Information, University of Arizona, Tucson, AZ, USA*
OLEGUER CAMERINO • *Laboratory of Human Movement Observation, INEFC-University of Lleida, Lleida, Spain*
MAURIZIO CASARRUBEA • *Human Physiology Section "Giuseppe Pagano", Laboratory of Behavioral Physiology, Department of Experimental Biomedicine and Clinical Neurosciences, University of Palermo, Palermo, Italy*
MARTA CASTAÑER • *Laboratory of Human Movement Observation, INEFC-University of Lleida, Lleida, Spain*
GIUSEPPE CRESCIMANNO • *Human Physiology Section "Giuseppe Pagano", Laboratory of Behavioral Physiology, Department of Experimental Biomedicine and Clinical Neurosciences, University of Palermo, Palermo, Italy*
BERTRAND L. DEPUTTE • *E.N.V.A., Plélan le Grand, France*
BARBARA DIANA • *Department of Human Sciences for Education "Riccardo Massa", CESCOM (Centre for Studies in Communication Sciences), University of Milano-Bicocca, Milano, Italy*
MASSIMILIANO ELIA • *Department of Human Sciences for Education "Riccardo Massa", CESCOM (Centre for Studies in Communication Sciences), University of Milano-Bicocca, Milano, Italy*
GIUSEPPE DI GIOVANNI • *Department of Physiology and Biochemistry, Faculty of Medicine and Surgery, University of Malta, Msida, Malta; School of Biosciences, Cardiff University, Cardiff, UK*
ALFONSO GUTIÉRREZ • *Faculty of Education and Sport Science, University of Vigo, Vigo, Spain*
MICHAEL HASS • *Oklahoma State University Institute of Technology, Okmulgee, OK, USA*

Charles C. Horn • *Biobehavioral Oncology Program, University of Pittsburgh Cancer Institute, Pittsburgh, PA, USA; Division Gastroenterology, Hepatology, and Nutrition, Department of Medicine, University of Pittsburgh, Pittsburgh, PA, USA; Department of Anesthesiology, Center for Neuroscience, University of Pittsburgh School of Medicine, Pittsburgh, PA, USA*
Gudberg K. Jonsson • *Human Behavior Laboratory, University of Iceland, Reykjavik, Iceland*
Aaron S. Kemp • *Psychiatry and Human Behavior, University of California, Irvine, CA, USA*
Haesook Kim • *School of Nursing, University of California, Los Angeles, CA, USA*
Mohammed R. Lenjavi • *Psychiatry and Human Behavior, University of California, Irvine, CA, USA*
Magnus S. Magnusson • *Human Behavior Laboratory, University of Iceland, Reykjavik, Iceland*
David McNeill • *McNeill Lab, Department of Psychology, University of Chicago, Chicago, IL, USA*
Alister U. Nicol • *Bioinformatics Department, Babraham Institute, Cambridge, UK*
Gerardo Ortiz • *Centro de Estudios e Investigaciones en Comportamiento, Universidad de Guadalajara, Guadalajara, Mexico*
Linda R. Phillips • *School of Nursing, University of California, Los Angeles, CA, USA*
Jean-Sébastien Pierre • *UMR 6553 CNRS, University of Rennes 1, Beaulieu, France*
David Pincus • *Crean School of Health and Life Sciences, Chapman University, Orange, CA, USA*
Ivan Prieto • *Faculty of Education and Sport Science, University of Vigo, Vigo, Spain*
Liesbet Quaeghebeur • *Faculty of Philosophy, University of Antwerp, Antwerp, Belgium*
Anaïs Racca • *Laboratory of Experimental and Comparative Ethology, University of Paris 13 Sorbonne Paris Cité, Villetaneuse, France*
Vincent Roy • *PSY-NCA, EA4700, Laboratoire de Psychologie et de Neurosciences de la Cognition et de l'Affectivité, Université de Rouen, Mont-Saint-Aignan, France*
Curt A. Sandman • *Psychiatry and Human Behavior, University of California, Irvine, CA, USA*
Andrea Santangelo • *Department of Neuroscience, Psychology, Drug Research and Child Health, University of Florence, Florence, Italy*
Ryan Schuetzler • *University of Nebraska Omaha, Omaha, NE, USA*
Frank Schwab • *Wuerzburg University, Wüerzburg, Germany*
Anne Segonds-Pichon • *Bioinformatics Department, Babraham Institute, Cambridge, UK*
Monika Suckfüll • *The Berlin University of the Arts, Berlin, Germany*
Vilhjalmur Thorsteinsson • *The Marine Research Institute, Reykjavik, Iceland*
Gunnar G. Tomasson • *School of Science and Engineering, Reykjavik University, Reykjavik, Iceland*
Ana Lilia del Toro • *Posgrado en Ciencia del Comportamiento: Análisis de la Conducta, Universidad de Guadalajara-Mexico, Guadalajara, Mexico*
Paul E. Touchette • *Pediatrics, University of California, Irvine, CA, USA*
Dagmar Unz • *University of Applied Sciences, Würzburg-Schweinfurt, Schweinfurt, Germany*
David Wilson • *University of Oklahoma, Norman, OK, USA*
Diana Lynn Woods • *Azusa Pacific University, School of Nursing, Azusa, CA, USA*
Maria Yefimova • *School of Nursing, University of California, Los Angeles, CA, USA*
Valentino Zurloni • *Department of Human Sciences for Education "Riccardo Massa", CESCOM (Centre for Studies in Communication Sciences), University of Milano-Bicocca, Milano, Italy*

Authors Biography

Magnus S. Magnusson, Ph.D. is the creator of the T-pattern model and analysis and the author of the THEME™ Behavior Research software. He currently serves as a Research Professor at the University of Iceland and is the founder and director of its Human Behavior Laboratory. He has worked as an invited professor at the Psychological Laboratory of University René Descartes Paris V in the Sorbonne, where in 1995 a formal interuniversity collaboration convention was initiated with the T-Pattern model and analysis as the common reference. It has now been signed by 24 universities in Europe, the USA, and Mexico. Dr. Magnusson has authored and coauthored numerous papers and chapters in many areas of behavioral and neuroscience including research on humans, animals, and neuronal networks in living brains. In 2015 the *Journal of Neuroscience Methods* published a comprehensive review of his T-Pattern Analysis (TPA) for the analysis of the real-time structure of behavior and interaction, and a paper presenting results of TPA of interactions in neuronal networks in the olfactory bulb. He has given many invited talks at conferences and universities across the world.

Judee K. Burgoon, Ph.D. is Professor of Communication, Family Studies and Human Development at the University of Arizona, where she is Director of Research for the Center for the Management of Information and Site Director for the Center for Identification Technology Research, a National Science Foundation Industry/University Cooperative Research Center. Previously she held faculty appointments at the University of Florida, Michigan State University, and University of Oklahoma and was a Visiting Scholar at Harvard University. She has authored or edited 13 books and monographs and over 300 articles, chapters, and reviews related to interpersonal, group, nonverbal, and verbal communication. Her recent work has emphasized deception, computer-mediated communication, and human–computer interaction. Her research has been funded by the National Science Foundation, Department of Defense, Department of Homeland Security, and IARPA, among others. Her awards and honors include, from the International Communication Association, the Steven Chaffee Career Productivity Award, Robert Kibler Mentorship Award, and election as Fellow; from the National Communication Association, the Distinguished Scholar Award for a lifetime of scholarly achievement, the Mark L. Knapp Award in Interpersonal Communication, and the Charles Woolbert Research Award for Scholarship of Lasting Impact. A recent survey identified her as the most prolific female scholar in communication in the twentieth century.

Maurizio Casarrubea, Ph.D. has, since 2005, served as a Research Associate at the University of Palermo—Italy, Institute of Human Physiology "Giuseppe Pagano," Laboratory of Behavioral Physiology, Department of Experimental Biomedicine and Clinical Neurosciences. He graduated from the University of Palermo with degrees in Medicine and Surgery, a medical specialization in Sport Medicine, and a Ph.D. in Neurosensorial Physiophathology. Dr. Casarrubea currently teaches Human Physiology in the School of Dentistry and the School of Medicine and Surgery and has organized a laboratory of

Behavioral Physiology. He has authored or coauthored numerous research papers, book chapters, and conference proceedings related to the study of anxiety-related behavior, the relationships between anxiety and pain, and the behavioral effects produced by the administration of psychoactive drugs. In addition, he regularly presents the results of his experiments to various congresses of Neurosciences and Behavioral Sciences. His current research activity is funded by the University of Palermo and is focused on the application of multivariate approaches to the study of anxiety-related behavior in rodents. He is member of the Physiological Society of Italy (S.I.F.) and of the Technologies of Knowledge Interdepartmental Center (C.I.T.C.), University of Palermo—Palermo, Italy.

Part I

Human Behavior

Chapter 1

Time and Self-Similar Structure in Behavior and Interactions: From Sequences to Symmetry and Fractals

Magnus S. Magnusson

Abstract

This chapter concerns the temporal structure of behavior and interaction amongst individuals as diverse as brain neurons and humans. It suggests a view of behavior and interaction in terms of recurrent self-similar tree structures, T-patterns, which thus have the basic characteristics of fractals and exemplify translation symmetry through the similarity of the recurrence of each, a view that is the basis for the special pattern (T-pattern) detection algorithms implemented in the THEME software especially developed for T-pattern detection. Derived concepts are defined and illustrated with special T-pattern diagrams. Some comparison is made with standard multivariate statistics methods. The analysis of Big Data and Tiny Data using this particular recurrent hierarchical and multiordinal pattern detection approach is discussed, as well as the use of T-pattern Analysis (TPA) to detect experimental effects that often remain hidden to standard statistical methods.

Key words Behavior, Interaction, Time, Patterns, Detection, Hierarchy, Self-similarity, Symmetry, Fractals, Big data, Experimental effects

1 Introduction

The broadness of the concept of *pattern* is considerable as indicated by the fact that most modern mathematicians consider mathematics the science of patterns [1]. Risking to state the obvious, a decision to search for patterns is thus next to meaningless without further specification of the *kind* of pattern. Even the simplest kinds of patterns may be of much interest as they sometimes describe the most diverse and apparently complex phenomena in nature and mathematics, as does, for example, one of the most famous mathematical series, the Fibonacci number series, $xi := xi_{-1} + xi_{-2}$ where the next term of the series is simply the sum of the preceding two [2].

Here, the T-system composed of a particular kind of pattern, called T-pattern, with extensions and additions, are proposed as tools and building blocks for the discovery, analysis, and description

Magnus S. Magnusson et al. (eds.), *Discovering Hidden Temporal Patterns in Behavior and Interaction: T-Pattern Detection and Analysis with THEME™*, Neuromethods, vol. 111, DOI 10.1007/978-1-4939-3249-8_1,

of real-time structure in behavior and interactions. The research reported in this book shares a focus on discovering and analyzing repeated patterns through T-Pattern detection and Analysis (TPA) using the THEME™ software especially developed for that purpose and often allowing the detection of patterns hidden to observers whether unaided or using other available tools. A number of publications contain background information and details (Magnusson [3–8]) recently complemented by a comprehensive review of TPA for the analysis of the temporal structure of behavior and interactions [9]. The present chapter thus aims at complementing rather than repeating, but contains a fairly self-contained description regarding the main concepts, methods and tools applied in this book as well as some idea about their theoretical, methodological and historical roots. Some more recent additions and future perspective regarding the T-System are also noted.

2 Why a Hierarchical Pattern Detection Approach?

As it turns out, the relevance of the T-pattern goes beyond the human interaction situation for which it was initially designed, e.g., regarding the analysis of multi-neuron firing networks within populations of brain neurons ([10, 11] and Chap. 17, Nicol et al. in this book) and some DNA analysis [7]. Seeking answers as to why, leads to common aspects of natural and biological structures including that of the brain.

Modern science tells us that from the smallest to the largest in our universe, structure is mostly due to a balance between a relatively small number of pushing and pulling forces. Of extremely different sizes and complexity, from subatomic particles to galaxy clusters, the elements at each level can thus be seen as results of this universal pushing and pulling [12, 13]. Thus, rather than simply clumping, the balancing of such opposing forces leads to *hierarchical structured clustering* that in the known universe appears as a hierarchy of complexity levels and as self-similar patterns of patterns of patterns, etc. over multiple scales giving rise to the recently discovered fractal structure of the universe [14] (Fig. 1).

Astronomer E.L. Wright's view of the gradual increase of structure in the universe is interesting here:

> But dust grains is the interstellar medium are very unlikely to be spherical, and the most likely situation is one where dust grains can collide and stick, or collide and shatter, leading to a distribution of sizes and a random distribution of shapes. In order to model such a situation, I have made randomly shaped grains in a computer using a random collision process.

He adds:Fractal dust grains are created from the coagulation of smaller subunits, starting with *monomers*. Two monomers collide

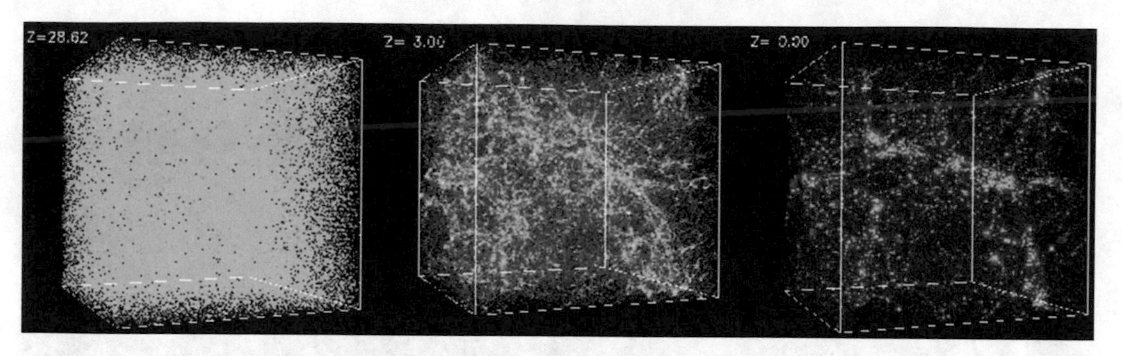

Fig. 1 This figure illustrates the gradual hierarchical and self-similar clustering in the universe, from almost homogenous energy near the Big Bang to gradually higher levels of hierarchical clustering until now. Adapted from http://cosmicweb.uchicago.edu/filaments.html

and stick to make a *dimer*. The figure above shows a process where two dimers collide to make a *tetramer*, two tetramers collide to make an *octamer*, and so on. This produces what I call CL fractals. (Quoted from his website http://www.astro.ucla.edu/~wright/dust/ where illustrative figures can also be found.) (Fig. 2)

Much of the brain's structure is also hierarchical and self-similar:

> …a nested hierarchy – smaller elements join together to form larger elements, which, in turn, form even larger elements, and so forth … many of the integrative aspects of brain function depend on this multi-scale structural arrangement of elements and connections. ([15], p. 41) (Fig. 3)

When trying to understand the world, discovering patterns that repeat is of primary importance for the survival of any individual and species as it allows some prediction of what its future may hold and needs to be dealt with. Science is also less concerned with the anecdotal.

Science searching for repeated patterns in nature, and mathematics being the science of patterns, suggests imagining and mathematically describing the abstract structure of the often dimly perceived repeated patterns of everyday life. Profiting from the power of computers, unimaginable throughout practically all of mankind's history, corresponding search algorithms can be developed. Obviously, creating a nouvelle kind of analysis of behavioral data must be based on knowledge about the structure of behavior, thus limiting hypotheses to what kinds of structure may realistically exist in behavior. Also necessary is knowledge about existing or obtainable data and about tools that are available or can be developed.

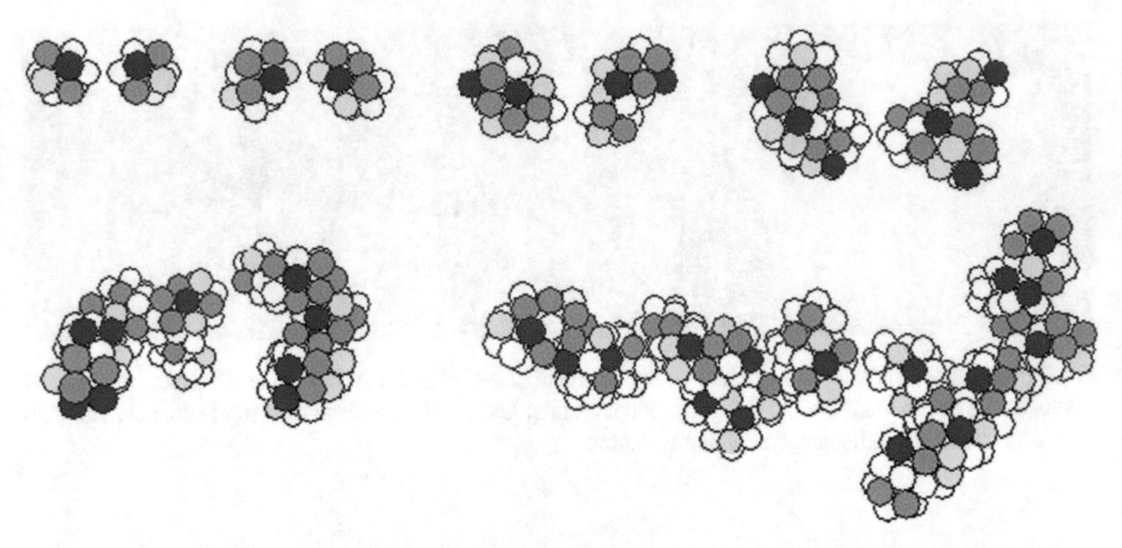

Fig. 2 This figure (adapted from E.L. Wright's website) illustrates how astronomer E.L. Wright of UCLA envisions the gradual hierarchical clustering of the universe as made of numerous different fragments rather than of a uniform kind of units. Given the multitude of different kinds of fragments, the number and kinds of different and gradually more complex combinations should quickly become, well, "astronomical"

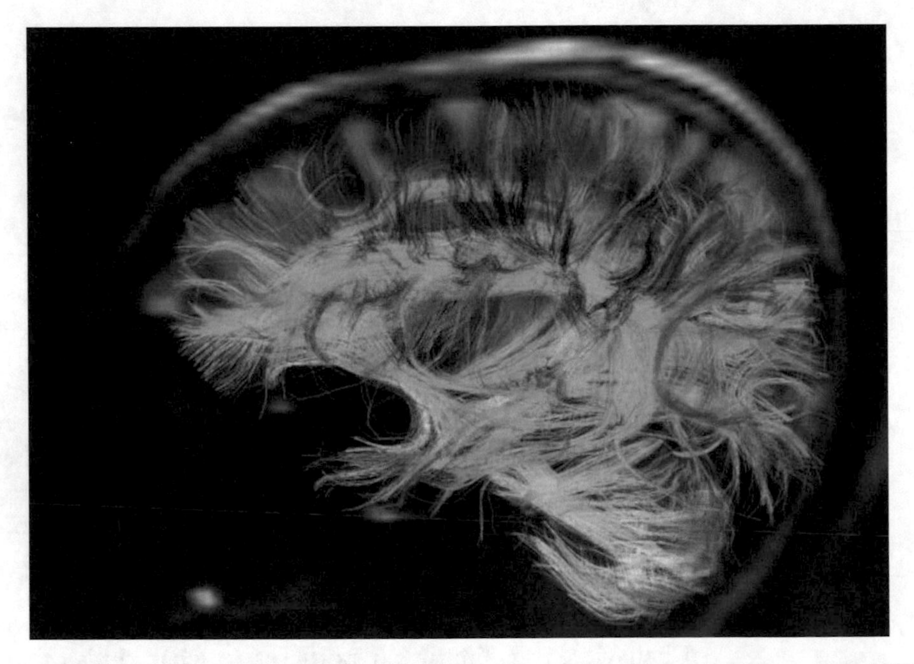

Fig. 3 A connectome of a human brain. The fractal- or tree-like structure is apparent. Figure from MGH Human Connectome Project Acquisition Team. In Neuroscience: Making connections, Jon Bardin, Nature, 21 March 2012. (License obtained from Nature via RightLink)

3 Looking for Ideas Regarding Hidden Patterns in Behavior

Behavior as repeated nonobvious or hidden *sequences* or *patterns* in time is a traditional viewpoint in behavioral science. Eibl-Eibesfeldt, sometimes called the father of human ethology (the biology of human behavior), thus opens his book in 1970 with these opening words:

> **Behavior consists of patterns in time**. Investigations of behavior deal with **sequences** that, in contrast to bodily characteristics, are **not always visible**. ([16]. Emphasis added.)

Repeated sequences or temporal patterns in behavior having some kind of nonrandom syntactic structure to be discovered is obviously a long established idea in, for example, Linguistics [17], Ethology [18] and in Radical Behaviorism, where repeated real-time probabilistic contingencies (temporal patterns) are also paramount [19].

Numerous other areas of behavioral science such as Anthropology, Social Psychology, and Cognitive Science are also concerned with the discovery and analysis of often nonobvious or hidden repeated behavioral patterns such as scripts, plans, routines, strategies, rituals, and ceremonies, which all are repeated patterns of patterns, etc., that is, they are hierarchical and syntactically constrained temporal patterns (see, as an early example [20]).

4 Speech and Nonverbal Behavior

Human speech and language had evolved in the context of other behavior for thousands of years (see, for example [21]) long before the first ideas of systematic analysis. Since then, centuries of analysis and inventions have led to modern Linguistics, recently including computational linguistics. In speech and writing we again find repeated hierarchical patterns of patterns varying greatly in complexity, content, function and meaning. From the simplest muscular movements in the vocal tract to phonemes, syllables, words, sentences and frequently repeated longer (vocal or written) patterns such as stories, poems, and massively repeated and standardized legal and religious verbal patterns, often of great length and complexity. These patterns are also made up of physical entities that do not simply clump in time and space, but form repeated hierarchical and syntactically constrained, spatially and temporally structured clusters. We thus see that they have nonrandom distances between their parts, which themselves have nonrandom distances between their parts at a lower scale, etc.

As the two major categories of human communicative behavior, verbal and nonverbal, are typically intertwined, a unified theory of both is a longstanding dream. Thus a leading linguist in his time,

Pike, begins his book "Language: in Relation to a Unified Theory of the Structure of Human Behavior" with a description of a party game—the repetition of a song—a timed sequence of words, but at each repetition, one more word is replaced by an equivalent gesture.

Finally, after further repetitions and replacements, there may be left only a few connecting words like *the*, and a sequence of gestures performed in unison to the original timing of the song. ([22], p. 1).

A little later ([22], p. 2) Pike continues:

> The activity of man constitutes a structural whole, in such a way that it cannot be subdivided into neat "parts" or "levels" or "compartments" insulated in character, content, and organization from other behavior. Verbal and nonverbal activity is a unified whole, and **theory and methodology should be organized or created to treat it as such**. {Emphasis added.}

After years of intensive collaborative efforts within a group including among others Birdwhistell [23] and Bateson analyzing speech and body movements, McQuown thus ends his foreword to their "Natural History of an Interview" (NHI) [24] with these words:

> It is expected that, as a result of such investigations, the frames for describing language, paralanguage and body-motion in English (and in other) language communities will be perfected, that the linguistic, paralinguistic, and body-motion markers of sentence-like units manifest in these communicative behavioral channels will be uncovered, and that the foundations of a general theory of the structure of human communicative behavior, as manifest through these channels, and such units, will eventually be worked out. It is hoped that the materials here presented may facilitate the first steps in this on-going process.

Bateson's chapter in NHI [24] starts with an overview of the impressive scientific scene in the first half of the twentieth century citing many of the principal contributors (e.g., Shannon, von Neumann, Ashby, Tinbergen, and Lorenz) ending with this interesting partial conclusion (p. 5):

> What has happened has been the introduction into behavioral sciences of very *simple, elegant, and powerful ideas* all of which have to do with the nature of communication in the widest sense of the word. The steps and sequences of logic have been coded into the causal sequences of computing machines and, as a result, the Principia Mathematica has become a cornerstone of science. {Italics added}.

Just as the study of phenomena from micro to cosmic scales depended on the invention of microscopes and telescopes, the invention of film and video recording has allowed the fleeting character of human behavior and interactions to be captured for unlimited inspection and the collection of high quality data. But paper-and-pencil behavioral data collection, whether directly or from film and video often overwhelmed the data processing possibilities of that time.

As computers became more easily available, existing statistical methods were implemented in software and attempts made to use them in behavioral research (e.g., [25–27]). But these statistical methods were created when minimizing calculation was essential and therefore did not fully exploit the new computational power. More importantly here, these methods were not developed for the discovery of complex multilevel, multiordinal, hierarchical real-time patterns within partly independent parallel processes such as human behavior and interactions. New models were thus needed concerning the kinds of behavioral structure allowing the development of adequate computational methods.

The present "beginning-from-scratch" or Cartesian mathematical and computational approach could also among other, be seen as a reaction to two common aspects of observational interaction research, at the time based on film and video recordings, (a) highly subjective qualitative interpretation and/or (b) nearly exclusive focus on the frequency and duration of behaviors. Only rarely were multivariate or sequential analyses applied and then mostly standard and simple ones allowing little chance of discovering much of the complex hidden real-time structure of behavior and interactions.

5 Towards a Fresh Neutral and Objective Look

Two world wars in less than half a century is probably more than enough to suggest that something is wrong and a fresh look at human behavior, interactions, and relationships was overdue. But how to obtain true understanding of that elusive and complex phenomenon? Experiments in unnatural laboratory environments meant that the results might not be valid in real life and watching the events of everyday life through glasses heavily tainted by theories and the collection and interpretation of data based on too much subjectivity might lead to a distorted view.

In his chapter in NHI, Bateson writes: "Our primary data are the multitudinous details of vocal and bodily action recorded on this film. We call our treatment of such data a "natural history" because a minimum of theory guided the collection of the data." ([24], p. 5). He also notes (p. 3) that Kurt Lewin [28] had already suggested a "mathematics of human relationships," thus starting on the hard scientific path away from subjectivity through mathematics.

Ethology and then human ethology promoted direct objective non-intrusive and open-minded observation of each species' behavior in its natural environment, which for modern humans is not easily defined otherwise than as human everyday environments.

In 1973, the Dutch ethologist Tinbergen's pioneering ethological research [29] earned together with K. Lorenz and K. von

Frisch the Nobel Prize in ethology/zoology "for their discoveries concerning organization and elicitation of individual and social behaviour patterns" (www.nobelprize.org/nobel_prizes/medicine/laureates/1973/) providing new inspiration for biologically oriented behavior research.

5.1 Towards T-Patterns and THEME

As a teenager, I had the good luck to find Tinbergen's then brand new and inspiring book "Animal Behaviour" [30] and then a few years later another by the French ethologist and psychiatrist Jacques Cosnier [31] and later I visited with Cosnier and his laboratory at the University of Lyon around 1975. Later, during my work in Paris (1983–1993), he directed the first (state) doctoral thesis where T-pattern Analysis was applied [32].

Until the 1970s, ethology was almost exclusively concerned with nonhuman behavior, but human ethological studies began in earnest and under that name with the work of Blurton-Jones [33] and McGrew [34]. At about the same time, most importantly here, Montagner, who earlier had analyzed social insect interactions using high-speed filming and frame-by-frame analysis [35], turned to extensive filming of children's interactions in their everyday kindergarten environments, followed by frame-by-frame analysis [36], and the many interesting results suggested great possibilities of further insight especially given powerful computational analysis. A visit with Montagner around 1975 at his laboratory resulted in his long standing invaluable support, a collaborative paper and two doctoral theses, applying TPA, that he directed [37–39].

Fortunately, well before the advent of the PC, I had read some new American literature about the computer revolution. This was just before I moved to Denmark and began my studies at the University of Copenhagen in the biological and animal behavior section of its Psychological Laboratory where ethology and behaviorism struggled with the difficult questions of behavioral research—and with each other, and with Chomsky's devastating assault on Skinnerian behaviorism still in fresh memory [40]. The triangular struggle between these three directions is partly indicated by the words of the ethologist Cosnier who wrote about "Chomsky's lacuna," a serious one from a biologist's point of view:

> ...if the competence is a part of the organizing system of the species, what we agree on, then speech activity should be placed within the general homeostatic system of the talking individual. ([31], p. 122, here translated from French).

Strongly inspired by recent research on social phenomena in animals and especially in social insects [35], I wrote in 1975 a B.A. thesis entitled "Social Organization and Communication in Social Insects and Primates (Humans Included)" [41] and began the theoretical, methodological and software development of my doctoral studies completed in 1983. Computational methodology including multivariate statistics and Artificial Intelligence was coming into focus and I experimented for a few years with multivariate statistics using the principal statistical program packages on the mainframe IBM computer of the North European University Computation Center (NEUCC) at the Technical University of Denmark, helped invaluably by my countryman, a statistician and computer statistics specialist there, Agnar Hoeskuldsson, but it turned out fairly soon that these methods were not fully adequate for the task at hand. So in 1978, I wrote a thesis entitled "The human ethological probabilistic structural multivariate approach to the study of children's nonverbal communication" [42] where I initiated the alternative approach that eventually led to the T-pattern, the T-System, and THEME.

Using the brand new smaller computers (PDP 11 and 8) of the Psychological Laboratory, I developed and programmed the first version of the (3000 lines) THEME software. I was helped getting started with Fortran IV programming by psychologist Strange Rosss, who was also the principal computer person of the laboratory, but I was soon on my own. My very first computer program, Theme, was running in 1980 on a PDP 8, based on my developing model of temporal structure in behavior. I first presented it with the first results of an analysis of toddlers' dyadic interactions, at an A.I. workshop at Uppsala University in Sweden. The title was "Temporal Configuration Analysis" [3] corresponding in essence to the present ("fast," see below) T-pattern Analysis. Now, after 35 more years of development, programming and application in my research mostly in Paris at the National Museum of Natural History 1983–1988 and the University of Paris (V, VIII and XIII) 1988–1993 and since at the Human Behavior Laboratory that I created at the University of Iceland in 1991, many new features have been added. Particularly useful was the special T-pattern diagram (see Fig. 12) that I developed at the Anthropology Laboratory of the Musée de l'Homme in Paris and at the Psychology Department of the University of Paris VIII around 1988–1989. Theme now also has a hundred times more lines (300 thousand), can process far larger data and is orders of magnitude faster. The research application has always involved collaboration with a number of researchers at a still growing number of universities mostly in Europe and the USA that since 1995 is based on a formal interuniversity collaboration convention started at the (Binet) Psychological Laboratory of University of Paris V in the Sorbonne and now includes 24 universities.

In around 1976, I met Paul Ekman in Copenhagen and then visited his laboratory in San Francisco. It was a great inspiration as his new FACS (Facial Action Coding System) provided high quality real-time data with possibilities of new intensive kinds of analysis unimaginable without the power of computers. A few studies have since combined the use of FACS and Theme, for example, in studies of face-to-face psychiatric and psychotherapeutic interactions [43, 44].

As I was travelling around the USA searching for behavior analysis software for human interaction research, Paul Ekman suggested that I visit with Kenneth Kay (who had just developed the CRESCAT behavior analysis software) and with Starkey Duncan, both at the University of Chicago. Duncan had also worked with the NHI group as he describes in his 1977 book:

> The underlying purpose of the materials presented in this work, as articulated by McQuown in his Foreword, was to initiate first steps in a process aimed at the development of "the foundation of a general theory of the structure of human communicative behavior... (p. 5)". Both in its conceptualization and its implementation, *The Natural History of an Interview* served as the foundation of the design of the present study, and indeed, provided much of the motivation for pursuing it. ([45], p. 136.)

Our collaboration and friendship started around 1977 and continued until his death in 2007. In 1973 Duncan had published a paper entitled "Toward a grammar for dyadic conversations" ([46], p. 29–46) and in 2005 he developed a kind of interaction grammar based on detected T-patterns [47].

6 The T-System

The T-system, developed since the seventies, is evolving as formally defined statistical (probabilistic) patterns of relations between sets of points on a discrete scale (mostly time and molecular sequences). Corresponding detection and analysis algorithms are still being created and implemented in the THEME™ software (copyright PatternVision Ltd; for full versions and free educational versions see patternvision.com).

With the exception of some consideration of DNA patterns [7], the focus has been on behavior in time considered in terms of point-events occurring in unit intervals on a discrete scale within an observation interval $[1, T]$. Here the event typically represents the beginning or ending of some behavior, such as running, sitting, smiling, looking, talking or swimming, each seen as a different kind of state with beginning- and endpoints (t^1, t^2; where $t^2 \geq t^1 \geq 0$) and an "ongoing" state interval $[t^1, t^2]$ with duration $(\geq 1) = (t^2 - t^1 + 1)$. The beginning (and ending) of each such state is considered without duration, but located within a unit interval on the discrete scale. Data are thus point events, and the series of occurrence points (unit intervals) of each type of event (called T-EventType) is called a T-series. Occurrences of each event-type (a behavioral category) must be recognizable as a repetition of the same behavioral event.

Note: A state may also have other associated event-types other than beginning and ending, for example, the point at min and/or max of some intensity or speed measure during the state interval, each giving rise to a different event occurrence series. Such series can be immediately included as can any point series within the observation interval (period).

6.1 The T-EventType

The correspondence between a behavior and its T-series is achieved simply by labeling the series with a description, a T-EventType string, which is defined using values (T-Items) from predefined nominal value classes, T-Classes. These event-type descriptions typically include a specification of an agent or thing (individual, group, ball, etc.) and the beginning or ending of some particular behavior. T-EventType definition strings (descriptions) thus typically resemble simple sentences such as, for example, "sue begins running fast" {sue,b,run,fast}, "ball ends rolling" {ball,e,roll} or "baby ends crying in bed" {baby,e,cry,bed}, "jack starts a company in electronics" {jack,b,startup,electronics}, "neuron 245 begins firing" {n245,b,spike} (*see* **Notes 1** and **2**) (Fig. 4).

6.2 T-Data

T-Data, the reference of all definitions in the T-System, is a collection of n (≥1) T-series, each representing the occurrences of an EventType within some observation period, the T-Period = [1, *T*], T being the length of the observation and thus the number of unit intervals (data points) where each T-event-type in T-Data occurs or not. The number of instances of an event-type divided by *T* gives its average probability of occurrence per unit time, used as a baseline probability by the detection algorithms. The number of occurrence data points in T-Data (discrete scale units with or without an occurrence) is thus n × *T*.

(*Note*: it is, of course, not important at what time the observation interval begins, that is, its offset time, but only its length; the number of occasions for each event-type to occur or not.

Note: Splitting time infinitely, that is, increasing the temporal resolution, does not, of course, increase precision, but up to a

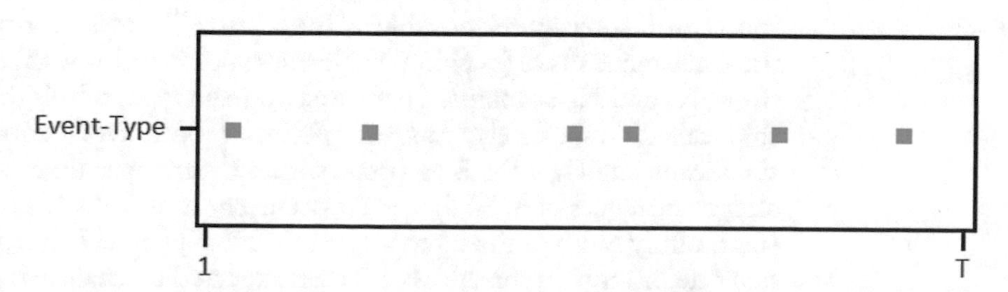

Fig. 4 This figure shows the basic type of data of the T-system, an event-type and its series of occurrence points (unit intervals) on a discrete scale within an observation interval [1, T]

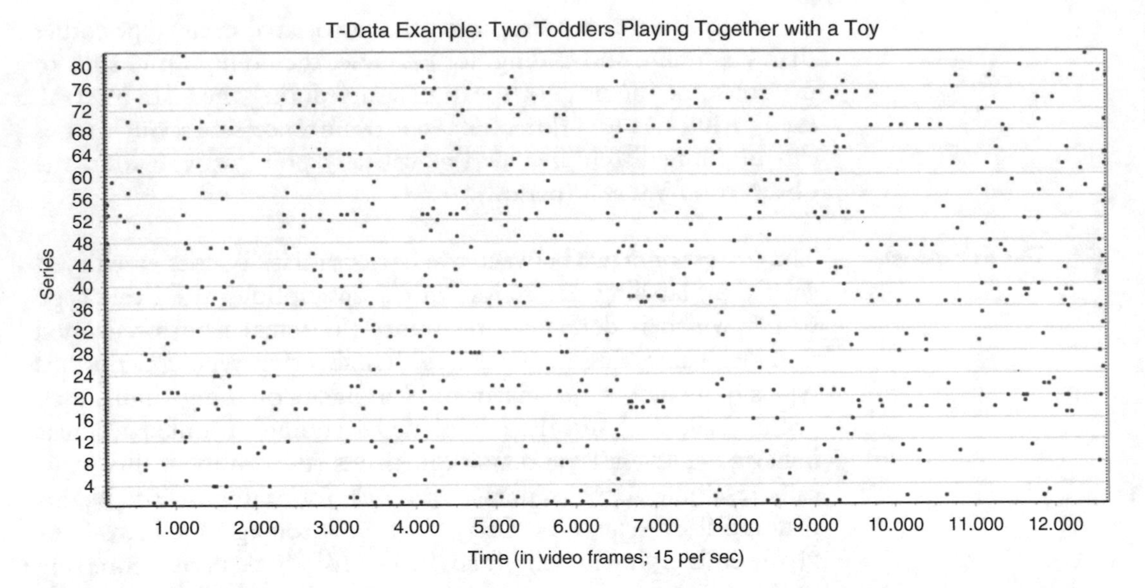

Fig. 5 Example of T-Data. This diagram shows the T-Data coded from a 13.5 min dyadic toddler's object interaction. The ethological categories used are mostly from the list of McGrew [34]. (Figure from THEME)

point, as it's the precision with which the events can be located on the discrete scale that really sets the limit.)

6.3 Multimodal and Multichannel Data

Each of the T-Series in T-Data may represent occurrences of very different phenomena, for example, anything from a bell ringing to a neuron firing, a person starting a company or a peek or special quirk in any analogous, for example, physiological measurement. Each T-Data may thus be multimodal and multichannel involving events of very different kinds and frequencies. Patterns found in such T-Data may therefore reveal relations across any channels and modalities (Fig. 5).

7 Critical Intervals: Fast and Free

Essential in the definition and detection of the T-Pattern structure is the *critical interval* relationship between two T-Series representing the occurrence times (positions) of event-types A and B with, respectively, N_A and N_B elements. There are two main types of this relationship called, respectively, *fast* and *free*, but both types concern a significant tendency for B to occur within a particular time window after occurrences of A. So for the N_A occurrences of A at t_i $\{i = 1 \ldots N_A\}$, significantly more of the intervals $[ti + d_1, ti + d_2]$ (short $[d_1, d_2]$) contain one or more occurrences of B than expected by chance assuming independent distributions of A and B and $[d_1, d_2]$ being the smallest such interval for the largest significant number of the A occurrences.

In the fast case, the lower limit, d_1 is fixed at A so the critical interval becomes $[d_1=0, d_2]$. The critical interval detection algorithm in this fast case thus begins with d_2 set to the greatest distance between A and B and then removing the longest remaining case until a significant interval is found or no more cases remain. The free critical interval is detected using the same method except that d_1 is first set equal to the shortest distance to a B and when the reduction of d_2 fails to find a CI, d_1 is set to the next shortest distance to a B. This procedure is repeated until a CI is found or d_1 cannot be further increased.

A special univariate fast critical interval, $[1, d_2]$ relates a series to itself, that is, the B series is the A series lag one (thus the lower limit of this critical interval is fixed at 1). It is used in the definition (below) of a special kind of T-patterns called T-burst (Fig. 6).

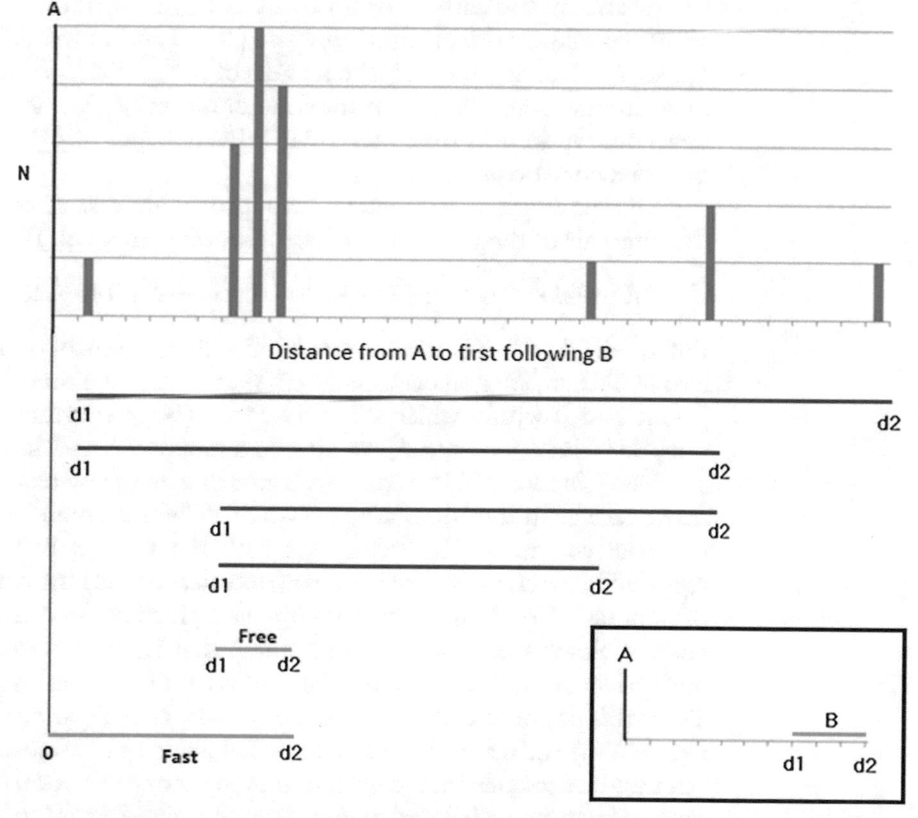

Fig. 6 For any two T-series, *A* and *B*, this illustrative example shows a frequency chart of distances measured from each element in *A* to the first concurrent or following in *B*. The *red lines* indicate some candidate critical intervals having failed the significance test. The green ones suggest significant cases, one free and the other fast (see text). The fast critical interval (CI) detection algorithm always has the occurrence time of A as the lower limit and starts trying with the longest distance from A to B, that is, [0, max(*B*−*A*)]. Testing each candidate interval, the candidate CI is narrowed as needed to the next shorter *A* to *B* distance until either significance is found or it fails, that is, [0, 0] has been passed. The current free critical interval detection algorithm differs from the fast in lifting the lower limit each time it fails until a CI is found or [max(*B*−*A*), max(*B*−*A*)] has been tested. Note that it is not possible to know only by inspecting the chart whether any of the shown intervals are significant as this requires knowing the number of occurrences of *A* and *B* as well as the observation interval [1, *T*] all used when testing the significance of each candidate interval (see Magnusson [5]). Note that each of the A to B distances may occur only once so that those corresponding to the CI may appear as a group or cluster rather than a peek as shown here

8 The T-Pattern

Somewhat intuitively stated, a T-Pattern is a recurring set of T-event-types occurring concurrently and/or sequentially with *significantly invariant* time distances between them, which can be noted in the following way:

$$Q = X_1 \approx dt_1 X_2 \approx \mathrm{d}t_2 X_3 \ldots X_i \approx dt_i X_{i+1} \ldots X_{m-1} \approx \mathrm{d}t_{m-1} X_m \quad (1)$$

where each of the X terms stands for some event-type in T-Data and the general term $Xi \approx \mathrm{d}tiXi_{+1}$ means that during occurrences of the pattern, consecutive terms Xi and Xi_{+1} are separated by a characteristic approximate time distance $\approx \mathrm{d}ti$, where characteristic also means statistically invariant. $X1.._{\mathrm{m}}$ is called the terminal sequence, $tX_1..m$ the terminal times sequence and m the length of Q. For Q occurring n times, the lengths of $[tX_1, tXm]j$; $j = 1..n$, are their durations and their sum the total duration of Q. The (behavioral) descriptions in the m terminal T-Event-types of Q are called its (behavioral) content.

An equivalent view replaces the approximate distances $\approx \mathrm{d}t$ by the intervals of their variation over their occurrences in T-Data:

$$Q = X_1[d_1, d_2]_1 X_2[d_1, d_2]_2 X_3 \ldots X_i[d_1, d_2]_i X_{i+1} \ldots X_{m-1}[d_1, d_2]_{m-1} X_m \quad (2)$$

Where $Xi\,[d_1, d_2]iXi_{+1}$ means, that within all occurrences of the pattern in T-Data, after an occurrence of Xi at t, there is a time window $[t + d_1, t + d_2]i$ within which Xi_{+1} will occur. (Note shorthand notation, the values of d_1 and d_2 are not the same for the m-1 intervals.)

Detecting long T-Patterns as a whole in data of any realistic size may overwhelm the computing power of PC's and even (the most) powerful computers. In accordance with the view of behavior as repeated hierarchical and self-similar (statistical fractal) the detection of complex T-patterns is bottom-up. For practical detection purposes a binary tree structure is assumed, that is, each nonterminal node having only two branches. Patterns are thus detected as pairs of T-EventTypes or already detected T-Patterns. Any T-pattern can be seen as a *left* and a *right* (concurrent or following) part. Equation (2) can thus be presented and detected as a binary tree of critical interval relations between a left and a right part, with the X terms in Eq. (2) as the terminal nodes of the binary tree of critical interval relations:

$$Q \rightarrow Q_{\mathrm{Left}}[d_1, d_2]Q_{\mathrm{Right}} \quad (3)$$

Recursively splitting left and right in this way thus leads to the presentation in Eq. (2), which is only considered a significant T-pattern if it has at least one such binary tree presentation where all CI's are significant.

For $Q_{\mathrm{Left}}\,[d_1, d_2]\,Q_{\mathrm{Right}}$ to be significant there must be a critical interval relation between the end times, $(., S)$, of Q_{Left} and beginning times, $(S, .)$, of Q_{Right}, that is,

Q_{Left} $[d_1, d_2]$ Q_{Right} is a T-Pattern only if $(., S_{left2})$ $[d_1, d_2]$ $(S_{right1}, .)$

Each instance of Q_{Left} ending at t, which is followed by the beginning of Q_{Right} within $[t+d_1, t+d_2]$ forms an instance of a new T-Pattern, Q, which begins where that Q_{Left} instance began and ends where that Q_{Right} instance ends. Such pairs of values form the Dseries for Q. The length (m) of Q is the sum of the lengths of Q_{left} and Q_{right}, that is, $m = m_{left} + m_{right}$.

8.1 Redundancy

As a number of significant binary critical interval trees may have the same Dseries and terminal sequence (e.g., ((AB)(CD)), ((((AB)C)D), and ((A(BC))D); where all occurrences of A, B, C, and D as parts of the patterns are the same and are thus considered equivalent, only whichever is detected first is kept and any others ignored. On the other hand, when two T-patterns, Q_a and Q_b, have identical Dseries (and thus the same number of occurrences) and the terminals of Q_a are the same and in the same order as in Q_b except that one or more are missing in Q_a, then Q_a is considered less complete and redundant and is dropped. (As their Dseries are identical, all continuations of Q_a, and Q_b would be the same, keeping both thus only adding further redundant detections.)

Note that as only two occurrences are needed for the definition and detection of critical intervals or T-Patterns, cyclical occurrence is not a defining aspect nor are other common aspects of time series such as trends (Fig. 7).

8.2 Statistical Validation

A Monte Carlo simulation method is used to evaluate the a priori probability of findings globally for all patterns detected in a given data set and for individual patterns. This is done through repeatedly randomizing and searching in the same data, always with the same search parameters and thus establishing a mean and standard deviation of the number of detected patterns of each length detected in randomized data and then comparing with the numbers detected in the initial data. The number of randomizations with searches is set such that a higher number only shows a negligible difference. Two different kinds of randomization are used, T-shuffling and T-rotation. T-shuffling replaces each T-series in T-data by a T-series with an equal number of random points within the T-period, thus totally removing any connection with the initial series except for the number of points. T-Rotation can be seen as folding T-data around a cylinder and then rotating each of its now circular T-series independently by a different random amount, thus barely altering its structure, but randomizing its relation to the other series. When detected T-patterns reflect synchronized cyclical relationships between T-series, the difference between detection in random versus real data may be much smaller for T-rotation than for T-shuffling. These methods can also be used to estimate the a priori expected number of occurrences of a single detected pattern by only implicating the event-types (T-series) included in the pattern (Figs. 8 and 9).

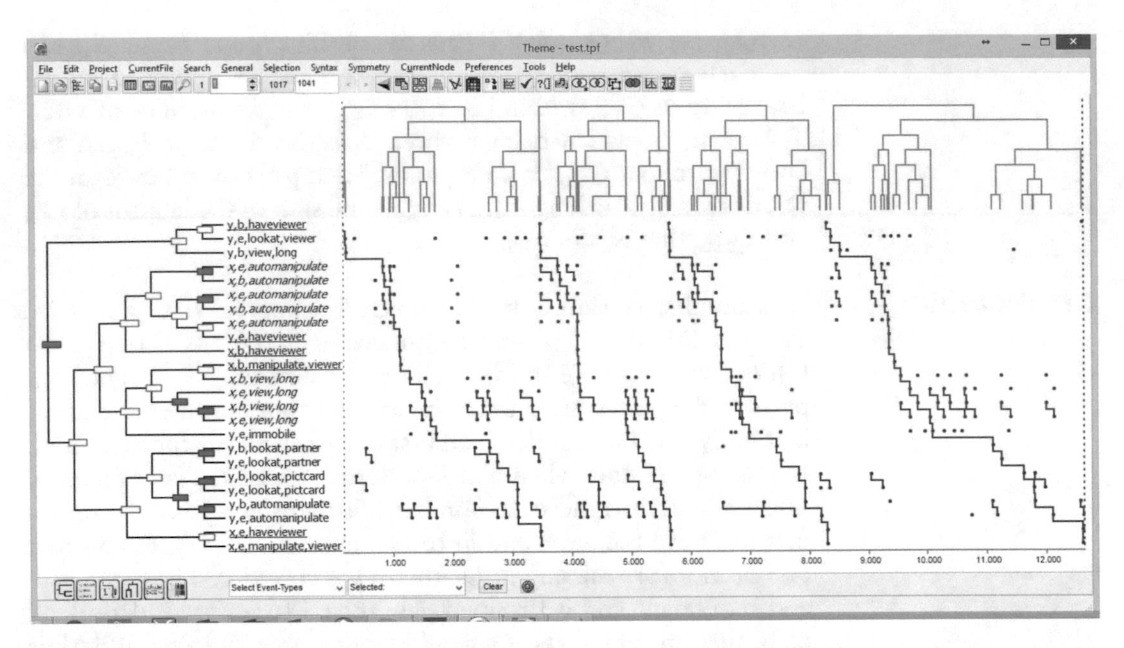

Fig. 7 This T-pattern diagram shows all the raw data (point) series involved in this T-pattern. The static tree on the left, shows the hierarchical level-by-level bottom-top detection and connection steps in the series part. At the top of the diagram are shown the dynamic versions of the static (left) tree corresponding to each occurrence of the pattern. These dynamic trees have the same hierarchical structure as the static (*left*) tree, but also show the real distances between the event-types (terminal nodes) at each occurrence of the tree (pattern) and point to the raw data point connections below. (THEME screen capture)

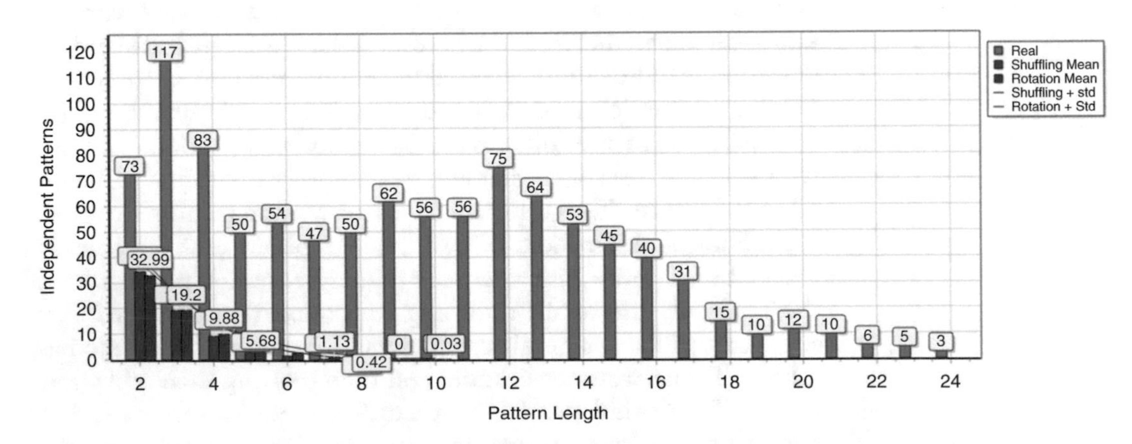

Fig. 8 This chart shows in green the number of different patterns of each length detected in the dyadic interaction of two toddlers playing with a picture viewer. In *blue* are the corresponding averages over 100T-shuffling randomizations and detections using the same search parameters and the *blue line* shows the average +1 standardization. In *red* are corresponding numbers for T-rotation, the other kind of randomization used. (Figure created with THEME™)

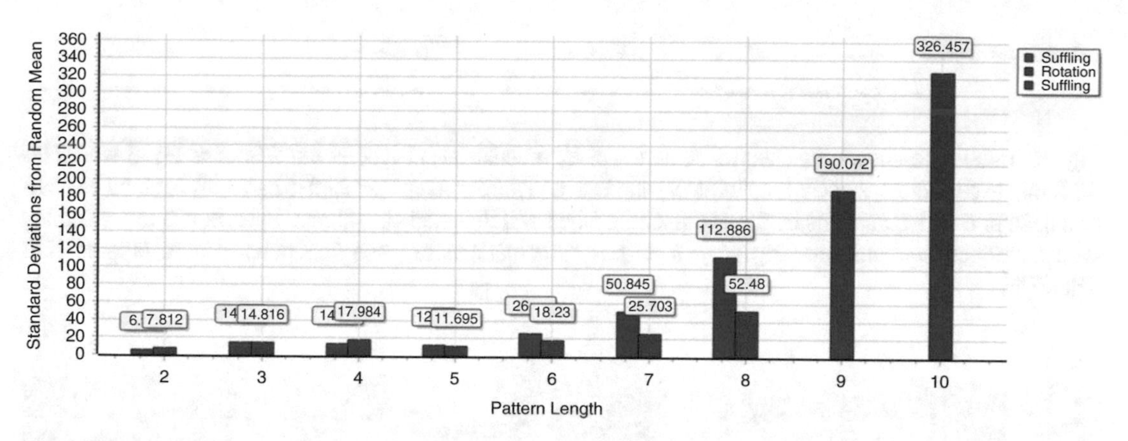

Fig. 9 For the 13.5 min dyadic object play interaction this shows for all detected T-patterns the standardized differences between the number of patterns detected in the data and the average number detected over 100 repetitions of, respectively, T-shuffling (*blue*) and T-rotation (*red*)

8.3 Derived and Related T-Concepts

The following sections concern primarily definitions of T-pattern extensions or aspects that are a part of the T-system.

8.4 T-Bursts

Bursts or bouts of some behavior occur when a number of its consecutive occurrences occur with shorter distances between them, that is, with a higher rate, than average. To capture these scale independent phenomena, a T-Burst is a special kind of (univariate) T-Pattern relating a series by a fast CI $[0, d]$ with itself lag-one. When such a fast relation exists in the series, consecutive points separated by a distance $\leq d$ are joined to form each T-Burst occurrence. Note that a T-Burst may be detected even if it occurs only once and possibly only has *three points*, which also defines the smallest T-Data that can be meaningfully analyzed with THEME™; a single T-Series with only three points within an interval $[1, T]$. Note, however, that the number of data points is T, the number of occasions for an event-type to occur or not.

T-bursts sometimes can highlight and make important use of the tendency of some behaviors to occur in bursts and sometimes with different effects from those expected from single occurrences (Figs. 10, 11, 12, and 13). For example, a rapid repetition of the same command may lead to a reaction different from a single one. As when simply saying "go" versus a rapid "go, go, go, go, go" indicating much urgency. An example where detecting T-bursts provides much increased predictability of following behavior is shown in Fig. 12.

8.5 Towards Grammars and Predictions

The T-patterns in a T-corpus describe different aspects of structure in T-data. Some T-corpus subsets may be related in various ways and may also contain somewhat redundant information. Summarizing in some way the information in a T-corpus may thus

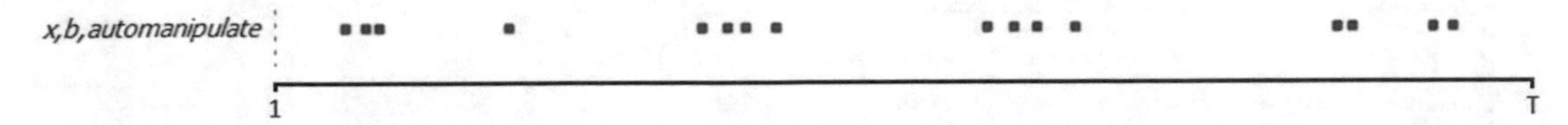

Fig. 10 This figure shows the occurrence series of a particular event-type in the T-Data of Fig. 5 where for 13.5 min two toddlers, x and y, take turns playing with a picture viewer. The point is here only to show an obvious case of a series contains bursts of the kind defined and detected as T-bursts. Here, bursts of x's fiddling with something without watching what it is (i.e., "automanipulate"; see Appendix). (Figure created with THEME™)

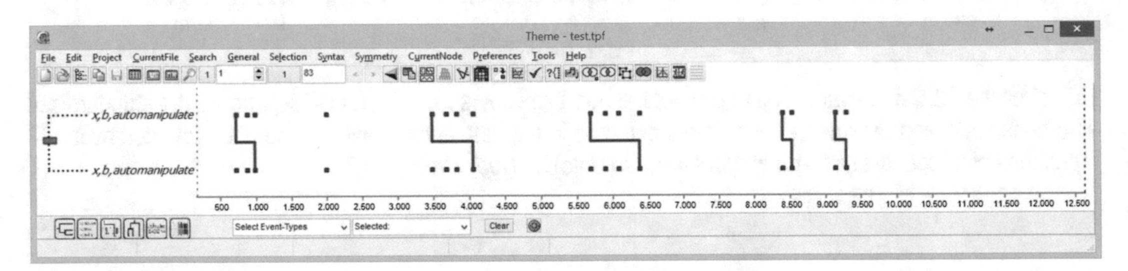

Fig. 11 This shows how Theme has identified and presented the bursts in x,b,automanipulate shown in Fig. 10. In Theme, a T-burst, a special relation between a series and itself, is shown as a special kind of T-pattern where the series is shown twice and the first and last point of each burst are connected. In this way, T-bursts are easily included in more complex T-patterns, which may or not themselves occur as T-bursts or contain other T-bursts. Such relations may reach any level found in the data. (THEME screen capture)

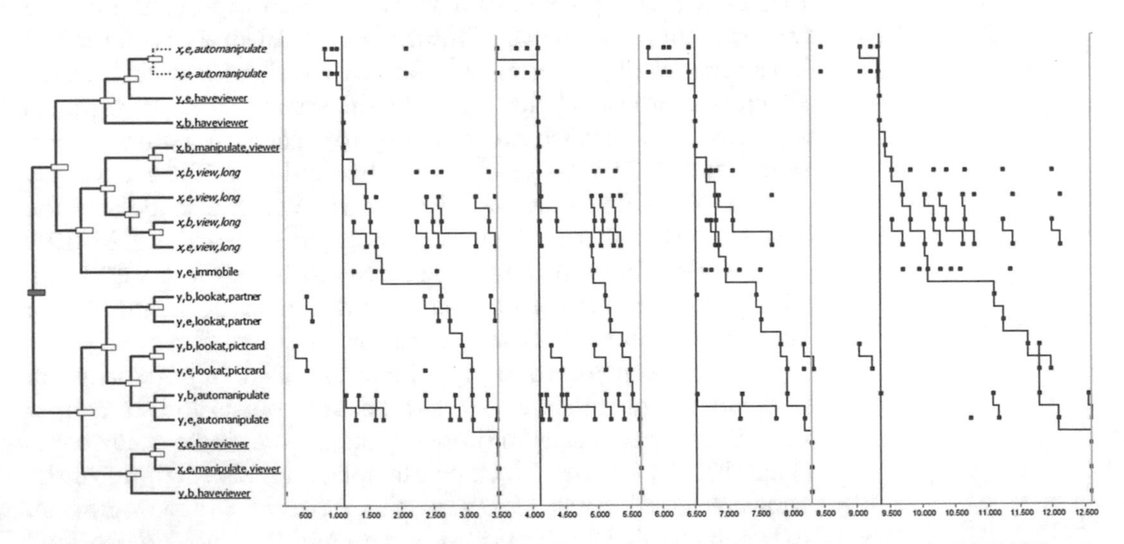

Fig. 12 This figure illustrates how a burst of some behavior (here x,e,automanipulate) can provide 100% prediction of a long following pattern while individual occurrence of that behavior cannot. Bursts may each contain a high number of repeated occurrences of the same behavior,so prediction gain based on first detecting bursts may increase from a very low level, for example, a few percent, to 100%, that is, from almost no predictive power to the maximum. The periods where y has the viewer begin at *green vertical lines* and end at *red vertical lines*. *X* thus has the viewer from *red* to *green x lines* (overall much longer than y). *X*'s burst of x,e,automanipulate thus strongly predict the switch of the viewer to her from y. Moreover, it can be seen that these bursts in x's behavior only occur, and almost all the time when the other, y, has the viewer. (Figure created with THEME™)

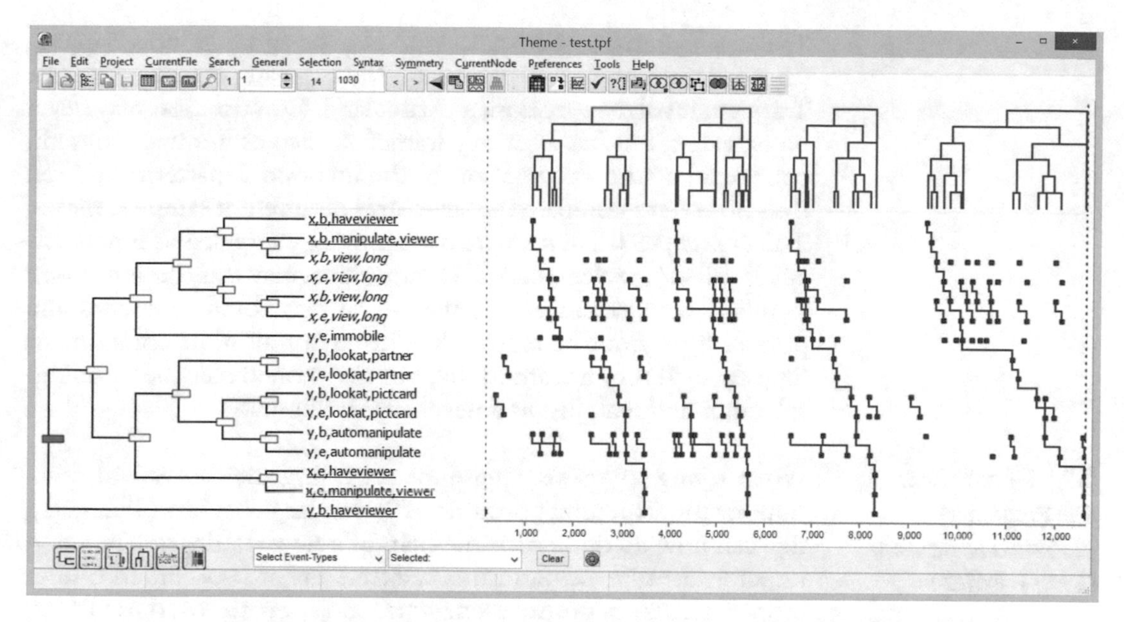

Fig. 13 This T-pattern diagram is similar to Fig. 7 except that it only shows the T-patterns while *x* has the viewer. (THEME screen capture)

be highly valuable and has already been done through the construction of flow-charts [47], which may also be seen as an important step towards the still evasive goal of automatic grammatical inference [8]. However, some concepts and tools facilitating this task have been developed and implemented in the Theme software. Among these, *T-composition* focuses on major splits or alternations within an observation period (T-period), while *T-markers* focus on key event-types or "organizers" [37, 48] or hubs, such that when they occur it is usually as parts of one or more T-patterns and thus indicate that these are ongoing. *T-predictors* (and T-retrodictors) concern predictive relations (that is, forward predictions, but with backward or "retro-diction" also possible) between T-event-types and/or T-patterns occurring together within more complex T-patterns.

8.6 T-Markers

An *x*% T-marker of T-pattern Q is behavior (event-type or pattern) such that *x*% of its instances are involved in occurrences of *Q*. In this sense, an instance of the marker indicates with *x*% probability that *Q* is ongoing. Thus, depending on the marker's position within the pattern, it predicts what may follows and/or precedes it. The focus here is thus on the occurrence of behavior indicating with some certainty that one or more particular T-patterns are ongoing.

8.7 T-Composition and Modularity

A stream of behavior, here as T-data, may be composed of modules without a hierarchical relation between them, some occurring in parallel and independently while others may be mutually exclusive. T-composition concerns the latter case and is a description of

T-data as alternating instances of two or more nonoverlapping (mutually exclusive) T-patterns. While most concepts in the T-system involve repetition, a particular T-Composition may never be repeated, but has a set of parameters such as number, content, complexity and total duration of the involved T-patterns and can thus be used to compare the structures of different samples, that is, T-data sets [49]. *Note*: (total duration/*T*) expressed as a percentage is called T-coverage. A T-Composition may thus itself not be a standard component at all, that is, just as some sentences and phrases it may not be repeated while any or all of its components may. As such they are anecdotal, but may nonetheless be meaningful and functional, just as a never repeated phrase.

8.8 T-Prediction: The Prediction of Particular Behaviors Using T-Patterns

When trying to answer questions regarding the causes and functions or the structural positions of particular behaviors (T-targets), this can now be done with the help of a new feature in Theme. A target event-type or pattern is selected and if it occurs in one or more patterns after other event-types and/or patterns, these 1–100 % T-predictors are automatically identified (see Fig. 14). The percentage is the forward or conditional probability of a particular T-target occurring after the T-predictor within a T-pattern given the number of occurrences of the predictor pattern in the data.

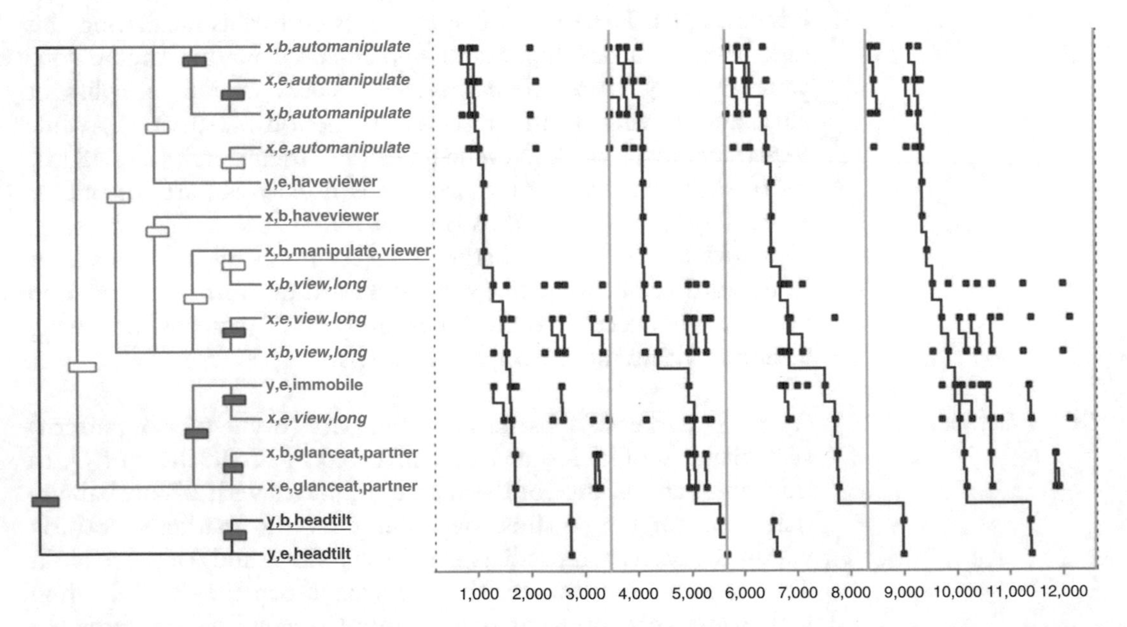

Fig. 14 This T-pattern diagram shows the result of using Theme to search through all the T-patterns detected in the particular T-data (see Fig. 5) for the behavior (event-type or pattern) that best predicts a head tilt in one of the two toddlers, *y*. The pattern shown in green was the best T-predictor found T-predicting 100% the occurrence of y's beginning and ending a head tilt as seen in this detected T-pattern. (Figure created with THEME)

The focus here is thus on the prediction (with retro-diction also possible) of the occurrence of particular behaviors and patterns within particular time distance windows relative to each other. Other structural concepts of the T-system are T-packets with +/- T-Associates and +/– T-zones, which have not been applied in the studies reported here, but are described elsewhere (e.g., [5, 8]).

8.9 Pattern Aspects for the Detection of Experimental Effects

As can be seen above, each T-Pattern occurrence has parameters and so do its occurrences taken together. The set of all T-Patterns detected in a T-Data, called a T-Corpus, also has such parameters and among those that most often have helped to distinguish between T-Corpuses, such as obtained from different subjects or (experimental) conditions, are the average and maximum lengths (m) and frequencies of their patterns. Some other useful descriptive parameters of a T-Pattern used, among other, in the chapters of this book are, for example:

- Number of actors: the number of actors in the terminal sequence (or string). Patterns with a single actor are called intra-individual, but those with more than one are called multi-individual or interactive patterns.
- Number of actor switches: the number of times through the terminal sequence $X_1..m$, the actor of Xi_{+1} is different from Xi.

8.10 T-Pattern Analysis (TPA) and Multivariate Statistics

For a quick view of the differences between TPA and standard multivariate statistical methods, it is useful to think of TPA as a multivariate statistical method with T-Data, of duration T, as a multivariate binomial data table with T rows and one column for each series (event-type). One essential differencewhich immediately sets TPA apart from most multivariate analyses is that the latter relates values within the same row and in special cases with a fixed row distance, while TPA automatically considers any distance ranges through its critical interval detection. Another difference is that TPA adds detected relations (patterns) to the data and thus becomes multiordinal, allowing that the repeated patterns and patterns of patterns, etc., which it detects, are entities occurring in time rather than abstract relations such as found through, for example, Factor Analysis, Multidimensional Contingency Table Analysis and Multivariate Regression Analysis of time series. Markov Chain analysis does not either have the equivalent of critical interval detection and the multilevel hierarchical and self-similar patterns of patterns branching structure. The same is true for Lag Sequential Analysis.

The kind of multivariate analysis that superficially most resembles TPA is Hierarchical Cluster Analysis as suggested by the similarity of its hierarchical tree (or nesting) diagrams and those of T-patterns, but the differences are essential as indicated by the multilevel point connections part of T-Pattern diagrams. (*Note*: that for

TPA, similarity does not mean belonging to the same class as it is not about classifying known entities or patterns, but rather gradually discovering them.) However, TPA could probably be considered as a special multiordinal hierarchical cluster analysis with critical interval detection providing a kind of similarity measure (and with an added evolution algorithm). A T-pattern detection algorithm of this nature has been implemented by Vishnevskiy et al. [50]. They also implement an improvement concerning optional pattern elements (decades overdue in Theme to be implemented when resources allow using a different and simpler method), but as they point out, at the price of, among other, some false detections and far more complex and slow statistical computations:

> Due to method's statistical roots, some patterns can be treated as noise. Also our method is computationally complex. The current version, implemented on MATLAB works approximately 100 times longer, than the algorithm, based on T-Pattern detection. One of the directions for future work is parallel implementation of the algorithm on multiprocessor computers or on Graphical Processing Units (GPUs).

This also serves as a reminder that among fundamental design criteria for the T-pattern detection algorithms in Theme have always been transparency and availability through use of mathematically simple statistics and programs for small and common computers, first PDP computers and then PCs since they appeared. Regarding speed, the results, presented in the first Theme paper in 1981 [3] demanded 6 h of computation on a PDP 8 (after countless debugging runs of similar duration) whereas Vishnevskiy's 100 times slower algorithm would have required some 600 h or nearly a month for each run. The latest T-pattern detection algorithm (in Theme 6) is about 50 times faster than the one they compared with, so the same task now takes around 15 s on a normal PC or 5000 times faster than Vishnevskiy's implementation.

8.11 Looking for New "Simple, Elegant, and Powerful Ideas"

As the T-Pattern concept has been applied, elaborated, and complemented to form the T-system, ideas in various areas have been explored and some are here noted as they seem particularly relevant and promising for the continued development: symmetry, self-similarity, and fractals.

8.12 Symmetry

Symmetry is still an unusual concept in behavioral research, but of paramount importance when considering patterns in other sciences as stated by Livio in his book about the young French mathematical genius Galois:

> …, symmetry is the paramount tool for bridging the gap between science and art, between psychology and mathematics. It permeates objects and concepts ranging from Persian carpets to the molecules of life, from the Sistine Chapel to the sought-after Theory of Everything. ([51], p. 2.)

And in the words of the eminent mathematician and popular science writer Ian Stuart:

> Symmetry is not a number or a shape, but a special kind of *transformation* – a way to move an *object*. If the object *looks the same* after being transformed, then the transformation concerned is a *symmetry*. For example, a square looks the same if rotated through a right angle. This idea, much extended and embellished, is fundamental to today's scientific understanding of the universe and its origins. ([52], p. ix. Emphasis added.)

A particularly relevant kind of symmetry is here the so called *translation* symmetry. If after being moved (in space or time), a pattern (anything) looks the same, it has translation symmetry under the operation of such movement, with exact sameness as a mathematical ideal, while in nature sameness can only be approximate. Thus T-Patterns may be seen as structures that show translation symmetry under recurrence (movement) in time or space. This also explains how a T-pattern with only two types of elements (A and B) each occurring only twice, can be highly significant, while, among other, no particular order permutation of the four elements can be significantly different from random. The focus on translation symmetry rather than on order (sequence, contingency etc.) alone, becomes possible using the information about the positions of the elements on a real scale (mostly discrete real-time). Thus even with A and B each occurring only twice, it is still possible to estimate how likely it is that they would, simply by chance, occur twice in the same observed order *and* with the observed distances between them, which may be very similar or identical. From this perspective, the number of data points is not four, but rather for each event-type as many as there are discrete time units in the observation period, where it either occurs or not. So for an observation period of length, T, there are here $2 \times T$ data points. This might suggest that given an almost infinite temporal resolution, that is, an almost infinitely small basic (discrete) time unit, infinite precision can be obtained, but this would require that the locations of event-types could be decided with the same nearly absolute precision, which, of course, is not realistic if only because real events have somewhat fuzzy limits, so it is not 100 % clear exactly when they begin (and/or end). However, for example, given a time location precision of 10^{-6} and an observation period of 10^{6} time units, where, for example, “A then B” occurs twice with exactly the same number ($>=0$) of time units separating them, it is possible to calculate how likely (extremely unlikely) this were to happen by chance were B distributed evenly, randomly and independently of A. The same is true for any particular variation from this exact distance between A and B. The question thus becomes whether the “A then B” structure is showing translation symmetry within some criterion of approximation. For T-Patterns, the degree of likeness is evaluated through the critical interval relation.

8.13 Fractals

With ancient roots, the mathematical concept of fractal is primarily a concept of the latter half of the twentieth century attributed to the mathematician Benois Mandelbrot [53] and is of great importance for the mathematical representation of natural phenomena. While purely mathematical fractals are infinite, natural (physical) ones are not and are thus called *pseudo-fractals.*

> Physical fractals typically display statistical self-similarity over scales differing by just a few factors of 10. Nonetheless, as Mandelbrot first observed, it's extremely useful to recognize the fractal properties of natural shapes. ([54], p. 293)

A multitude of well-known branching structures such as, for example, various natural trees or the arteries in lungs are examples of statistical (approximate) pseudo-fractal *objects.* As each branch of a T-Pattern, is also a T-pattern, it is approximately (statistically) *self-similar,* making the occurrences of a particular T-pattern those of a particular recurring *statistical pseudo-fractal object.* It may be noted that self-similar suggests (translational) symmetry under the operation of simply changing the scale, as the pattern again is approximately the same only smaller (or bigger), just as a triangle remains a triangle independently of variation in size and angles, making it look different in some ways but not regarding its triangle-ness. This is true for detected T-patterns, but staying within the limits set by its critical interval relations, the variation can appear quite big over the occurrences of a particular T-pattern (see Fig. 15).

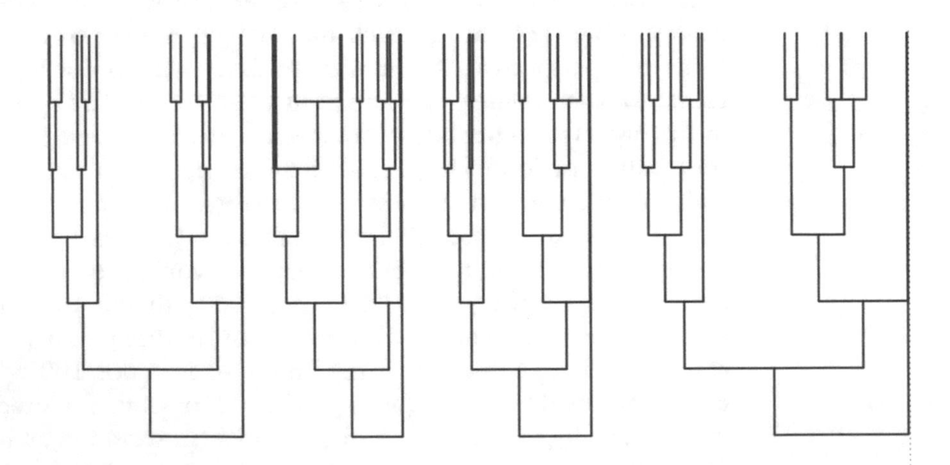

Fig. 15 This figure shows (upside down) only the dynamic T-pattern tree structure shown in Fig. 7. The four distinct pattern occurrences can be seen as consecutive pictures taken with some intervals between them of the same tree in some wind making the branches swing, a case of translation symmetry. Generally the bigger branches can move more and their distances to others change more. Small branches near the top of the same big branch will stay relatively close, while the distances between those at the top of different big branches may vary much more. This explains why search algorithms that assume all distance variations (using an interval size fixed a priory) between all parts to be the same when a large pattern repeats will normally fail if the pattern has a fractal or tree structure and recurs in some "wind". (Figure created with THEME)

9 Discussion

The various terms of the T-System and in particular the T-pattern concept, have here been defined in a fairly formal manner, but are sometimes better communicated and explained by reference to the apparently obvious everyday observation that the instances of various behaviors sometimes occur "together," concurrently and/or sequentially, as pairs, triplets, quadruples, and combinations of these, etc. with predictable time distances between the consecutive n-tuples ($n>=1$), and that these approximately timed combinations occur more often than expected if each part occurred independently of the others (e.g., a greeting, a dinner, or a frequent verbal expression in terms of its main parts.) The repeated patterns are thus typically hierarchically composed of patterns that also occur independently and possibly as parts of different T-Patterns as, for example, in speech and text where even meaningless components such as sub-meaning vocal tract movements or, in text, vertical and horizontal segments and curves, combine into spoken or written phonemes or letters that again combine into meaningful words and phrases. Seeking objective information, the task here is to start by obtaining information about the locations of repeated simple elements, possibly below the level of meaning or function, and then searching for repeated higher order patterns, which may reach the level of "meaning" or function, while some that don't in any obvious way, may still be useful, for example, for diagnostic purposes or for the detection of any experimental effects.

9.1 Physical and Temporal Connectomes

We meet mathematics again as a unifying language in the modern study of the neuronal connectome of the brain firmly based on Graph Theory, the mathematics of networks and providing already this interesting estimation:

> Given the estimated number of neurons ($\approx 10^{11}$) and number of synaptic connections ($\approx 10^{15}$) in the human brain, a complete map of the connectome at the micro scale would be extremely sparse – fewer than one in a million (less than one ten thousandth of a percent) of all synaptic connections actually exist. ([15], p. 41)

Similarly, one wonders what tiny proportion of all mathematically possible temporal T-patterns (connections) actually exists in neuronal (and human) behavior and interactions, but it seems likely to be even smaller.

9.2 Neuronal vs. Human T-Patterns

TPA finds somewhat more complex patterns in human interactions than in the neuronal interactions ([10, 11], and Nicol et al in this volume). One possible reason is that while all the neurons registered are probably at the same organizational level, the human behavior categories commonly used, for example, anything from eyeblinks to locomotion, are probably not. As the neuronal connection networks of brains are nested hierarchies with connection hubs at various levels [15], including the activation moments of hubs at various nesting levels might give rise to T-Patterns more like those found in human behavior.

9.3 Big Data

Automatic data collection from social media, mobile devices and other kinds of electronic interactions, has made colossal data volumes common place and is leading to a new field of research: social physics [55]. Consequently, much attention is directed towards mining the gigantic data volumes constantly piling up at a rate never seen before in human history. TPA was not developed for such data mining and has to date not been applied in that area. The biggest data analyzed with TPA to date are minuscule in comparison, that is, data collected automatically from microchips implanted in living brains simultaneously recording the firing of hundreds of individual neurons with a temporal resolution of a few millionth of a second during trials lasting tens of seconds and producing hundreds of data files each with tens of thousands of spikes from hundreds of neurons. So even if TPA with Theme was rather developed to deal with the analysis of tiny data it has been applied successfully to such fairly substantial neuronal data [10, 11], which are, however, dwarfed by modern big data.

9.4 Tiny Data

Given the current attention to data mining in big data it may be forgotten that statistical analysis of tiny data, possibly only a few instances of a few types of events, can also be a great challenge as most statistical methods were not developed for such data. On the other hand, the T-Pattern model and algorithms were, from the very beginning developed for meaningful quantitative and structural analysis of tiny data, for example, from dyadic interactions lasting only a few minutes, where a single complex interaction pattern possibly occurring only twice can be a highly significant T-Pattern detectable by TPA, but beyond the reach and intended use of available statistical methods. It is worth noting that some types of events will by nature remain be very rare and thus only provide tiny data, such as, for example, tsunamis, the melt down of nuclear power plants, major economic system crashes, earthquakes and volcanic eruptions or the collapse of states and political blocks.

9.5 Social Physics

In 2007, Buchanan writes in his book "The Social Atom":

> A few years ago, I worked as an editor for *Nature*,… I noticed then that the authors of some of the research papers … were seriously seeking to find mathematical regularities in the human world of the kind known to physics; … finally beginning to take Schelling's way of thinking seriously. Since then, an explosion of modern research in what I would like to call "social physics" has convinced me that we stand at an important moment in history. We're now witnessing something akin to a "quantum revolution" in the social sciences. ([56], p. x).

It seems that the approach described and applied in this book may fit in this evolution towards a more mathematical or "social physics" approach to the study of everyday behavior and interactions, while attempting to understand in similar mathematical terms the behavior and interactions of organisms and other biological entities from proteins in Cell City (Note 4), to neurons in brains, to humans in cities and societies.

In our current extremely unequal "world of the 1 %" that nobody seems to want, "social physics" may help dealing with such unwanted "mathematical" twists in nature:

> Even when everyone is content with racial integration, but just not wanting to be a part of an extreme minority (for example, 30 %), this leads to segregation even though no single individual would have wanted this". ([56], p. viii). {Buchanan here referring to Schelling's [57] famous segregation (simulation) experiment.}

9.6 Breaking the Ice

In the first years of TPA with Theme, in the early 1980s the ice to break was quite thick especially among some statisticians, who were often automatically considered experts when numbers were involved. The analysis of data in standard matrix format (and rarely involving time series) such as obtained from, for example, questionnaires, were in great demand and close to the minds of most statisticians with excellent adequate tools in easily available statistical packages.

In continuation of the already mentioned collaborations and support, one important ice-breaker was a series of doctoral thesis (see [9]) directed by my long-time (mostly 1983–1993) collaborator Janine Beaudichon director of the (Binet) Psychological Laboratory in the old Sorbonne while also vice-president of the University of Paris V. Another icebreaker was an invitation in 1993 from our collaborator Teresa Anguera, then director of the Department for Methodology for the Behavioral Sciences at the University of Barcelona, to publish in the European Journal of Psychological Assessment, where she was then a guest editor [4]. A lesson learned from these earlier years is that it is not always sure that a paper containing some mathematics and computation is best understood any statistician or computer scientist, among other some knowledge of behavioral science may also be required as well as some general openness to new approaches. Another lesson learned is that easily available tools may sometimes disproportionally influence the choice of analytical approach and even research, a fate that TPA will hopefully never befall.

9.7 From DNA in Cells, to Cells in Brains, to Brains in Cities

Findings reported in this book and in earlier publications (e.g., [9, 58]) indicate that T-patterns are salient features in the temporal organization of behavior and interactions at very different time scales with time units varying from approximately the order of s^{-6}

to s^5 (days) among organisms from brain neurons ($\geq m^{-6}$) to a humans [10, 11, 59]. It seems fairly obvious, even if no empirical analyses confirm this yet, that T-patterns exist at far larger scales such as well-known and named patterns repeated throughout history (uprisings, wars, elections, great games and strikes) all apparently sharing the essential T-pattern characteristics of repeated structured hierarchical self-similar clusters in time and/or space.

Ending on an even more speculative note, there is apparently a structural similarity of behavioral T-patterns and the various repeated patterns, such as genes, on the physical DNA molecule [7]. Even more surprisingly there also seem to be functional analogies between various behavioral T-patterns in human societies and genes and proteins within the cell (sometimes called "Cell City," *see* **Note 4**). Thus to meet particular intra-cellular needs particular genes (DNA T-Patterns) may for a while be expressed more frequently leading to more copies of the corresponding protein molecules and in human cities to facilitate some processes, the frequency may be increased of some T-patterns such as talks, sessions and dinners.

In any case, TPA, that is, the T-system and corresponding software, may hopefully continue to find some constructive use for researchers in various fields and with some luck be scrutinized and improved possibly beyond recognition by talented individuals with fresh energy and time.

10 Notes

1. The following is an example of T-Class vs. T-Item (or nominal variable vs. nominal value) correspondence table as used in the Theme software. Class names begin in column 1 and its items in column >1. (Here classes are also shown in upper case and items in lower case.) Such tables are provided for each set of data files as simple .txt files, called vvt.vvt, where vvt stands for "variable value (correspondence) table".

```
ACTORS
    bill
…
 jack
B_E {for beginning or ending}
b
e
POSTURE
    stand
    sit
    …
    lie
LOCOMOTION
```

walk
crawl
...
order
HEAD
headtilt
...
nod
toungout
HANDS
reach
manipulate
...
lay
automanipulate
REACH
up
down
LOOKS
lookat
...
glanceat
FOCUS
viewer {a toy}
...
partner
DURATION
short
long
EXTENT
full
OTHER
haveviewer

2. The following is a part of a T-Data file as entered into Theme, that is, as a simple tab delimited .txt file, here based on the full version of the variable vs. value correspondence table of the VVT file in Note 1. The *Time* variable, always an integer, indicates the time in the basic time unit of the time scale used (e.g., microsecond, second, hour, or day). The *Event* variable indicates the type of event observed at the time unit (unit interval) on the left.

Note that while each event-type either occurs or not at any discrete time point (unit interval), any number of different event-types may occur at the same discrete time point. Hundreds of such files can be analyzed in a single Theme run, either independently, in various combinations corresponding to levels of experimental variable or all combined. The minimal size of data that may be meaningfully analyzed by Theme is a single T-data file with a minimum of one T-series with a minimum of three occurrences of a single T-event-type. Such data may allow the detection of a single T-burt occurring once, but a T-Data with a single series with two occurrences of each of two event-types may allow the detection a two occurrences of a T-pattern.

Time	Event
155	: {means observation begins}
155	y,b,haveviewer
155	y,b,kneel
155	y,b,manipulate,viewer
161	y,b,lookat,viewer
211	y,e,lookat,viewer
214	y,b,view,long
318	y,e,view,long
319	y,b,lookat,viewer
369	y,e,lookat,viewer
370	y,b,lookat,pictcard
424	y,b,sit
424	y,e,kneel
519	x,b,walk
541	x,e,walk
….	
12603	y,b,haveviewer
12611	y,b,manipulate,viewer
12612	y,b,glanceat,partner
12615	y,e,automanipulate
12622	y,e,glanceat,partner
12624	y,b,sit
12624	y,e,kneel
12628	y,b,glanceat,partner
12639	y,e,glanceat,partner
12639	y,e,sit
12640	& {means observation ends}

3. Regarding Cell City see, for example, http://www.open.edu/openlearn/profiles/guest-218

Appendix: Meaning of Codes

Toddlers X and Y playing with a viewer

Categories mostly those of McGrew [34]

Auto manipulate: touching something without looking at what is being touched.

View,long: viewing something for more than 3 s.

Glance at partner: a short glance at the other.

x or y, b or e, haveviewer: the viewer begins or ends being on the side of, respectively, x or y.

Immobile: being completely still, silent and not moving at all

Headtilt: tilting the head clearly to one side; a category much studied by Montagner [24] with results providing much inspiration for the present methodological development of T-patterns and Theme. It seems to function often as a part of effective begging behavior and/or be soothing for the receiver.

References

1. Devlin K (1997) Mathematics: the science of patterns: the search for order in life, mind and the universe. Scientific American Library
2. Posamentier AS, Lehmann I (2007) The fabulous Fibonacci numbers. Prometheus Books, New York, NY
3. Magnusson MS (1981) Temporal configuration analysis: detection of a meaningful underlying structure through artificial categorization of a real-time behavioral stream. Workshop on artificial intelligence Uppsala University. Part of a 1983 thesis at the Psychological Laboratory, University of Copenhagen. Communications. http://hbl.hi.is
4. Magnusson MS (1996) Hidden real-time patterns in intra- and inter-individual behavior: description and detection. Eur J Psychol Assess 12(2):112–123
5. Magnusson MS (2000) Discovering hidden time patterns in behavior: T-patterns and their detection. Behav Res Meth Instr Comput 32(1):93–110
6. Magnusson MS (2004) Repeated patterns in behavior and other biological phenomena. In: Oller DK, Griebel U (eds) Evolution of communication systems: a comparative approach. The MIT Press, Cambridge, pp 111–128
7. Magnusson MS (2005) Understanding social interaction: discovering hidden structure with model and algorithms. In: Anolli L, Duncan S Jr, Magnusson MS, Riva G (eds) The hidden structure of interaction: from neurons to culture patterns. Volume 7. Emerging communication: studies in new technologies and practices in communication. ISBN: 1-58603-509-6, approx. p 300
8. Magnusson MS (2006) Structure and communication in interaction. In: Riva G, Anguera MT, Wiederhold BK, Mantovani F (eds) From communication to presence. Ios Press, Amsterdam, pp 127–146
9. Casarrubea M, Jonsson GK, Faulisi F, Sorbera F, Di Giovanni D, Benigno A, Crescimanno G, Magnusson MS (2015) T-Pattern analysis for the study of temporal structure of animal and human behavior: a comprehensive review. J Neurosci Methods 239:34–46
10. Nicol AU, Kendrick KM, Magnusson MS (2005) Communication within a neural network. In: Anolli L, Duncan Jr S, Magnusson MS, Riva G. ISBN: 1-58603-509-6, approx. p 300
11. Nicol AU, Segonds-Pichon A, Magnusson MS (2015) Complex spike patterns in olfactory bulb neuronal networks. J Neurosci Methods 8:4. doi:10.1016/j.jneumeth.2014.09.016
12. Rees M (1999) Just six numbers: the deep forces that shape the universe. Weidenfeld & Nicolson, London
13. Schumm BA (2004) Deep down things: the breathtaking beauty of particle physics. Johns Hopkins University Press, Baltimore, MD
14. Baryshev Y, Teerikorpi P (2002) Discovery of cosmic fractals. World Scientific Publishing, Singapore
15. Sporns O (2012) Discovering the human connectome. The MIT Press, Cambridge, MA
16. Eibl-Eibesfeldt I (1970) Ethology, the biology of behavior. Holt, Rinehart and Winston, Austin, TX
17. Chomsky N (1957) Syntactic structures. De GruyterMouton, Mouton
18. Bateson PPG, Hinde RA (eds) (1976) Growing points in ethology. Cambridge University Press, Cambridge
19. Skinner BF (1969) Contingencies of reinforcement: a theoretical analysis. Appleton Century-Crofts, New York, NY
20. Miller GA, Galanter E, Pribram KH (1960) Plans and the structure of behavior. Henry Holt & Co., New York, NY
21. McNeill D (2012) How language began: gesture and speech in human evolution. Cambridge University Press, Cambridge
22. Pike KL (1967) Language in relation to a unified theory of the structure of human behaviour. Mouton, The Hague
23. Birdwhistell RL (1970) Kinesics and context: essays on body motion communication. Ballantine Books, New York, NY

24. McQuown N (ed) (1971) The natural history of an interview. Microfilm collection of manuscripts on cultural anthropology. Microfilm Department, Joseph Regenstein Library, University of Chicago, Chicago, IL, pp 95–98
25. Colgan PW (ed) (1978) Quantitative ethology. John Wiley & Sons, New York. NY
26. Monge PR, Cappella JN (eds) (1980) Multivariate techniques in human communication research. Academic, New York, NY
27. Bakeman R, Gottman JM (1986) Observing interaction: an introduction to sequential analysis. Cambridge University Press, Cambridge
28. Lewin K (1935) A dynamic theory of personality. McGraw Hill Book Company Inc., New York, NY
29. Tinbergen N (1963) On the aims and methods of ethology. Z Tierpsychol 20:410–433
30. Tinbergen N (1966) Animal behaviour. Time-Life International Publ, New York, NY
31. Cosnier J (1971) Clefs pour la psychologie. Segher, Paris
32. Plety R (1985) Ethologie de l'interaction chez des enfants du 1er cycle de l'enseignement secondaire au cours d'un apprentissage des mathématiques en groupe dans la résolution de problèmes. Doctoral thesis (thèse d'état ès sciences), University Claude Bernard Lyon 1
33. Blurton Jones N (ed) (1972) Ethological studies of child behavior. Cambridge University Press, Cambridge
34. McGRew WC (1972) An ethological study of children's behavior. Academic, New York, NY
35. Montagner H (1971) Les communications interindividuelles dans les sociétés de guêpes. Journal de Psychologie normale et pathologique : intercommunications chez les animaux, nos. 3–4. Presses Universitaires de France, Paris, pp 281–296
36. Montagner H (2012) L'enfant et la communication: comment gestes, attitudes, vocalisations deviennent des messages. Collection: Enfances, Dunod 2012. EAN13:9782100577309, p 320
37. Montagner H, Magnusson M, Casagrande C, Restoin A, Bel JP, Nguyen Hoang P, Ruiz V, Delcourt S, Gauffier G, Epoulet B (1990) Une nouvelle méthode pour l'étude des organisateurs de comportement et des systèmes d'interaction du jeune enfant. Les premières données La Psychiatrie de l'Enfant 33:391–456
38. Filiatre J-C (1986) Contribution à l'étude des systèmes de communication intra et interspécifiques chez un canide, Canis familiaris. Doctoral thesis, Faculty of Science and Technology, University of Franche-Comté, Besançon, France
39. Casagrande C (1995) Organisation des interactions sociales dyadiques de nourrissons de 4/5 mois. Contribution a une nouvelle méthode d'étude. Doctoral thesis in biological and fundamental sciences, psychology. University of Franche-Comté, Besançon. www.theses.fr/1995BESA2058
40. Chomsky N (1959) Review of skinner, verbal behavior. Language 35:26–58
41. Magnusson MS (1975) Social organization and communication in social insects and primates (humans included). B.A. thesis (in Danish) at the Psychological Laboratory, University of Copenhagen, p 80. http://hbl.hi.is
42. Magnusson MS (1978) The human ethological probabilistic structural multivariate approach to the study of children's nonverbal communication. Thesis (in Danish), University of Copenhagen's Silver Medal, University of Copenhagen Library. http://hbl.hi.is
43. Haynal-Reymond V, Jonsson G, Magnusson M (2005) Non-verbal communication in doctor-suicidal patient interview. In: Anolli L, Duncan S Jr, Magnusson MS, Riva G (eds) The hidden structure of interaction: from neurons to culture patterns. Ios Press, Amsterdam
44. Merten J, Schwab F (2005) Facial expression patterns in common and psychotherapeutic situations. In: Anolli L, Duncan S, Magnusson MS, Riva G (eds) The hidden structure of interaction: from neurons to culture patterns. Ios Press, Amsterdam
45. Duncan SD Jr, Fiske DW (1977) Face-to-face interaction: research, methods and theory. Lawrence Erlbaum Associates, Hillsdale, NJ
46. Duncan SD Jr (1973) Toward a grammar for dyadic conversations. Semiotica 9:29–46
47. Hardway C, Duncan S Jr (2005) Me first! Structure and dynamics of a four-way family conflict. In: Anolli L, Duncan S, Magnusson MS, Riva G (eds) The hidden structure of interaction: from neurons to culture patterns. Ios Press, Amsterdam
48. Magnusson MS, Beaudichon J (1997) Detection de "marqueurs" dans la communication referentielle entre enfants. In: Bernicot J, CaronPargue J, Trognon A (eds) Conversation, interaction et fonctionnement cognitif. Presse Universitaire de Nancy, Nancy, pp 315–335
49. Bonasera SJ, Schenk AK, Luxenberg EJ, Tecott LH (2008) A novel method for automatic quantification of psychostimulant-evoked route-tracing stereotypy: application to *Mus musculus*. Psychopharmacology (Berl) 196:591–602
50. Vishnevskiy VV, Vetrov DP (2010) The algorithm for detection of fuzzy behavioral patterns.

In: Spink AJ, Grieco F, Krips OE, Loijens LWS, Noldus LPJJ, Zimmerman PH (eds) Proceedings of measuring behavior 2010. Eindhoven, The Netherlands, p 166

51. Livio M (2006) The equation that couldn't be solved: how mathematical genius discovered the language of symmetry. Simon & Schuster, New York, NY
52. Stewart I (2007) Why beauty is truth: the history of symmetry. Basic books, first trade paper edn
53. Mandelbrot BB (1982) The fractal geometry of nature. W. H. Freeman and Company, New York, NY
54. Kautz R (2011) Chaos: the science of predictable random motion. OUP, Oxford
55. Pentland A (2014) Social physics. The Penguin Press, New York, NY
56. Buchanan M (2007) The social atom: why the rich get richer, cheaters get caught, and your neighbor usually looks like you. Bloomsbury, New York, NY
57. Schelling TC (1971) Dynamic models of segregation. J Math Sociol 1:143–186
58. Anolli L, Duncan S, Magnusson MS, Riva G (2005) The hidden structure of interaction: from neurons to culture patterns. Ios Press, Amsterdam
59. Hirschenhauser K, Frigerio D, Grammer K, Magnusson MS (2002) Monthly patterns of testosterone and behavior in prospective fathers. Horm Behav 42:172–181

Chapter 2

Interactive Deception in Group Decision-Making: New Insights from Communication Pattern Analysis

Judee K. Burgoon, David Wilson, Michael Hass, and Ryan Schuetzler

Abstract

Interpersonal deception is a dynamic process in which participating individuals adjust and adapt their behaviors as the deception proceeds. Using THEME, we demonstrate that deceptive communication in group settings is highly patterned. We further examine patterning behavior using the strategy-focused lens of Interpersonal Deception Theory (Buller and Burgoon, Commun Theory 6(3):203–242, 1996). Correlation and regression analyses suggest that (1) deceivers tend to be strategically assertive as they carry out deception in group settings, and (2) individuals suspicious of deception tend to engage in probing behavior, ostensibly attempting to confirm their suspicions. Our findings demonstrate the value of analyzing deceptive behavior in terms of patterning to gain greater insight into the complex deception process.

Key words Deception, Group decision-making, Nonverbal communication, Pattern analysis, THEME

1 Introduction

Interpersonal deception is a complex and dynamic interaction between two or more people in which messages are knowingly sent by a deceiver in order to foster a false belief or conclusion by the receiver [1]. Deceptive interchanges, like other forms of interpersonal message exchange, are an iterative process of sending, receiving, and updating messages in response to the messages and feedback of interlocutors. During a deceptive exchange, deceivers attempt to manipulate the interchange in order to accomplish their goals while evading detection and may do so by resorting to a variety of different strategies, such as obfuscating, becoming reticent, or distancing themselves from what they are saying [2] or alternatively, taking a more assertive and persuasive approach in order to win over their targets [3, 4]. In turn, potentially deceived interactants, if made suspicious, may adopt strategies of their own to uncover the truth. For example, suspectors might make their suspicions manifest by probing for more information, or they might

Magnus S. Magnusson et al. (eds.), *Discovering Hidden Temporal Patterns in Behavior and Interaction: T-Pattern Detection and Analysis with THEME™*, Neuromethods, vol. 111, DOI 10.1007/978-1-4939-3249-8_2,

quietly monitor the target of their suspicion but maintain an impassive exterior that masks their own skepticism. Because interpersonal communication is an adaptive and dynamic process, the emergent communication between deceivers and their targets may take a variety of forms and evolve as the deception progresses. The strategic, adaptive, and dynamic nature of communication thus poses significant challenges to its analysis. Simple aggregate measures or analyses conducted in the opening seconds of an interaction may be highly misleading about the trajectory of a deceptive interchange, and what happens early may bear little resemblance to what happens later. This analysis becomes even more complex when applied to group interactions, inasmuch as the web of relationships and message exchanges increases exponentially the potentialities for message transmission and receipt. Senders may produce one-to-one or one-to-many messages to recipients who may or may not transmit feedback to the sender and who may opt to remain a passive observer or who may take up the sending role. These complexities may account for why knowledge of deception in group interaction is sparse [5]. Because of this challenge, the current investigation attempted to uncover regularities in deceptive communication episodes by applying the pattern discovery and analysis tool THEME [6, 7]. Specifically, we applied it to group interactions in which both deception and suspicion were present to determine whether this analytical tool would deliver more insight into deception in interpersonal and group interactions.

THEME is a commercially available program designed to detect subtle temporal patterns in a set of data. It can be used to discover and analyze any data that is composed of discrete events that are arranged according to some temporal indicator. This flexibility has allowed the software to be used in such diverse contexts as team interactions [8], behavior of autistic children [9], and family conflict [10]. The current investigation, in which one deceiver attempted to deceive two other group members, one of whom was induced to be suspicious and one of whom remained naïve to the deception, offered an excellent test bed to examine the utility of THEME to uncover patterning of interaction in nonverbal behavioral data identified by trained observers. The patterning uncovered by this analysis grants a novel angle from which to view the strategic and dynamic nature of group deception interactions.

In what follows, we first articulate the theoretical underpinnings and research questions guiding the investigation, as informed by Interpersonal Deception Theory (IDT; [1]). Next, we briefly describe the experiment and behavioral observation procedures from which we derive the coded data used by the THEME software. We then describe the THEME analysis approach before turning to our report of the results of our analysis within the context of our IDT lens. We end with a discussion of our findings, including some limitations of our approach and opportunities for future deception research using THEME.

It is important to note at the outset that both the participant sample size and the nonverbal behaviors that are coded are limited and far short of the ideal for either. Nevertheless this corpus—as one of the first derived from lengthy group interactions—offers the temporal scale and unconstrained discourse among group members to permit a first glimpse of THEME's ability to uncover interesting regularities in truthful, deceptive, and suspicious behavioral patterns.

1.1 Theoretical Background

As already noted, with rare exceptions (e.g., [5, 11, 12]), little prior research has considered deception in groups or its detection in groups [13], the lion's share of work having focused on dyadic deception [14]. For this investigation, our focus was on group interactions in which one deceiver attempts to deceive two other group members. This type of interaction is particularly well-informed by the perspective that deception is interactive and dynamic [1], since group communication processes are typically much more complex than dyadic interactions and require deceivers to manage the perceptions and communication of two or more receivers at once [11]. The demands on deceivers to manage this process, especially in the face of a suspicious interactant, are therefore likely to call forth diverse schemes for successfully achieving their ends while allaying suspicions. It is well understood that human communication tends to be highly patterned, and these patterns have been the subject of much empirical effort (e.g., [15–17]). In addition to turn-taking and similarly obvious communication patterns, interactions also follow patterns that are sometimes subtle and imperceptible to the casual observer [18]. These patterns, perhaps especially those which cannot be easily observed, could provide important insights, especially under the assumption that behavioral patterns are fluid over the course of a deceptive interaction [19–22].

IDT [1, 23] was proposed as a "merger of interpersonal communication and deception principles designed to better account for deception in interactive contexts" ([1], p. 203). The theory places deception in the context of interactive communication, and proposes that deceivers exhibit both strategic and nonstrategic behavior. Strategic behavior in this context refers to intentional, deliberate activities that are not necessarily manipulative but rather goal-directed. Thus, deceivers may orchestrate and adapt their deception to put forward the most successful self-presentation, to allay receiver suspicions and to achieve their desired ends (such as to persuade another to accept their advocated position). Additionally, they also display inadvertent indicators of discomfort, true emotional state, cognitive taxation, and attempted behavioral control that result in impaired communicative performance. These are what IDT regards as nonstrategic behaviors.

The notion that deception is strategic is not unique to IDT. The original four-factor theory of Zuckerman, DePaulo and Rosenthal

[23] recognizes that deceivers engage in behavioral control. The later self-presentational perspective of DePaulo [24] and the alternative models of deception advanced by Vrij [25] all acknowledge, implicitly or explicitly, that deceivers are goal-oriented and adapt their communication to achieve their deceptive goals. However, the strategic nature of deception is particularly salient in the context of an ongoing interaction. The deceiver must monitor the reactions of the other interactant(s) in order to guide the downstream interactions to further convince them, fill in exposed holes in the deception, and so on.

Though IDT broadly addresses many facets of interactive deception, the focus of our research here is on two interaction-focused factors that should be particularly evident in our pattern analysis approach: *behavior patterns associated with deception* and *behavior patterns associated with suspicion.*

1.2 Behavior Patterns Associated with Deception

We have noted that during deceptive interchanges, deceivers may display both strategic and nonstrategic behavior. Considerable empirical work has identified that certain kinds of behaviors such as adaptor gestures (scratching, face-rubbing, fidgeting, and the like) are signs of discomfort or nervousness [26] and are unlikely to be displayed intentionally. Understudied are the ways in which deceivers might behave strategically. In the context of group deception, this strategic behavior might be manifested as the deceiver attempting to control the interaction, either dominating the conversation or, at a minimum, soliciting desired responses with carefully constructed speaking actions that initiate patterns of communication. Research has shown that the initiation and control of conversation are dominant strategies, as is using highly expansive and expressive gestural patterns that convey strength [27]. Thus, initiating patterns by adopting the speaker role and displaying patterns that entail illustrator gestures would constitute strategic activity. Other forms of strategic behavior are also conceivable. For example, deceivers might opt for a more submissive "lying low" strategy after introducing deception, letting other group members initiate lines of discussion and further develop the deceptive ideas without his or her intervention. One way this pattern would manifest is through the use of backchannel head movements—nods, shakes, and other head movements—while in the listener role that encourage the speaker to continue talking [28]. Because there is no clear-cut prediction of what strategies a deceiver might adopt, we left as a research question whether and in what form strategic patterns might emerge. If no repeated patterning involving deceivers were to emerge, or if patterns only entailed adaptor behavior, that would be suggestive of a lack of strategic activity on their part.

RQ1: To what extent and in what ways do deceivers exhibit strategic communication patterns when engaged in group deception?

1.3 Behavior Patterns Associated with Suspicion

Research related to IDT has shown that receivers often become aware of the presence of deception, even if they do not label it as such; when their suspicions are piqued, their own behavior often telegraphs their suspicions to senders [2, 29]. In turn, IDT contends that senders perceive suspicion when receivers signal disbelief, uncertainty, or a desire for more information, and such suspicion, whether perceived or actual, prompts senders to increase their own strategic activity to mitigate such suspicion [1].

To ensure that suspicion would be invoked in the current experiment and have some degree of homogeneity in its cognitive representation, we experimentally induced suspicion but did not indicate who might be engaging in deceit or even how certain such a possibility might be. Just as deceivers are active rather than passive interlocutors who may engage in deliberate, strategic actions of their own, suspectors possess a range of alternative strategies for attempting to unmask duplicitous team members. Sensing that deception is taking place may spur them to probe for further information or evidence to confirm or disconfirm that suspicion. Employing this strategic information-seeking behavior [30], they might engage in verbal or nonverbal behavior that encourages suspected deceivers to talk more and in so doing to betray their duplicity verbally through what they say or nonverbally through "tells" of malicious intentions. In the context of patterned interactions, this probing behavior might manifest as an increase in interdependent interaction, i.e., patterns that include other members of the group. However, suspectors might also adopt a cagier strategy of increased watchfulness while remaining on the interaction sidelines, allowing other group members to lead the conversation and elicit diagnostic indications of a suspect's hidden agenda. Thus, we posed as a research question,

RQ2: To what extent do suspicious individuals exhibit regularities in their interaction patterns and what is the complexion of those patterns?

2 Method

2.1 Overview

The experiment consisted of Reserve Officers' Training Corps (ROTC) cadets participating in a simulated operations task called StrikeCom, a serious multi-player strategy game developed by the University of Arizona Center for the Management of Information [31]. StrikeCom is designed to investigate factors affecting teamwork and communication and to elicit the kinds of interdependencies and interaction patterns that typify decision-making in scenarios that resemble real-world conditions, incorporating the kinds of decisions military personnel must make in searching out and destroying enemy weapons caches. Although the game was introduced to participants as an exercise in collaboration, in reality it was intended to test for the effects of deception and suspicion on group processes and outcomes.

2.2 Sample

Participants ($N=42$; 31 males, 11 females) were undergraduate ROTC cadets recruited from the university's local ROTC chapter who participated as part of their curriculum in a simulation of an air operations center decision-making activity. Participants were randomly assigned to one of three roles—defined by the intelligence assets they would control—to search territory in an enemy country, to locate caches of missiles, and to destroy the weapons. As an incentive to boost motivation, cadets were informed they would be observed by their superior officer during performance of the task and would receive a special reward if they were the most successful team in destroying all of the missiles. The sample ranged in age from 18 to 24 years ($M=19.69$, SD = 1.52) and included 33 Caucasian, 7 Hispanic/Latino, and 2 Asian/Pacific Islander individuals.

2.3 Independent Variables

The independent variables of deception and suspicion were delivered through the random role assignment. Those assigned to the role of *Space* (e.g., satellite imagery), were instructed to introduce faulty information that would steer the group *away from* the missile silos. They were to act as though they had friends and family members still living in the vicinity of the missile sites and their true allegiance was to the enemy cause. Their task was also to evade detection. They were thus induced to become *deceptive*. Those assigned to *Intel* (e.g., spies on the ground) were induced to be *suspicious* by being told that other team members might or might not introduce inaccurate and misleading information into the team deliberations. The third member of each team controlled *Air* (e.g., surveillance planes) and was naïve to the deception and suspicion inductions. Air served as the control group against which to compare the other two manipulations. This within-group design permitted examining the dynamics resulting from having both deception and suspicion present in the group. Table 1 summarizes the three roles, their associated manipulations, and the different instructions given to each participant.

2.4 Experimental Procedure

Participants reported to a research site and were randomly assigned to three-person groups and to one of three roles within their group: Space, Air, or (Human) Intel. Through a succession of five turns, teams were to utilize their respective assets to locate targets hidden on a computer-displayed map overlaid with a 6 × 6 grid. After sharing search results with each other, they were to arrive at a consensus strike plan to destroy the missile launchers (see Figs. 1 and 2). Instructions were delivered by video on the same computers to be used during the simulation. Prerecorded instructions assured that they were uniform for all groups and players. Participants completed a practice run through the simulation individually before undertaking the team version so that all participants were familiar with the game.

Table 1
Independent variable manipulations

Role	Manipulation	Excerpts from manipulated instructions
Air	Naïve	Standard game instructions. Served as control group
Intel	Suspicion	"In group deliberations, there is always the possibility of faulty or inaccurate information. You are strongly advised to discuss your teammates' search results and recommendations in detail…This faulty or inaccurate information can come from ANY source internal to this simulation, such as the players"
Space	Deception	"Although most of us typically think that 'honesty is the best policy,' there are times when being truthful is not in the best interest… Your real task is to deceive your team members about the true location of the enemy camps… You must prevent your teammates from discovering the true location of the enemies and must convince your team to destroy empty territory"

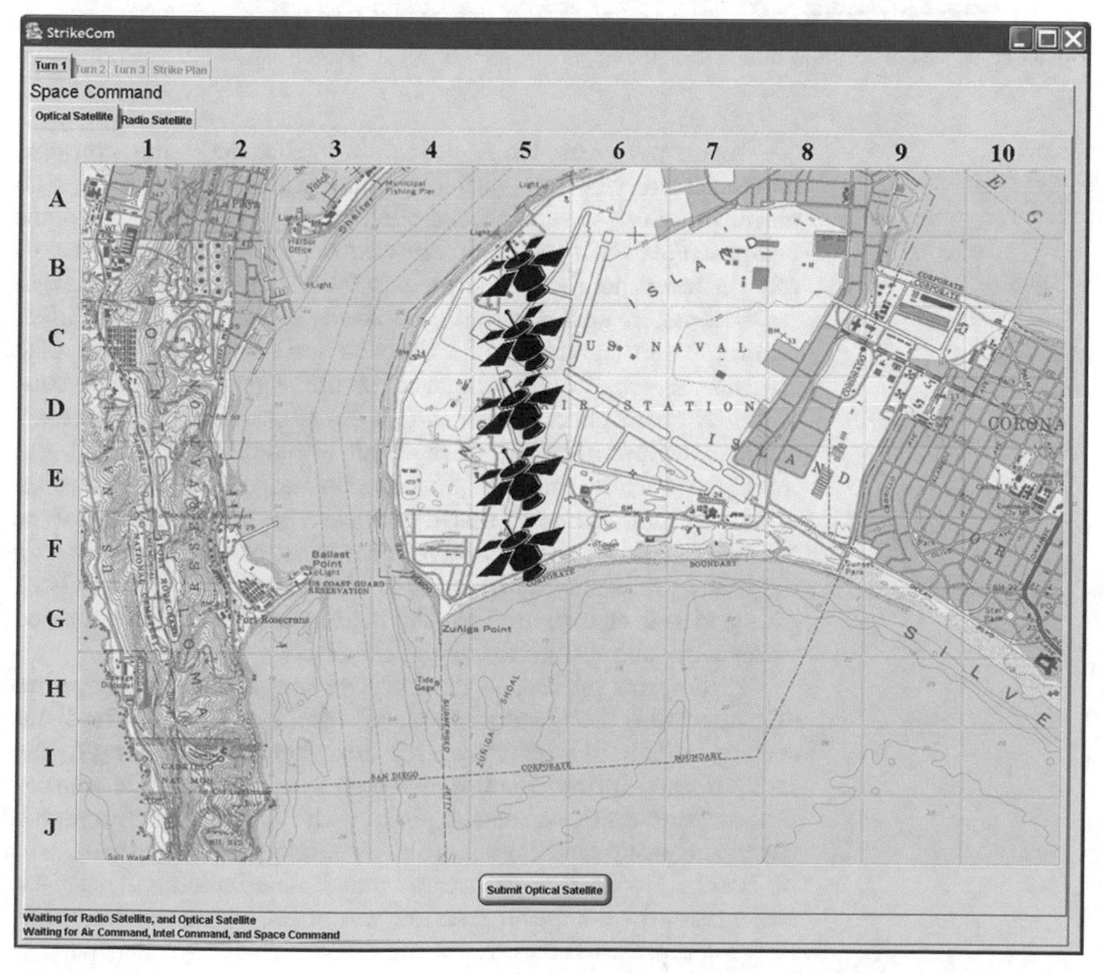

Fig. 1 Sample of 10×10 StrikeCom grid and map showing search selection for satellite intelligence

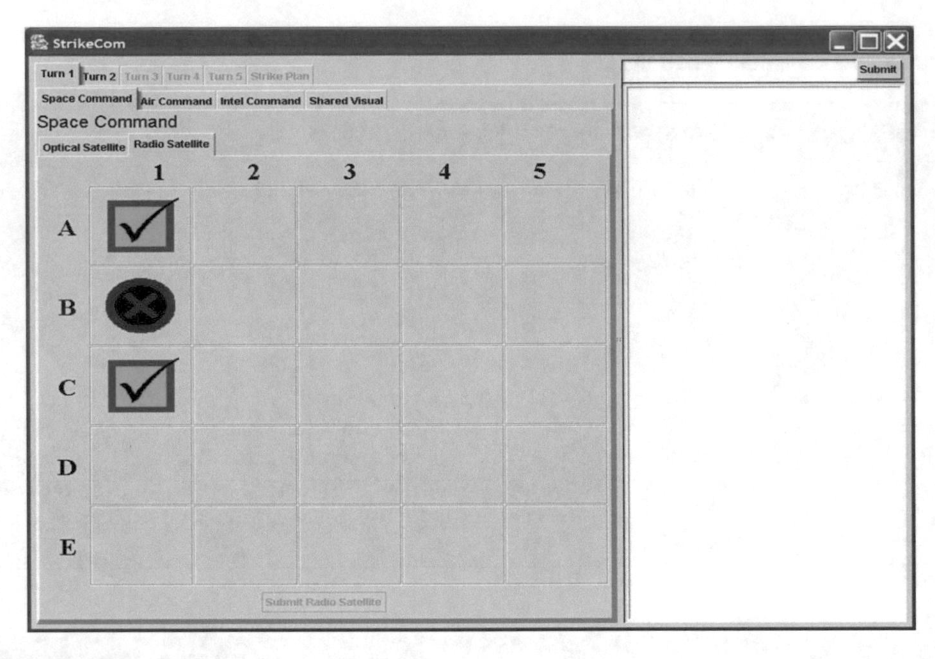

Fig. 2 Sample return of results after a search turn

Teams were seated at round tables so that they could communicate face-to-face but with computer screens in front of each player on which were displayed the map and grid. Privacy screens prevented players from seeing one another's computer; they had to rely on team members' oral reports for the information returned by their assets. Teams would complete a search turn, share information, plan their strategy for the next search and then undertake the next search. Results displayed to other group members consisted of green checks, indicating a grid location was apparently clean, a yellow question mark signifying uncertainty about the intelligence from that area or a red icon suggesting a high probability (but not certainty) that weapons were present. Players were informed in advance that different information assets had varying degrees of reliability and that their single asset could only check one grid cell at a time. This instruction introduced the needed uncertainty about the accuracy of the assets.

Participants proceeded through five search rounds, presenting and discussing their ostensible search results after each round and strategizing about where to search next, before proceeding to the strike round, during which they were to arrive at a single consensual decision on three areas to strike with bombs. After five search turns, the group developed a strike plan, choosing the territories to be bombed to destroy the weapons caches. After submitting their strike plan, the computer returned a report of their success in hitting the three grid areas where the weapons had been located.

The search and strike discussions averaged 25 min (range = 15–51), thus providing much longer and richer interaction than is typical of interpersonal communication experiments and far longer than the brief utterances that typify most deception studies [14].

2.5 Self-Report Measures

Participants first completed a pretest survey to assess demographics and prior computer experience. Following the simulation, they completed a posttest survey asking about their experiences playing the game and their personal motivation during the game. Manipulation checks were also collected to assess whether those in the Space role (deceivers) misreported information during the game and those in the Intel role (suspectors) actually were suspicious of their team members. Following the posttest survey, the participants were fully debriefed and discussed the task as part of their ROTC curriculum on decision-making and the impact of new technologies on such decision-making.

2.6 Nonverbal Behavioral Observation

Video recordings of the groups' discussions during the exercise were manually time-coded using a software tool called C-BAS [32]. The kinesic nonverbal behaviors that were coded for analysis are summarized in Table 2. Using C-BAS, trained coders watched each session's video, pressing and holding a key each time a specific behavior began, then releasing the key when the behavior ended. This allowed the software to record both frequencies and durations of each behavior in very granular (down to individual frames at 30 frames-per-second) increments, thus providing fine-grained data of behavior during the interactive sessions.

Table 2
Nonverbal behaviors coded for THEME analysis using C-BAS

Behavior	Description
Illustrator gestures	Movements used to accompany speech. They can be used to clarify speech and aid in listener understanding [34]
Adaptor gestures	A broad category of kinesic behaviors used to "satisfy physical or psychological needs" [26]. These adaptors in our classification schema refer to adaptors *using the hands*
Lip adaptors	Adaptor behavior limited to the mouth: pursing, licking, or biting lips, tongue-showing, and other related mouth movements indicating concentration, consternation, confusion, or nervous activity [26]
Speaker head movements	Head movements used during speech to illustrate and complement what is being said, to punctuate speech, or signify tense [35, 36]
Backchannel (listener) head movements	Nods and shakes while listening, which provide visual feedback and/or encouragement to the speaker, showing that they are being heard and understood [28]

Table 3
Summary statistics for frequencies of nonverbal behaviors included in THEME analysis ($N = 14$)

	Air		Intel		Space	
Behavior	***M***	**SD**	***M***	**SD**	***M***	**SD**
Illustrator gestures	53.0	46.2	63.4	26.3	47.1	39.6
Adaptor gestures	50.6	53.5	33.7	40.9	38.1	36.7
Lip adaptors	17.7	18.7	26.9	18.9	38.2	25.6
Speaker head movements	22.3	21.2	20.3	20.2	22.3	17.8
Backchannel (listener) head movements	52.1	39.5	41.0	23.8	46.5	36.7

Note: Means and standard deviations are aggregated across all 14 interactive sessions

The means and standard deviations for each of the coded behaviors are summarized in Table 3. They reveal that participants engaged in quite a few illustrator gestures. Head movement was also quite frequent, especially backchannel nods (listener head movements). Although the amount of speaker head movement did not differ by role, such movement was featured differentially in the initiation patterns that emerged. Thus, aggregate data may mask an important behavior, while THEME analysis can expose its importance. One other noteworthy statistic is the frequency of lip adaptors. Those in the space (deceiver) role used more than twice as many as those in the air (naïve) role. Although this is evidence of nonstrategic behavior by deceivers, it does not obviate the presence of strategic behavior, as the analyses to follow reveal.

These frequency data were the input for our THEME analysis, described in the next section. Each behavior's beginning and end was also marked.

2.7 THEME Analysis

The THEME software program is designed to discover and analyze patterns among discrete events within time-oriented data [7, 6]. The software uses specially coded data to build a timeline of events, then searches for patterns within the timeline that meet very stringent statistical significance thresholds and other criteria. Each occurrence of an event is signified by a text-based code that specifies the type of event that occurred along with a timestamp of when it occurred. In our dataset, a sample event type that occurred would be coded as "Space,b,SpeakingNods," signifying that the individual in the Space role began a speaking nod at the associated time. These coded data are aggregated by the software into a time sequence that includes each distinct event type and the times at which that event type occurred.

2.7.1 Pattern Search Parameters

THEME allows the researcher to specify various parameters when searching for patterns. We used a maximum search level of four to find complex patterns yet stay within operational bounds of our system. Variability in group discussion length necessitated adjusting the pattern recognition parameters for each session. To overcome potential bias because of this variability, we chose to dynamically weight the minimum occurrences according to the length (in seconds) of each interaction session. In this way the minimum occurrences threshold was set lower for datasets with shorter interactions and higher for datasets with longer interactions, with a maximum threshold of ten occurrences. This dynamic threshold assured that we were only analyzing truly recurrent events, but that we did not artificially deflate the number of patterns discovered for datasets from shorter interactions.

2.7.2 The T-Pattern

A t-pattern is defined as a set of event types that occur either concurrently or sequentially with significantly invariant time differences between the pairs [6]. When THEME detects a t-pattern, a string of event types is provided. This string reports the pattern structure observed. An example t-pattern discovered in our dataset is the following:

((Intel,b,BackchannelNods (Intel,b,Left,Adaptors Space,b,Left, Adaptors)) (Space,e,SpeakingNods Intel,b,BackchannelNods))

This pattern shows an interaction between the Intel and Space roles, with Intel demonstrating backchannel nods and adaptors while the Space role is speaking (indicated by the end of speaking nods during this pattern). A representation of where this pattern appeared during an interaction is shown in Fig. 3.

It is possible, especially given the large number of events in each session, that some or all of the patterns discovered by THEME are the result of chance. To ensure that this is not the case, THEME also compares the results of the pattern discovery process to randomized results. To accomplish this, THEME randomizes the order of the event types to produce a new sequence of events. Pattern discovery is then performed on the randomized data using the same parameters previously selected, and the resulting discovered patterns are compared against those patterns which were discovered using the original data. This procedure is repeated several times to get an average number of patterns from the random data. As shown in Fig. 4, which is produced from the coded data from one of the team discussions, the communication patterns from the original dataset occur much more frequently than can be attributed to chance. In the figure, which represents the counts of different lengths of discovered patterns, we see substantially more observed patterns of lengths three, four, and five in the real data than in the randomized data (which are represented by the smaller bars only visible at patterns of lengths two, three, and four). Take,

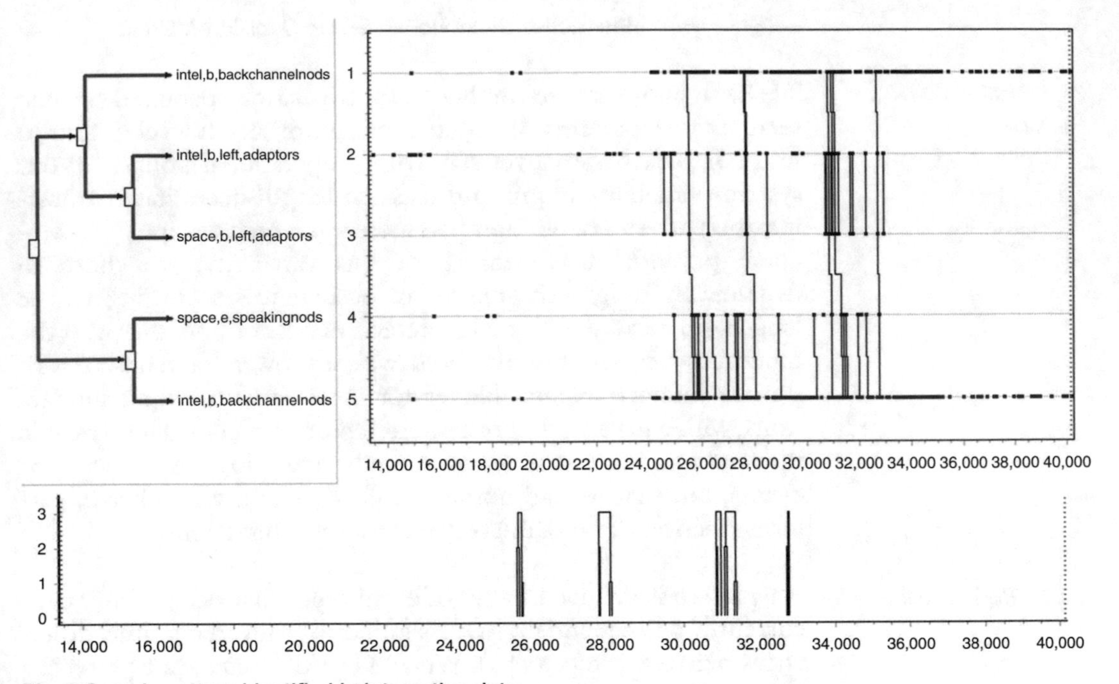

Fig. 3 Sample pattern identified in interaction data

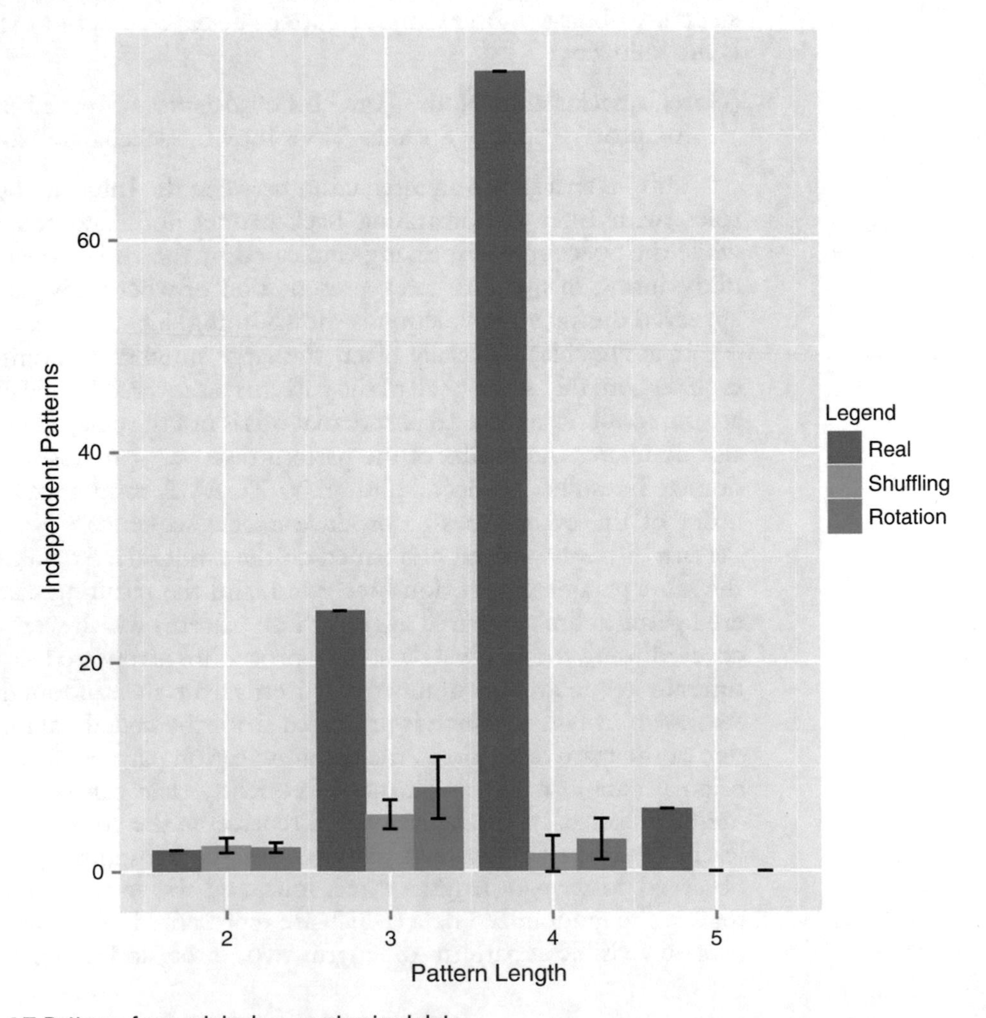

Fig. 4 T-Patterns from original vs. randomized data

for example, the bars representing patterns of length four. In the real, unshuffled dataset (the red bar), THEME discovered 76 independent patterns that included exactly four different or repeating event types. The randomized data produced an average of three patterns of this length.

3 Results

3.1 Manipulation Checks

Prior to testing hypotheses, manipulation checks were conducted to confirm that those in the Space role actually deceived, those in the Human Intel were actually suspicious, and that their suspicion level differed from those in the naïve Air role. On the rating of "I was accurate in reporting my asset's information to the group," those in the Space role ($M=2.36$, $SD=2.10$) were significantly less accurate than those in the Intel ($M=6.86$, $SD=0.36$) and Air ($M=6.43$, $SD=1.09$) roles, $F(2, 39)=45.29$, $p<0.001$, $\eta^2=0.70$. Space participants largely misled their group, although the large standard deviation indicates that not all were equally compliant. Less accurate reporting also tended to be related to a poorer team game score ($r=0.22$, $p=0.08$, one-tailed)—an indication that Space's deception produced an objectively quantifiable impairment of team performance.

Suspicion failed to differ by role, $F(2, 39)=0.89$, $p=0.47$, $\eta^2=0.04$, in part because Space also reported being suspicious of Air, making both Intel and Space somewhat more suspicious than Air of their team members. However, suspicion ratings were quite low, indicating that these cadets, who were acquainted with one another, were largely unswayed by the suspicion manipulation. These self-report results warrant caution in over-interpreting findings related to suspicion. Interestingly, suspicion was strongly and inversely related to measures of interaction, relationship and task communication quality ($r=-0.48$, $r=-0.67$, and $r=-0.41$, respectively; $p<0.001$). It was also negatively related to team performance: Those with poorer game scores reported being more suspicious ($r=-0.25$), implying that suspicion may have registered at a more subconscious level and resulted in reduced acceptance of team members' reports of their assets' findings.

3.2 Overview of THEME Analysis

With the THEME software configured as described previously, we conducted the pattern recognition analysis on each of the coded interaction sessions. THEME provides a plethora of pattern-related data which can be used in analyses. THEME discovered many patterns in the interactions, ranging from a low of 48 to over 1600 patterns in a given group. These data were then exported to Microsoft Excel® for the full analysis of the pattern data. Using Excel®, we filtered the patterns according to a number of different criteria described below (for example, excluding all patterns in which a certain role was not a participant, or limiting results to

Table 4
Summary of pattern discovery using THEME

Session	Session length (min)	Unique patterns	Total patterns	Actor switches per pattern M	Actor switches per pattern SD	Actors per pattern M	Actors per pattern SD	Pattern level Max	Pattern level M	Pattern level SD	Pattern length Max	Pattern length M	Pattern length SD
1	17	305	2593	0.70	0.70	1.59	0.56	5	3.34	0.73	3	1.96	0.43
2	25	239	2576	0.71	0.71	1.59	0.54	5	3.27	0.77	3	1.90	0.47
3	34	408	5715	0.60	0.72	1.47	0.50	5	3.27	0.69	4	1.88	0.37
4	15	416	2307	1.39	1.01	2.10	0.70	6	3.55	0.80	4	1.98	0.46
5	28	349	4468	0.54	0.67	1.45	0.51	5	3.27	0.67	3	1.89	0.35
6	51	146	3530	0.29	0.53	1.26	0.44	5	2.84	0.65	3	1.72	0.48
7	24	101	966	0.43	0.59	1.38	0.49	4	2.99	0.69	2	1.76	0.43
8	19	903	7126	1.32	0.90	1.86	0.47	5	3.64	0.64	3	2.01	0.33
9	17	1603	8804	1.43	0.89	2.11	0.62	5	3.63	0.56	3	1.98	0.20
10	21	71	567	0.21	0.48	1.20	0.43	5	2.93	0.70	3	1.75	0.47
11	33	76	967	0.45	0.60	1.42	0.55	4	2.78	0.67	2	1.64	0.48
12	29	198	2567	0.82	0.73	1.74	0.61	4	3.15	0.65	2	1.85	0.35
13	17	356	2436	0.73	0.82	1.56	0.55	6	3.55	0.77	4	2.02	0.44
14	15	48	304	0.40	0.57	1.40	0.57	6	3.17	1.02	3	1.81	0.64

only those initiated by a particular event). The results of the pattern discovery process are summarized in Table 4.

THEME produces a vast amount of potential data for exploration. We narrowed our focus to the issues raised by IDT and the research questions set forth earlier. For the purposes of this exploratory investigation into the patterned behavior of deceptive group interactions, we conducted three types of analysis. First, we conducted a descriptive analysis. Second, we completed an exploratory, zero-order correlation analysis. Third, we conducted a hierarchical regression analysis.

3.3 Descriptive Statistics

The 14 sessions analyzed produced significantly varying metrics. One session, for example, produced 416 patterns, while another produced only 76. Rather than skew our results comparing, for example, the number of patterns in which one particular role participated (227 vs. 38 patterns), we converted the counts to percentage of total patterns in the session, which yielded a better comparison of relative differences (55 % vs. 50 %). Such adjusted percentage-based comparisons were used for all pattern analyses in

Table 5
Means and standard deviations of behaviors and general pattern statistics ($N = 14$)

		Air		Intel		Space	
Variable	**Description**	***M***	**SD**	***M***	**SD**	***M***	**SD**
% Illustrator gestures	Percent of total illustrator gestures in a session performed by this role	20.6	15.6	32.6	12.0	46.9	23.1
% Adaptor gestures	Percent of total hand adapt or behaviors in a session performed by this role	27.4	15.4	43.9	16.1	28.7	12.7
% Lip adaptors	Percent of total lip adaptor behaviors in a session performed by this role	36.0	20.6	29.0	12.0	35.0	17.5
% Speaker head movements	Percent of total speaking head movements in a session performed by this role	34.2	21.4	30.4	16.0	35.4	19.5
% Listener head movements	Percent of total listener head movements in a session performed by this role	42.4	29.7	26.6	17.4	31.0	18.5
% of speaking activities	Percent of total speaking activities in a session performed by this role	29.8	16.9	31.2	11.4	39.1	19.4
% Total patterns	Percent of total session patterns in which this role participated	46.2	21.2	41.3	23.2	54.1	19.3
% Total patterns solo	Percent of total session patterns in which this role was the sole actor	21.3	13.5	25.8	18.6	27.9	17.5
% Total patterns initiating	Percent of total session patterns which this role initiated with some behavior	34.5	17.0	26.7	13.4	38.8	19.0
% Total patterns with switches	Percent of total session patterns in which this role participated and in which there was at least one switch between actors	25.0	16.8	25.9	22.8	28.2	16.4
Pattern length (complexity)	Number of consecutive event types (length) of patterns in which this role participated	1.42	0.7	1.31	0.8	1.67	0.6

Note: Each percentage represents the average percent of a session's patterns in which the specified role participated with the specified behavior

this report. Table 5 summarizes means and standard deviations of the metrics used in our analyses, both for the percentages of raw behaviors (i.e., without considering patterned behavior), and for the percentages of patterned behavior as calculated using THEME.

The descriptive statistics produced some noteworthy observations. First, group interactions were lengthy, averaging almost 25 min, indicating that groups were, for the most part, engaged in the task and devoted a fair amount of time to discussion. This is in sharp contrast to many laboratory group tasks that are only a few minutes in length.

Second, the number of unique patterns was quite high ($M = 373$, $SD = 417$) but the large standard deviation indicates that there was substantial heterogeneity across groups. The same was true of the total number of patterns ($M = 3209$, SD = 2522). Another interesting finding was the number of patterns that were repetitive (not unique) for each group. Calculated as a percentage of the total number of patterns, the groups averaged 88.7 % redundancy in patterning; that is, their interactions showed a high degree of structure and repetition. Some of this can be attributed to the fact that the beginning and ending of some behaviors would constitute a pattern. However, patterns went beyond simple beginnings and ends of single behaviors and tended to be interactive. Pattern level averaged 3.2 behaviors, and number of actors per pattern averaged 1.6, indicating that the patterns more often than not involved two actors.

3.4 Bivariate Correlation Analysis

Next, we examined intercorrelations among pattern features such as the number of patterns that each role *initiated*, the number of patterns in which the role was simply *included*, the average complexity (length of patterns) of each role's patterns, the number of patterns which *excluded* a given role, the number of patterns in which a given role was the *sole* actor (i.e., self patterns rather than interactive patterns), and various measures of interactivity in patterning, such as the average number of turn-switches between actors in a given pattern (i.e., when event types from one actor triggered event types from another actor), or the number of patterns which included one, two, or all three actors. These metrics represent a fraction of the pattern-related data that can be produced by the THEME software.

The interdependent nature of group interaction data meant that independent analyses could not be conducted between roles. Instead, we conducted analyses on patterns in which the Air, Space, or Intel player participated. For the measure of complexity, which concerned the length of patterns, Space—the deceiver—was involved in the longest patterns ($M = 1.67$, SD = 0.6), whereas Intel—the suspicious role—participated in the shortest patterns ($M = 1.31$, SD = 0.8), and Air—the naïve role—fell in between ($M = 1.42$, SD = 0.7). Thus, rather than being reticent or removed from the interactions, Space's behavior was part of the most complex patterns.

For diversity of patterns, we used the total percentage of group patterns in which the individual was involved. Space was involved in the largest percentage of patterns ($M = 54.1$ %, SD = 19.3 %), with Intel involved in the least ($M = 41.3$ %, SD = 23.2 %), and Air again in the middle ($M = 46.2$ %, SD = 21.2 %). This further suggests active engagement by Space, whereas Intel acted as an observer during interactions between Space and Air.

We also examined the percentage of patterns in which a role was the sole participant. These patterns constituted self-patterning. Once again, Space was most active ($M = 27.9$ %, SD = 17.5 %), Intel was second ($M = 25.8$ %, SD = 18.6 %), and Air was the least likely to be in a pattern alone ($M = 21.3$ %, SD = 13.5 %). This implies that Space's communication was not random but rather produced systematic and repeated patterns.

A summary of the zero-order correlation analysis is given in Table 6. We highlight several relationships that relate to our research questions. First, deceivers could opt for two distinctly different strategies—either opting for reticence, in which the deceiver largely lets other group members do the talking and occasionally hitch-hikes on their misstatements, uncertainties, and erroneous conclusions, or a persuasive deception strategy, in which they take a more active and assertive role in misleading the group. The former better fits a flight or defensive response pattern whereas the latter better fits a fight or offensive stance. The results, though showing both types of patterns, more often fit the latter characterization in that there was an extremely high correlation, $r(12) = 0.92$, $p < 0.01$, between the deceiver (Space) initiating communicative patterns and the deceiver starting patterns with speaking actions (either illustrators or speaking nods). Put differently, nearly all interactions initiated by the deceiver were done so with him or her presumably controlling the situation by initiating lines of talk and guiding the conversation. Importantly, neither the Air nor the Intel role had a significant correlation between these two factors, $r(12) = 0.46$, $p = 0.10$ and $r(12) = 0.23$, $p = 0.43$, respectively. Across all groups the deceiver tended to initiate more patterns with speaking activities than did the other members of the group. These findings inform RQ1. Deceivers were more inclined to tilt in favor of active, persuasive deception than more passive, opportunistic deception.

Given IDT's proposition that suspectors also behave strategically and may telegraph their suspicions through their behavior, we might also expect some effects to emerge in the behavior of the group members other than the deceiver. This is the focus of RQ2. In our case, the Intel role was given a suspicion manipulation in which he or she was notified that others in the group may not act in the best interest of the group. Accordingly, we might expect to see probing activities from the Intel role as he or she attempted to uncover the deception and confirm suspicions. The zero-order correlation analyses reveal several interesting findings in this regard. The Intel role tended to be highly interactive in the patterning. For instance, nearly all patterns that included Intel also included a switch between actors, $r(12) = 0.96$, $p < 0.01$. Patterns that included a single actor were very negatively correlated with involvement from Intel, $r(12) = -0.84$, $p < 0.01$, while patterns with two or three actors were very positively correlated with involvement from Intel, $r(12) = 0.81$, $p < 0.01$ and $r(12) = 0.62$, $p = 0.02$, respectively. Thus, when Intel was implicated in patterned behavior, he or she tended to draw another actor into the pattern, which could be interpreted as investigative behavior. Perhaps more intriguing, however, is the tendency of Intel's interactive behavior to focus on the Space role.

Table 6
Select intercorrelations of session-level patterning behaviors ($N = 14$)

Percent of session patterns	Speaking behavior			Interactivity				Specific roles					
	Began by air talking	Began by intel talking	Began by space talking	With at least one switch	With one actor	With two actors	With three actors	No air	No intel	No space	With air	With intel	With space
Started by air	0.46	0.17	**−0.65**	−0.04	−0.05	−0.11	0.36	**0.76**	**−0.66**	0.29	**0.74**	**−0.57**	−0.52
Started by intel	−0.17	0.23	−0.48	**0.74**	**−0.57**	**0.65**	0.20	−0.06	0.27	**−0.61**	−0.29	−0.07	**0.89**
Started by space	−0.30	−0.31	**0.92**	−0.49	0.45	−0.37	−0.46	**0.59**	−0.40	0.35	−0.46	**0.91**	−0.17
with air	**0.55**	0.34	**−0.63**	0.03	−0.19	0.04	0.45	**−0.56**	0.37	**0.58**	–	0.01	**−0.74**
With intel	−0.08	0.26	−0.51	**0.96**	**−0.84**	**0.81**	**0.62**	**0.66**	**−0.59**	**0.59**	–	–	−0.11
With space	−0.52	**−0.58**	**0.74**	0.03	−0.09	0.15	−0.08	0.45	0.11	−0.53	–	–	–
No air	−0.39	−0.32	−0.06	**0.68**	**−0.58**	**0.72**	0.08	–	–	–	–	–	–
No intel	0.27	−0.07	0.10	−0.52	0.16	−0.16	−0.10	−0.47	–	–	–	–	–
No space	0.35	0.48	−0.53	**0.61**	**−0.56**	0.42	**0.65**	−0.12	−0.32	—	—	—	—

Note: Correlations in bold are significant at $p < 0.05$

Patterns that included Intel, but were not necessarily initiated by Intel, were highly correlated with patterning that *excluded* Air (and often included Space), $r(12) = 0.66$, $p = 0.01$, while patterns initiated by Intel tended to be highly correlated with patterns that *included* Space, $r(12) = 0.89$, $p < 0.01$. These results reveal that far more of Intel's patterning involved Space than Air, which could be an indication that Intel was engaging with Space because of his or her suspicion of deception. These findings shed light on the probing behavior of suspicious individuals. Rather than adopt a passive observer role, Intel was actively engaged in threaded, interdependent communication with Space and far more so than with Air.

3.5 Multiple Regression Analysis

While our exploratory zero-order analysis yielded several interesting insights, simple correlations must be interpreted with caution inasmuch as they merely indicate an association, without measuring or controlling for possible alternate explanations. To assess more adequately the strategic nature of deception in our group setting and possible joint influences on group conduct, we chose to also employ an exploratory multiple regression analysis.

A first analysis focused on the deceiver, specifically on the extent to which the deceiver tended to control and manage the interaction in order to guide the conversation as desired, and so is directly related to RQ1. For this analysis, the criterion measure was the number of patterns initiated by performing speaker-related actions (illustrators or speaking nods). The chief predictor variable of interest was whether the role (i.e., deception condition) affected this strategic speaking patterning. Included as covariates were gender, age, ethnicity, and a measure of general computer experience. We also controlled for differences in motivation between the different people assigned to the different roles. Motivation was a general measure of self-reported monitoring of others and vigilance for anything being amiss. Individuals who were more motivated to attend to others' communication could be expected to be either more strategic or more active in terms of speaking (or both), and controlling for this possibility allowed us to better understand the effect of the deception manipulation on strategic patterning behavior. Table 7 summarizes the descriptive statistics, including reliabilities where appropriate, for each of the variables included in the regression model.

The multiple regression was conducted hierarchically using SPSS version 20. Because of the small sample size and the exploratory nature of this investigation, we used a more relaxed criterion of $p < 0.10$ for entry into the model, and $p < 0.20$ for removal. Predictors were added in steps, with stepwise regression as the entry method at each step, beginning with inclusion of the demographic variables (gender, age, dummy-coded ethnicity categories) and computer experience. We then added motivation in step 2 to additionally control for differences in participant engage-

Table 7
Means, reliabilities, and intercorrelations among variables ($N = 42$)

Variable	*M*	SD	α	1	2	3	4	5	6	7	8	9
1. Male	—	—	—	—								
2. Hispanic / Latino	—	—	—	0.12	—							
3. Asian/Islander	—	—	—	−0.12	−0.10	—						
4. Age	19.69	1.52	1.00	0.24	0.22	0.10	—					
5. Computer experience	3.53	0.94	0.71	0.52	0.04	0.11	−0.18	—				
6. Individual motivation	4.59	1.36	0.69	−0.06	0.06	−0.13	0.06	0.06	—			
7. Intel dummy code	—	—	—	0.08	0.09	0.08	0.08	0.23	−0.01	—		
8. Space dummy code	—	—	—	−0.04	−0.04	0.08	0.01	−0.12	−0.07	—	—	
9. Speaking Initiation	14.47	13.70	1.00	0.03	−0.21	0.17	0.07	0.18	0.20	−0.19	0.42	—
10. Relational quality	5.92	0.62	0.79	0.00	−0.21	−0.02	0.05	−0.30	−0.06	−0.12	−0.05	−0.41*
11. Interaction quality	5.99	0.66	0.73	0.01	−0.07	0.17	−0.18	−0.12	−0.07	0.10	−0.05	−0.48*
12. Task quality	6.21	0.66	0.59	0.11	0.02	0.10	−0.01	−0.02	−0.15	0.15	−0.26	−0.44*

Note: Correlations with absolute value of 0.31 and greater are significant at $p < 0.05$

ment and effort. Lastly, step 3 included dummy variables for Intel and Space (with Air as the reference group because it best served as a control group for comparing the two manipulated roles). The only significant predictor in the model was the Space role, $\beta = 0.41$, $F(1,40) = 8.45$, $p = 0.006$, $R2 = 0.17$, adjusted $R2 = 0.15$. Regardless of demographics or motivation, the Space (deceptive) role initiated significantly more patterns with speaking actions than did the reference group (the Air role), and Intel did not differ from Air. These patterning data increase our understanding of the strategic deception processes. Relative to a naïve group member, deceivers tended to initiate more observed patterns, whereas suspicious group members did not differ significantly from the naïve group member.

One of the ways in which a group member can eschew active participation and encourage others to talk instead is through the use of backchannel head movements. We also explored this measure as a criterion, specifically considering the number of patterns that began with this behavior. The multiple regression model produced three predictors. In step 1, computer experience entered as a predictor; in step 2, gender entered as a predictor; and in step 3, age entered. The total three-variable model was significant, $F(3,38) = 4.55$, $p = 0.008$, $R2 = 0.26$, adjusted $R2 = 0.21$. Those with less computer experience, males, and younger students were more likely to start interaction patterns with backchanneling. The role did not factor into this pattern.

On the other hand are indications of nonstrategic behavior such as adaptor behavior. We examined two criterion measures, percent of total session patterns beginning with adaptor gestures and patterns beginning with lip adaptors. In the model analyzing adaptor gestures, three predictors emerged: ethnicity, the Intel role and the Space role. The total three-variable model was significant, $F(3,38) = 5.67$, $p = 0.003$, $R2 = 0.31$, adjusted $R2 = 0.25$. Those of Hispanic descent exhibited more adaptors, whereas those in Intel and Space roles exhibited fewer patterns beginning with adaptors than those in the Air role. For lip pursing, only a single predictor emerged: the Intel role. Those in the Intel role initiated more patterns beginning with lip pursing than those in the Air role, $F(1,40) = 8.01$, $p = 0.007$, $R2 = 0.17$, adjusted $R2 = 0.15$, possibly as a telltale sign of uncertainty or concentration.

4 Discussion

Deception in group contexts is complex, and relatively little work has been conducted to understand the intricate communicative processes within group deception. Using IDT [1] as a guiding framework, we addressed two research questions regarding the strategic behavior of deceptive individuals and the tendencies of deceived individuals who are suspicious of their deception. Using

THEME [7, 6], we extracted and analyzed the interactional patterns evident in 14 different experimental group sessions in which both deception and suspicion were introduced. In a two-stage approach, we first used zero-order correlation analysis to investigate tendencies within the patterning behavior. We then used regression analysis to more accurately model the strategic speaking behavior of the deceptive group member.

4.1 Answering RQ1: The Actively Strategic Deceiver

RQ1 attempted to explain the strategic behavior of deceptive individuals, following the logic of IDT [1]. Though IDT states that deceivers will be strategic as they carry out their deception, there are at least two ways in which such strategic behavior might surface. The first possibility is a "lying low" strategy in which the deceiver strategically withholds information or allows group members to entertain false possibilities in order to achieve the desired outcome. Such behavior would be shown in our pattern analysis by, for example, a lack of participation in patterned interactions or a tendency toward backchannel encouragement (i.e., backchannel nods).

The second possible strategy a deceiver might employ would be a more active—perhaps even manipulative—role in guiding the group interaction, leading the group astray by actively working to engender false assumptions or conclusions. This type of behavior would be supported in our dataset by the deceiver being more engaged in interactional patterns, particularly with behaviors associated with active speaking actions (e.g., illustrators and/or speaking nods).

The exploratory correlation analysis supports the latter of these two possibilities. The patterning behavior of the deceptive Space role was found to be generally correlated with presumably strategic initiation of interactional patterns, more so than that of the other two roles. The regression analysis furthered this line of discovery, focusing on the strategic actions of the deceiver, in which the dependent variable was operationalized as the extent to which interactional patterns were initiated by speaking behaviors (which could be expected from someone who is trying to strategically manipulate the direction of the interaction). Our analysis revealed that, even after controlling for salient individual characteristics that might affect patterned speaking behaviors, the deceptive role initiated significantly more interactional patterns with speaking behaviors than did the other two roles.

While these findings are very preliminary, they indicate that deceivers in group settings may attempt to actively and strategically guide the interaction in order to successfully achieve their deception. The deceivers in our sample tended to be actively engaged in interactional patterns, and they tended to initiate patterns with speaking-related behaviors, as compared to the suspicious or naïve group members.

4.2 Answering RQ2: The Investigative Suspector

RQ2 focused on the behavior of the suspicious individual. IDT proposes that individuals who become aware of deception will attempt to uncover or, at a minimum, confirm the deception taking place. Like the behavior of deceivers, however, suspectors' behavior might also take different forms. One strategy might be a less active, watchful approach in which suspectors carefully examine the other members' behaviors without initiating much interaction in order to identify the deceiver. A second, more active strategy would be one in which the suspector engages in probing behavior, initiating interactions with other group members and attempting to discover the deceiver.

The Intel role that was made to be suspicious during the interaction produced substantial evidence that indicated probing behavior. The interactions with the suspicious role tended to show signs of interactivity and, most tellingly, a focus on the deceptive Space role in that interaction. These results intimate that suspicious individuals in groups where deception has been introduced may actively engage rather than passively observe the other members of the group in order to uncover the deceiver.

It is important to note that while some of the above conclusions could be drawn from other forms of analysis such as summary statistics of the individual coded behaviors, such analyses would in no way reveal the interdependencies and evolutionary character of the interaction patterns that are evident in our pattern analysis. For example, although the speaking behavior of the Intel role could be inferred somewhat from summary statistics, simply counting the number of speaking behaviors for each role and comparing those totals would not reveal their importance to specific roles and their embeddedness in overall patterns of behavior associated with a given role, and the tendency of the Intel role to be interactive and focused on the deceiver role would be overlooked. Put differently, it is the structure and relatedness among behaviors that would not be evident in simple summary statistics, and the strength of the THEME pattern analysis approach lies in detecting and analyzing those structures.

Moreover, a particular value of THEME is revealed in the study of interactions between the group members, which interactions are modeled in patterns using THEME. Using THEME to detect subtle, even imperceptible patterns during these group interactions, we were able to uncover tendencies of deceivers and the other group members as they interact with one another. These tendencies could be aggregated and studied in an objective way, and the findings extracted therefrom expand our understanding of the intricacies of deception in group settings. Further opportunity exists to better understand the patterns that surface during deceptive group communication.

Also offering new insights were the results on pattern length. The THEME software was able to discover and analyze patterns that ranged in length from two to as many as six different or repeated behaviors. As we have demonstrated, these patterns occur much more frequently than those that could be expected to occur by chance. That deceptive behavior is so intricately patterned suggests the need to drill deeper into the full gamut of patterns exhibited by each dyad or group. Such exploratory work would lead to more insights into how variable each dyad or group is in the enactment of deception and whether these patterns that are imperceptible to the unaided observer are associated with judges' intuitive notions of suspicious or deceptive communication.

4.3 Future Use of THEME

The current analysis only begins to exploit some of the potential of THEME analysis in uncovering the intricacies of human interaction. For example, IDT posits that interactions are dynamic and iterative, which implies that behaviors and their patterning may change over time. In the case of interpersonal deceit, there is evidence that behavioral displays do not remain constant; what is displayed early in an interaction is not the same as what is displayed later [21, 33]. Thus, one might expect that behavioral patterns themselves are subject to change. Some that are present at one juncture may disappear later, while other new patterns may emerge. It is also possible that deceivers, being focused on attaining their ulterior ends, may lose the routinized character of their communication patterns and show greater heterogeneity than truth tellers, whose patterning might show greater stability over time. THEME can answer this question by analyzing whether more patterns emerge prior to or following a particular juncture in the interaction. Other forms of statistical analysis only reveal whether the trajectories of particular behaviors change over time but not whether entire patterns of behavior change their frequency and complexion over the course of an interaction. THEME has the potential to reveal whether interdependence deepens or attenuates over time as well as whether patterns distribute themselves relatively equally over time or are more bunched early or late in an interaction. Questions such as whether more strategic or nonstrategic behavior is evident early or late in an interaction can also be answered by THEME, thus going beyond superficial understanding of group interaction to unpack some of its complexities.

4.4 Limitations

As with any study, our work is not without limitations. As stated, our analyses have been highly exploratory in nature. While we believe the discovered patterning behavior to be both valid and novel, the relationships observed among these behaviors and with other measured variables remain tentative. As such, there exist several opportunities for future research to further explore the tendencies observed in this work. Our correlation and regression analyses also suffered from small sample sizes ($N = 14$ and 42, respectively), and the risk of error due to sampling error should be considered while interpreting our results. These (and other) limitations notwithstanding, we believe that our findings provide a convincing argument regarding the value of pattern discovery and analysis afforded by the THEME software.

5 Conclusion

We have demonstrated the use and value of a novel pattern discovery and analysis tool in studying the complex interactions that take place in the context of group deception. The interactional patterns discovered, together with their general tendencies as uncovered in our exploratory analyses, provide additional insight to the knowledge base of group deception. Our hope is that the findings in this paper will spur other analysis and discovery in this fertile area of research.

Acknowledgements

Portions of this research were supported by funding from the National Science Foundation (Grant # 0725895 and #1068026). The views, opinions, and/or findings in this report are those of the authors and should not be construed as an official US government position, policy, or decision.

References

1. Buller DB, Burgoon JK (1996) Interpersonal deception theory. Commun Theory 6(3):203–242
2. Buller DB, Burgoon JK (1994) Deception: strategic and nonstrategic communication. In: Daly JA, Wiemann JM (eds) Strategic interpersonal communication. Erlbaum, Hillsdale, NJ, pp 191–223
3. Burgoon JK (2005) Measuring nonverbal indicators of deceit. In: Manusov V (ed) The sourcebook of nonverbal measures: going beyond words. Lawrence Erlbaum Associates, Hillsdale, NJ, pp 237–250
4. Dunbar NE, Jensen MJ, Burgoon JK, Bessarabova E, Bernard DR, Robertson K, Kelley K, Adame B, Eckstein J (2010) Strategies for persuasive deception in CMC. Paper presented at the 43rd Hawai'i international conference on system sciences, Kauai
5. Burgoon JK, George JF, Kruse J, Marett K, Adkins MA (2014) Credibility assessment in meetings: deception. In: Nunamaker JF, Romano N, Briggs R (eds) Advances in MIS: collaboration science, technologies, processes and applications. M.E. Sharpe, Armonk, NY, pp. 109–127
6. Magnusson MS (2006) Structure and communication in interactions. In: Riva G, Anguera MT, Wiederhold BK, Mantovani F (eds) Communication to presence: cognition, emotions and culture towards the ultimate communicative experience. Ios Press, Amsterdam, pp 127–146
7. Magnusson MS (2005) Understanding social interaction: discovering hidden structure with model and algorithms. In: Anolli L, Duncan JS, Magnusson MS, Riva G (eds) The hidden structure of interaction: from neurons to culture patterns. Ios Press, Amsterdam, pp 4–22
8. Koch SC, Müller SM, Schroeer A, Thimm C, Kruse L, Zumbach J (2005) Gender at work: eavesdropping on communication patterns in two token teams. In: Anolli L, Duncan JS, Magnusson MS, Riva G (eds) The hidden structure of interaction: from neurons to culture patterns. Ios Press, Amsterdam, pp 266–279

9. Plumet M-H, Tardif C (2005) Understanding the functioning of autistic children. In: Anolli L, Duncan JS, Magnusson MS, Riva G (eds) The hidden structure of interaction: from neurons to culture patterns. Ios Press, Amsterdam, pp 182–192
10. Hardway C, Duncan JS (2005) "Me First!" Structure and dynamics of a four-way family conflict. In: Anolli L, Duncan JS, Magnusson MS, Riva G (eds) The hidden structure of interaction: from neurons to culture patterns. Ios Press, Amsterdam, pp 210–221
11. Marett LK, George JF (2004) Deception in the case of one sender and multiple receivers. Group Decis Negot 13(1):29–44
12. Fuller CM, Twitchell DP, Marett K (2012) An examination of deception in virtual teams: effects of deception on task performance, mutuality, and trust. IEEE Trans Prof Commun 55(1):20–35
13. Frank MG, Feeley TH, Paolantonio N, Servoss TJ (2004) Individual and small group accuracy in judging truthful and deceptive communication. Group Decis Negot 13(1):45–59
14. DePaulo BM, Lindsay JJ, Malone BE, Muhlenbruck L, Charlton K, Cooper H (2003) Cues to deception. Psychol Bull 129(1):74–118. doi:10.1037/0033-2909.129.1.74
15. Leavitt HJ (1951) Some effects of certain communication patterns on group performance. J Abnorm Soc Psychol 46(1):38–50
16. Dawson RO (1987) Guide to the micro galaxy: investigation into communication patterns in small groups. Program Learn 24(3):230–233
17. Perlow LA, Gittell JH, Katz N (2004) Contextualizing patterns of work group interaction: toward a nested theory of structuration. Organ Sci 15(5):520–536
18. Grammer K, Kruck KB, Magnusson MS (1998) The courtship dance: patterns of nonverbal synchronization in opposite-sex encounters. J Nonverbal Behav 22(1):3–29
19. Buller DB, Comstock J, Aune RK, Strzyzewski KD (1989) The effect of probing on deceivers and truthtellers. J Nonverbal Behav 13:155–169
20. Burgoon JK, Buller DB, White CH, Afifi WA, Buslig ALS (1999) The role of conversational involvement in deceptive interpersonal communication. Pers Soc Psychol Bull 25:669–685
21. Stiff JB, Corman SR, Snyder E, Krizek RL (1994) Individual differences and changes in nonverbal behavior: unmasking the changing faces of deception. Commun Res 21:555–581
22. White CH, Burgoon JK (2001) Adaptation and communicative design: patterns of interaction in truthful and deceptive conversations. Hum Commun Res 27:9–37
23. Zuckerman M, DePaulo BM, Rosenthal R (1981) Verbal and nonverbal communication of deception. In: Berkowitz L (ed) Advances in experimental social psychology, vol 14. Academic, New York, pp 1–59
24. DePaulo BM (1992) Nonverbal behavior and self-presentation. Psychol Bull 111(2):203–243
25. Vrij A, Edward K, Roberts KP, Bull R (2000) Detecting deceit via analysis of verbal and nonverbal behavior. J Nonverbal Behav 24(4): 239–263
26. Burgoon JK, Guerrero LK, Floyd K (2010) Nonverbal communication. Pearson, Boston
27. Burgoon JK, Dunbar NE (2006) Nonverbal expressions of dominance and power in human relationships. In: Manusov V, Patterson M (eds) The Sage Handbook of nonverbal communication. Sage, Beverly Hills, CA
28. Duncan JS (1974) On the structure of speaker-auditor interaction during speaking turns. Lang Soc 3(2):161–180
29. Burgoon JK, Buller DB, Dillman L, Walther JB (1995) Interpersonal deception. Hum Commun Res 22(2):163–196
30. Ramirez A, Walther JB, Burgoon JK, Sunnafrank M (2002) Information-seeking strategies, uncertainty, and computer-mediated communication: toward a conceptual model. Hum Commun Res 28(2):213–228
31. Twitchell DP, Wiers K, Adkins M, Burgoon JK, Nunamaker JF (2005) StrikeCom: a multi-player online strategy game for researching and teaching group dynamics. In: 38th annual Hawaii international conference on system sciences. IEEE Computer Society, Hawaii
32. Meservy TO (2010) C-BAS 2.0. Final report to the Center for Identification Technology Research
33. Hamel L, Burgoon JK, Humpherys S, Moffitt K (2007) The "when" of detecting interactive deception. Paper presented at the annual meeting of the National Communication Association. IL, Chicago
34. Holler J, Beattie G (2003) Pragmatic aspects of representational gestures. Gesture 3:127–154
35. Birdwhistell R (1970) Kinesics and context: essays on body motion communication. University of Pennsylvania Press, Philadelphia
36. Chovil N (2004) Measuring conversational facial displays. In: Manusov V (ed) The sourcebook of nonverbal measures: going beyond words. Lawrence Erlbaum Associates, Hillsdale, NJ, pp 173–188

Chapter 3

Imposing Cognitive Load to Detect Prepared Lies: A T-Pattern Approach

Valentino Zurloni, Barbara Diana, Massimiliano Elia, and Luigi Anolli

Abstract

One of the most well-documented claims in the deception literature is that humans are poor detectors of deception. Such human fallibility is exacerbated by the complexity of both deception and human behavior. The aim of our chapter is to examine whether the overall organization of behavior differ when people report truthful vs. deceptive messages, and when they report stories in reverse vs. chronological order, while interacting with a confederate. We argue that recalling stories in reverse order will produce cognitive overloading in subjects, because their cognitive resources are already partially spent on the lying task; this should emphasize nonverbal differences between liars and truth tellers. In the present preliminary study, we asked participants to report specific autobiographical episodes. We videotaped them as they reported the stories in chronological order or in reverse order after asking to lie about one of the stories. We focused in analyzing how people organize their communicative styles during both truthful and deceptive interactions. In particular, we focused on the display of lying and truth telling through facial actions. Such influences on the organization of behavior have been explored within the framework of the T-pattern model. The video recordings were coded after establishing the ground truth. Datasets were then analyzed using Theme 6 beta software. Results show that discriminating behavioral patterns between truth and lie could be easier under high cognitive load condition. Moreover, they suggest that future research on deception detection may focus more on patterns of behavior rather than on individual cues.

Key words Deception, Cognitive load, T-Pattern microanalysis, Theme, Detection, Nonverbal cues

1 Introduction

Deception is an articulated and complex communication act aimed at influencing the beliefs of others [1–3]. There are many kinds of deception, such as lies, fabrications, concealments, misdirection,

---

This chapter is dedicated to the memory of Professor Luigi Anolli.
Professor Anolli made an effective contribution to the introduction and development of communication psychology in Italy. Focusing on the miscommunication field, he closely examined deceptive communication in its different aspects. Within the communication domain, Professor Anolli also gave special attention to nonverbal communication. As well, he focused on new methodological devices of analysis. In Italy, he introduced the use of "Theme" software for the recognition of hidden patterns in human interaction. His contribution and his effort in the methodological and theoretical approach we have embraced were crucial for the realization of this study.

Magnus S. Magnusson et al. (eds.), *Discovering Hidden Temporal Patterns in Behavior and Interaction: T-Pattern Detection and Analysis with THEME™*, Neuromethods, vol. 111, DOI 10.1007/978-1-4939-3249-8_3,

bluffs, fakery, mimicry, tall tales, white lies, deflections, evasions, equivocation, exaggerations, camouflage, and strategic ambiguity. These forms of deception are common, everyday occurrences in interpersonal, group, media, and public contexts. However, one of the most well-documented claims in the deception literature is that humans are poor detectors of deception. Such human fallibility is exacerbated by the speed, complexity, volume, and global reach of communication and information exchange that current information technologies can afford now, enabling the collection of massive amounts of information—information that must be sorted, analyzed, and synthesized. The interactions and complex interdependencies of information systems and social systems make the problem even more difficult and challenging [4].

A recent meta-analysis reveals that although people show a statistically reliable ability to discriminate truths from lies, overall accuracy rates average 54 % or only a little above chance [5]. Moreover, the average total accuracy rates of professional lie catchers (56 %) are similar to that of laypersons. What is the reason for the near-chance performance of human lie detection? Do liars behave consistently with people beliefs about deceptive behavior?

2 Cues to Deception

Three recent, comprehensive meta-analyses [6–8] which considered above 130 studies published in English examining 158 different cues to deception, including a large variety of nonverbal, verbal, and paraverbal cues, as well as certain content features of deceptive and truthful messages, reveal that many conflicting results have been found.

DePaulo, Lindsay, Malone, Muhlenbruck, Charlton, and Cooper [6] examined around 100 different nonverbal behaviors. Significant findings emerged for 21 behaviors. The cues were ranked in terms of their effect sizes. The highest effect sizes were found in the cues that have not often been investigated, such as changes in foot movements, pupil changes, smiling. Concentrating on the cues that were investigated more often, the largest effect size was found for pupil size (dilation). Compared to truth tellers, liars' pupils look more dilated. Gaze aversion is not a valid indicator of deception. The simple heuristic that liars are more nervous than truth tellers is not supported because many cues of nervousness, such as fidgeting, speech disturbances or blushing, are not systematically linked to deception. However, liars appear tenser, have a tenser voice, have their chin in a higher position, press their lips more, and have less pleasant looking faces. They also sound more ambivalent, less certain and less involved, and make more word and sentence repetitions.

In different studies it was further found that liars make fewer illustrators, fewer hand and finger movements, and fewer leg and foot movements than truth tellers. DePaulo and colleagues [6] showed that the effect for illustrators was small, whereas the effect for hand and finger movements was somewhat more substantial. In those studies where a difference was found, liars typically showed fewer leg and foot movements than truth tellers. However, the effect for leg movements was not significant in this meta-analysis, perhaps because a "no difference" was found in many individual studies. Conversely, hand and finger movements appear to have the strongest relationship with deception. Vrij, Winkel, and Akehurst [9] found that 64 % of their participants showed a decrease in hand/finger movements during deception, whereas 36 % showed an increase of these movements during deception. Finally, most researchers that obtained an effect for shifting position reported that liars change their sitting position more than truth tellers do. In most studies where shifting position was measured, however, no difference between truth tellers and liars was found. It is therefore probably best to conclude that shifting position is not related to deception [8].

Even if research on this topic revealed only a few, and usually weak, relationships between nonverbal cues and deception, police training packages often include nonverbal and paraverbal behaviors as important cues to deception. Lay people and professionals alike have believed that these behaviors are particularly helpful in catching a liar [10, 11]. Although there seems to be no reliable cues, a closer look reveals that some studies have found reliable differences between liars and truth tellers under certain conditions, which may account for the overall inconsistency in findings. An explanation for the contradictory findings obtained across individual studies might be that studies differ regarding experimental method, the type of sample, or the operationalizations used to measure the nonverbal behaviors of interest. A host of moderator variables may blur the association between behavioral cues and deception.

These studies reveal that a sign equivalent to Pinocchio's growing nose has not been found [8]. In fact, most of the nonverbal cues do not appear to be related to deception. However, for some cues a weak relationship with deception emerged. Why only a few and rather weak relationships between nonverbal behavior and deception have been found? Could it be that more nonverbal cues to deception exist than we are aware of, but that researchers so far have overlooked some diagnostic nonverbal cues to deception? And could it be that a combination of nonverbal cues, rather than individual cues, would reveal diagnostic indicators of deceit?

3 The Imposing-Cognitive-Load Approach to Deception

Nonverbal cues to deception are typically faint and unreliable, as confirmed by the studies above mentioned. A contributing factor is that the underlying theoretical explanations for why such cues occur, like nervousness and cognitive load, also apply to truth tellers [12]. In order to explain deceptive communication, Zuckerman, DePaulo, and Rosenthal [13] pointed out that liars show signs of deceit as an outcome of experiencing emotions or cognitive load. However, those assumptions may lead to opposite behaviors. For example, arousal typically leads to an increase in eye blinks whereas cognitive load typically leads to a decrease in eye blinks. Moreover, the emotional approach predicts an increase in certain movements (signs of nervous behavior), whereas the cognitive load approach predicts a decrease in movements during deception as a result of neglecting the use of body language.

Different researches have recently examined whether liars actually do experience these processes more than truth tellers [14, 15], with controversial results. Studies in the past have focused on eliciting and amplifying emotions [16] for example by asking questions, but it is uncertain whether this procedure will necessarily raise more concern in liars than in truth tellers. For instance, Inbau, Reid, Buckley and Jayne [17] pointed out that liars feel more uncomfortable than truth tellers during police interviewing. Conversely, DePaulo and colleagues [6] comprehensive meta-analysis regarding how liars actually behave, and Mann, Vrij, and Bull's [18] analysis of behaviors shown by suspects during their police interviews gave no support for the assumption that liars, above all, appear nervous. DePaulo and colleagues [6] stressed that experiencing emotions is not the exclusive domain of liars. Truth tellers may also experience them and, as a result, may also display nonverbal cues associated with emotion. Moreover both liars and truth tellers gain from being believed, and will attempt to appear convincing [6].

Conversely, only a few efforts focused on unmasking the liars by applying a cognitive lie detection approach [16, 19]. Vrij and his colleagues [12, 16] have recently suggested that lying can be more cognitively demanding than truth telling. First, formulating the lie may be cognitively demanding. Except when the liar deceives by omission (e.g., when he omits to give the addressee some information that he/she thinks or knows is relevant to the addressee's goals), he needs to invent a story and must monitor its fabrication so that it is plausible and adheres to everything the addressee knows or might find out. Moreover, liars must remember what they have said to whom in order to maintain consistency.

Furthermore, liars are typically less likely than truth tellers to take their credibility for granted [12]. As such, liars will be more inclined than truth tellers to monitor and control their demeanor in order to appear honest to the investigator. Such monitoring and controlling is cognitively demanding. They may also monitor the interlocutor's reactions carefully in order to assess whether they appear to be getting away with their lie, and this too requires cognitive resources.

Liars also have to suppress the truth while they are fabricating, and this is also cognitively demanding. While activation of the truth often happens automatically, activation of the lie is more intentional and deliberate, and thus requires mental effort. A single deceptive act can be governed by a plurality of intentions, embedded in each other and hierarchically organized. Such is the case of a prepared (packaged) lie, in which different layers of communicative intentions are at work [20]: (a) a hidden (covert) intention (the speaker intends to deceive the addressee by manipulating the information); (b) an ostensive (overt) intention (the speaker intends to convey the information manipulation to the addressee). This second intentional layer is, in its turn, twofold: (b1) informative intention (the speaker wants to give the addressee the manipulated information as if it were true); (b2) "sincerity" intention (the speaker wants the addressee to believe that what he has said is true, in order to respect the Sincerity Rule of Searle [21]: "I want you to believe that I believe what I am saying to you"). Therefore, deceptive communication appears to require at least a second-order intentional system and in certain cases (especially in prepared lies) a third-order intentional system.

Lying is not always more cognitively demanding than truth telling [22]. Perhaps the reasons given as to why lying is more cognitively demanding could give us insight into when it is more cognitively demanding. For example, lying is likely to be more demanding than truth telling only when liars are motivated to be believed. Under those circumstances it can be assumed that liars take their credibility less for granted than truth tellers and hence will be more inclined than truth tellers to monitor their own behavior and/or the interlocutor's reactions. Moreover, for lying to be more cognitively demanding than truth telling, liars must be able to retrieve their truthful activity easily and have a clear image of it. Only when liars' knowledge of the truth is easily and clearly accessed suppressing the truth will be difficult for them. Obviously, truth tellers also need to have easy access to the truth for the task to be relatively undemanding. If truth tellers have to think hard to remember the target event (e.g., because it was not distinctive or it occurred long ago), their cognitive demands may exceed the cognitive demands that liars require for fabricating a story.

To sum up, the cognitive load of the deceiver (a) arises as a function of the entity and gravity of the deceptive contents, and (b) depends on the context's significance. This raises the distinction between *prepared* and *unprepared lies*. The former is cognitively planned in advance and examined by the deceiver at least in its main aspects, while the latter is spontaneously said, often by an answer to an unexpected question, without any mental planning. Anolli and colleagues [20] introduced a further distinction between *high-* and *low-content lies.* A deceptive act is high-content when it concerns a serious topic, is said in an important context, and is characterized by the presence of notable consequences and effects for the deceiver, for the addressee or even for other people. High-content lies may request previous planning, since they are generally foreseen and prepared. In this case, the liar has to elaborate his best communicative way to convince the partner. He has to be careful in the deceptive message planning, paying attention to its internal consistency and compatibility with the partner's knowledge. Moreover, he has to be as spontaneous as possible in the deceptive message execution in order to be believable [20].

In order to discriminate more effectively between truth tellers and liars, a lie catcher could exploit the different levels of cognitive load that they experience. If liars require more cognitive resources than truth tellers, they will have fewer cognitive resources left over. Cognitive demand can be further raised by making additional requests. Liars may not be as good as truth tellers in coping with these additional requests.

One way to impose cognitive load is by asking speakers to tell their stories in reverse order [12, 16]. The underlying assumption is that recalling events in reverse order will be particularly debilitating for liars—whose cognitive resources have already been partially depleted by the cognitively demanding task of lying—because (a) it runs counter to the natural forward-order coding of sequentially occurring events [23, 24], and (b) it disrupts reconstructing events from a schema [25]. This is analogous to the finding in the cognitive-attention literature that information processing in the primary task is slower in dual-task conditions than in single-task conditions [26].

In an experiment, half of the liars and truth tellers were requested to recall their stories in reverse order [16], whereas no instruction was given to the other half of the participants. More cues to deceit emerged in the reverse-order condition than in the control condition. Observers who watched these videotaped interviews could distinguish between truths and lies better in the reverse-order condition than in the control condition. For example, in the reverse-order experiment, 42 % of the lies were correctly classified in the control condition, well below what is typically found in nonverbal lie detection research, suggesting that the lie detection task was difficult. Yet, in the experimental condition, 60 % of the lies were correctly classified, more than typically found in this type of lie detection research.

4 Discovering Hidden Patterns in Deceptive Behavior

Since no diagnostic cue to deception occurs, it could be that a diagnostic pattern does arise when a combination of cues is taken into account [8]. Senders are able to arrange a set of different signaling systems to communicate and make their communicative intentions public, like language, the paralinguistic (or suprasegmental) system, the face and gestures system, the gaze, the proxemics and the haptic, as well as the chronemics. Among others, Anolli [27] argued that each of these communicative systems bears its contribution and participates in defining the meaning of a communicative act in an autonomous way. However, the generative capacity of each signaling system should be connected to produce a global and unitary communicative action, with a more or less high consistency degree.

Meaning is not connected with a unique and exclusive signaling system but is generated by the network of semantic and pragmatic connections between different systems. Such a process is ruled out by the so-called principle of *semantic and pragmatic synchrony* [27], according to which meaning is originated by a nonrandom combination of different portions of meaning, each of whom produced by a given signaling system. Thus, the meaning of a word, an utterance or a gesture hinges upon its relations to every piece of meaning arising out of each signaling system within the same totality.

Although it sounds reasonable to suggest that looking at a combination of nonverbal and verbal behaviors will lead to more accurate classifications of liars and truth tellers than investigating nonverbal and verbal behaviors separately, researchers rarely investigate both types simultaneously. Vrij and his colleagues [28, 29] examined participants' nonverbal and verbal behavior and obtained the most accurate classification of liars and truth tellers when both the nonverbal and verbal behaviors were taken into account. On the basis of a combination of four nonverbal behaviors (illustrators, hesitations, latency period, and hand/finger movements) they correctly classified 70.6 % of participating truth tellers and 84.6 % of liars, whereas any of these behaviors separately resulted in much more disappointing findings.

Ekman and colleagues found similar patterns of nonverbal behaviors during deception. Up to 80 % of truths and lies could be detected when a trained observer paid attention to micro-facial expressions [30]. When the tone of voice was taken into account in addition to micro-facial expressions, 86 % of the truths and lies could be detected [31]. Other studies have shown that between 71 and 78 % of correct classifications were made when the researchers investigated a cluster of behaviors [28, 32, 33]. In other words, more accurate truth/lie classifications can be made if a cluster of nonverbal cues is examined rather than each of these cues separately.

Clustering nonverbal cues to deception brings out some issues that remain unresolved at present. First of all, which kind of behavior should be clustered? Currently, different researchers examine different clusters of behavior, and it cannot be ruled out that a cluster that is effective in pinpointing lying in one situation or group of participants is not effective in another situation or group of participants [8]. Second, is there a criterion for grouping different behaviors into the same pattern? For example, the temporal distance between one cue and another is decisive. How much time must elapse between two cues to consider them as part of the same pattern? Finally, liars sometimes deliberately attempt to appear credible to avoid detection. For example, able liars, or people who are informed about the operating method on nonverbal behaviors, can successfully employ countermeasures to conceal nonverbal cues to deception. If it is known amongst terrorists, spies and criminals which lie detection tools will be used to catch them, they may learn more about these tools and attempt to beat them. If they succeed in doing so, the tools are no longer effective.

Of course, people can easily control only those patterns that are manifest and have a macroscopic nature, easily readable from the outside time by time. However, patterns in behavior are frequently hidden from the consciousness of those who perform them as well as to unaided observers [34]. As Eibl-Eibesfeldt [35] argued, "behavior consists of patterns in time. Investigations of behavior deal with sequences that, in contrast to bodily characteristics, are not always visible". Order alone is not a valuable criterion to detect hidden recurrent behavior patterns because deceptive strategies are characterized by a large complexity and by a great variability in the number of behaviors occurring between the liar and his interlocutor.

One approach is to include the element of time in the analysis of nonverbal deceptive behavior. Neural networks have been used frequently over the last few decades for static pattern recognition and pattern recognition in time. Discovering the real-time multilayered and partly parallel structure of even the most common dyadic verbal and nonverbal interactions remains a formidable challenge where clues to deception may be hidden to unaided observers in anything from the tiniest of details to intricate aspects of temporal structure.

A mathematical approach that may be particularly suitable for defining and discovering repeated temporal patterns in deceptive behavior is T-pattern sequential analysis. T-pattern detection was developed by Magnus S. Magnusson [34, 36, 37] to find temporal and sequential structure in behavior. The algorithm is implemented in the software THEME [38]. T-pattern analysis focuses on determining whether arbitrary events *e1* and *e2* in a symbolic string of {*ei*} events sequentially occur within a specified time interval at a rate greater than that expected by chance. In this way, it detects repeated patterns of intra- or inter-individual behavior coded as events on one-dimensional discrete scales. This kind of analysis has been used in a wide variety of observational studies, including microanalysis of *Drosophila* courtship behavior [39], cooperative behavior between humans and dogs when constructing an object [40], patient-therapist communication in computer assisted environments [41], and analysis of soccer team play [42]. The common feature in all of the above studies is the need to identify repeated behavioral patterns that may regularly or irregularly occur within a period of observation.

Since human behavior is highly organized, in this chapter, rather than concentrating on individual nonverbal cues of lies, our aim is to observe and describe the overall organization of behavior patterns which repeat over time. In particular, we focused on the display of lying and truth telling through facial actions. Research on single cues to deception has highlighted that facial displays are not reliable indicators of deception. They have great communicative potential (e.g., eye contact is used to persuade others) and, as a result, people are practiced at using and, therefore, controlling them [8]. We were interested in verifying the presence of significant and systematic differences in the organization of facial patterns when participants tell the truth or lie, both in the condition of imposed cognitive load and that of no cognitive load manipulation, in order to detect regularities in the temporal patterns of individuals belonging to the same condition.

5 The Present Study

5.1 Method

5.1.1 Participants

Participants were 12 students, all females, aged from 20 to 26 (mean age 23.25 ± 2.4), all native to the same geographic area. They were recruited at the faculty of Psychology from the University of Milano-Bicocca. After being recruited, all participants gave their informed consent both to audio and video recording.

5.1.2 Materials and Setting

The present study was carried at the University of Milano-Bicocca in an audio-isolated laboratory room equipped with four cameras, set to video-record participants' full-lengths and close-ups. The cameras were connected to a two channel quad device (*split-screen* technique).

5.1.3 Experimental Procedure

Participants were asked to report two autobiographical episodes, the first being a simple night out with friends, the second being about the last time they went to a restaurant with relatives. They were asked to lie about several elements of one (and only one) of the two reports; 15 min were given to recall the events (for the truth condition) and prepare a story (for the untruthful report), considering that they then would have to tell these reports to a confederate who didn't know which one was untruthful. In a first condition, participants had to report the story in chronological order, while in the second, they had to report it in reverse order (starting from the end of the story and going back to the beginning). Each story had to last around 5 min, marked by audio signals.

5.1.4 Experimental Design

A 2 × 2 experimental design was carried out. Included independent variables were:

- Content (1 = truth; 2 = lie).
- Cognitive load (1 = normal—chronological order; 2 = induced—reverse order).
- Nonverbal cues (Fig. 1) were the measured dependent variables.

5.1.5 Manipulation Check

To establish the ground truth and verify cognitive load manipulation and motivation, we asked the subjects to complete a questionnaire after the experiment was finished. Later, we watched the video recordings with the participants and asked them when they lied (veracity status).

5.2 Data Analysis

The coding grid was built basing on literature review of facial displays (gaze and facial micro-movements) in lie detection [6, 8, 43, 44] (Fig. 1).

We coded 2 min of each report, using a frame-by-frame video coder (Theme coder). Each behavior occurrence was considered as

ITEM	DESCRIPTION	ITEM	DESCRIPTION
b	begin of the event type	au20	Lip Stretcher
e	end of the event type	au23	Lip Tightener
		au24	Lip Pressor
ACTOR		au25	Lips part
actor1	participant	au28	Lip Suck
		au41	Lid Drop
FACE		au43	Eyes Closed
au1	Inner Brow Raiser	au45	Blink
au2	Outer Brow Raiser		
au4	Brow Lowerer	***GAZE***	
au5	Upper Lid Raiser	gazeup	
au6	Cheek Raiser	gazedown	
au12	Lip Corner Puller	gazeleft	
au14	Dimpler	gazeright	
au17	Chin Raiser		

Fig. 1 Coding grid

a punctual event (no duration). The occurrences of each behavior (event types) form the "T-dataset".

To assess inter-rater reliability of the T-dataset, Cohen's Kappa was calculated on 10 % of the same video materials independently coded by two coders, using a "blind" coding procedure. Although differing through categories, inter-coder reliability was found to be good to satisfactory (ranging from 0.79 to 0.91; $p < 0.05$). When disagreements were identified or the agreement was not perfect, the specific cases were discussed and agreed by both coders.

Datasets were then analyzed using Theme 6 XE3 beta software. A first analysis was conducted, aimed to explore the effects of the independent variables on the organization of nonverbal behavior in each condition, identifying their most significant patterns. We had a theme project file for each condition (normal/truth, normal/lie, reverse/truth, reverse/lie). Individual datasets were joined together through the "concatenate into a multi sample file" function.

We then conducted a second analysis, exploring the specific qualities that characterize truthful and untruthful behaviors within the "normal" and "reverse" conditions through the aforementioned concatenate into a multi sample file function.

5.2.1 Detection Procedure

Theme detects statistically significant time patterns in sequences of behaviors. The term T-pattern stands for temporal pattern; it is based on the timing of events, relative to each other. T-pattern detection [36, 37, 45] was developed for finding temporal and sequential structure in behavior. The algorithm implemented in the software detects repeated patterns of intra- or inter-individual behavior coded as events on one-dimensional discrete scales. T-patterns are repeated occurrences of two or more event types such that their order (allowing also for concurrence) is the same each time and the distances between them are significantly fixed as defined by a statistically defined and detected critical interval relationship. A minimal T-pattern consists of two event types. An event type is a category of observable behavior whereas an event is an instance of behavior occurring at a particular time unit without a duration [36]. The search naturally stops when no more patterns can be found and the result of such procedure is the discovery of patterns occurring more often than chance and embedded within the raw data stream in a nonintuitive manner.

By means of relevant options, the program allows the experimenter to set specific search parameters. Parameters set in present study were: critical interval type (free), significance (0.0001), minimum occurrence (5), lumping factor (0.90), minimum % of samples (100), packet base type (off), types of randomizations (shuffling and rotation), number of runs per type (20).

5.2.2 Selection Procedure

In the first analysis, complex T-patterns were considered, since they are regarded as the most interesting, due to their potential meaning and the (low) likeliness to be detected using other definitions and algorithms [36, 38]. However, instead of ignoring less complex patterns (e.g., simple two event type patterns), we selected those presenting a demonstrated behavioral relationship, basing our choices on literature review.

In the second analysis we used the function of Theme "selection – multi-sample file selection—statistical". Theme select patterns that appear significantly more often in samples selected than in the multi-sample file as a whole. The level of significance was 0.05.

5.3 Results

5.3.1 First Analysis

We compared the number of patterns detected (Fig. 2) and the mean of their lengths and number of levels (Fig. 3) among the four experimental conditions.

Qualitative Assessment

Behavioral Patterns

Within the first condition (truth when cognitive load is normal), most relevant patterns included a combination of different eyes movements with blinking (au45) and eyes closed (au43), a combination of different action units (au1 + au4; au1 + au2), and spontaneous smile (au6 + au12).

Analogous patterns were detected in the second condition (lie when cognitive load is normal). In particular, significative patterns included a combination of different eyes movements with blinking (au45) (e.g., Fig. 4), chin raised (au17) and spontaneous smile (au6 + au12), and a combination of different action units (au1 + au2).

Within the third condition (truth when cognitive load is induced), lots of patterns included a combination of different eyes

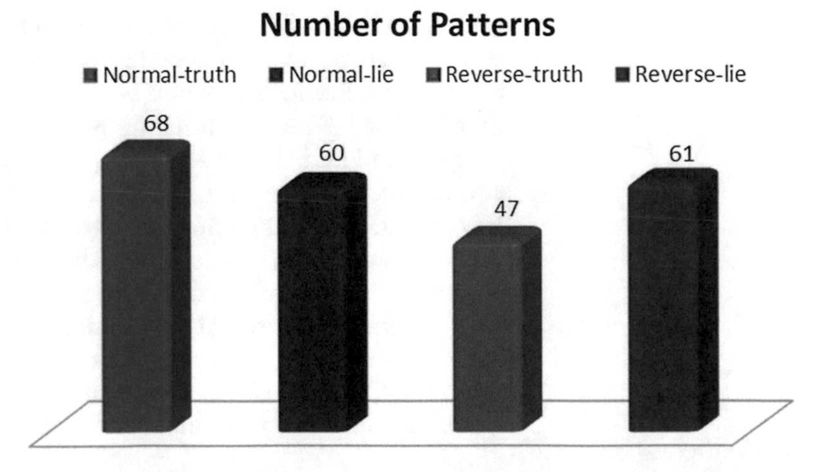

Fig. 2 First analysis: number of patterns in different conditions

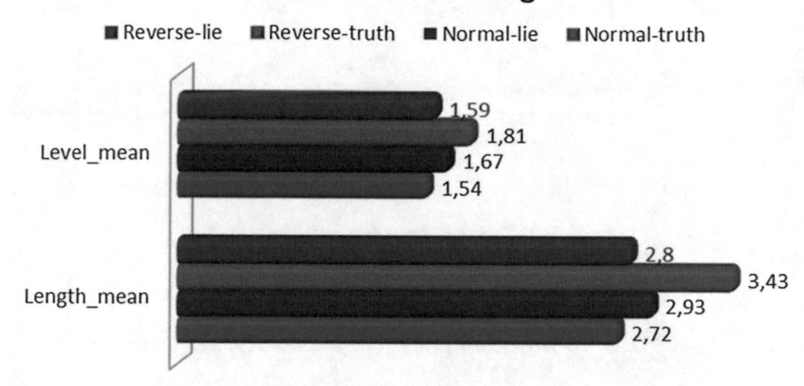

Fig. 3 First analysis: patterns' level and length in different conditions

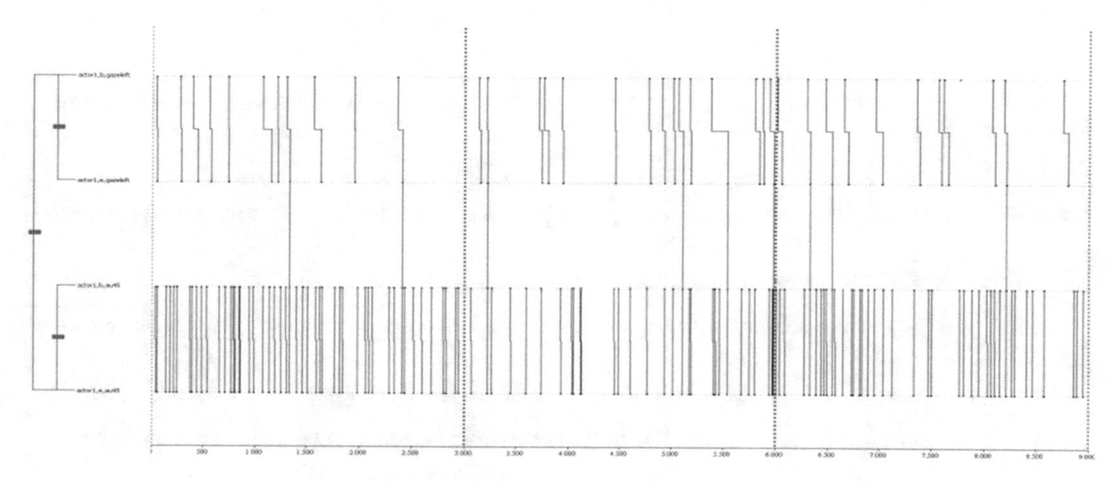

Fig. 4 A T-pattern in normal-lie condition: combination of eyes movements with blinking (au45)

movements with blinking (au45), and spontaneous smile (au6 + au12) (Fig. 5). Most patterns were composed of the repetition of a single event type.

In the fourth condition (lie when cognitive load is induced), most relevant patterns included a combination of eyes movements with blinking (au45) and eyes closed (au43), a combination of different action units (au17 + au23), and fake smile (au12 + au25) (Fig. 6). Lots of patterns were composed of one repeated event type.

5.3.2 Second Analysis

Qualitative Assessment

We compared the number of patterns detected (Fig. 7) and the mean of their lengths and number of levels (Fig. 8) between two experimental conditions (the two levels of the independent variable "cognitive load").

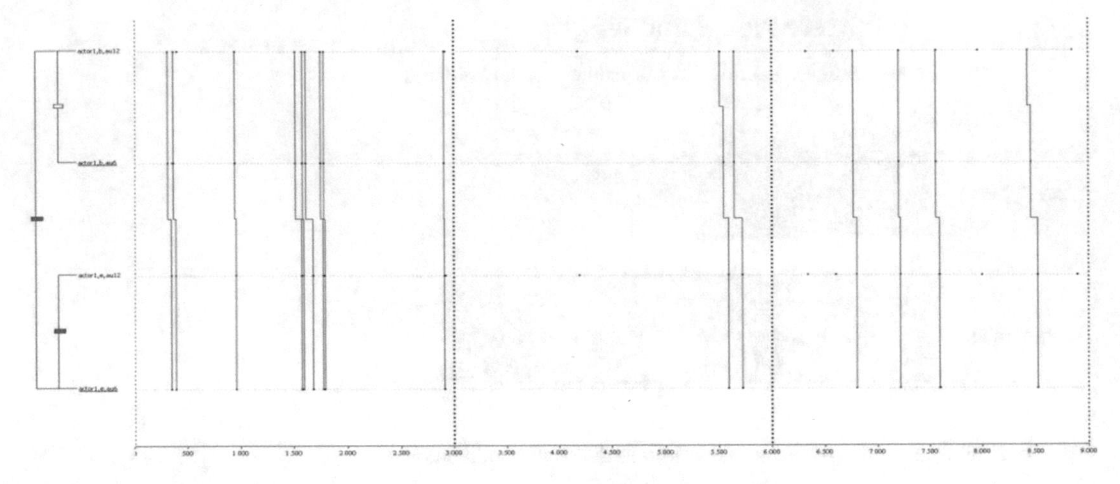

Fig. 5 A T-pattern in reverse-truth condition: spontaneous smile (au6 + au12)

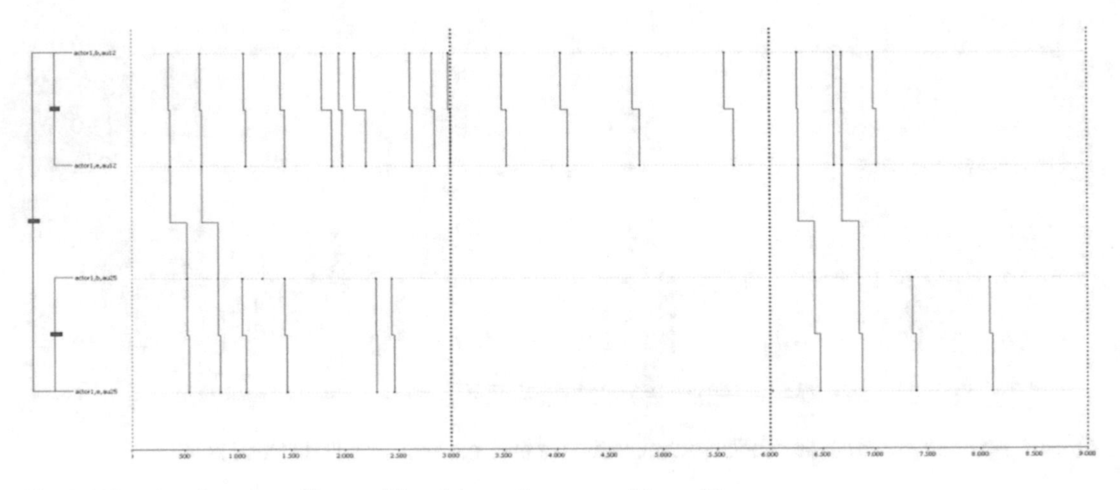

Fig. 6 A T-pattern in reverse-lie condition: fake smile, e.g., au12 + au25

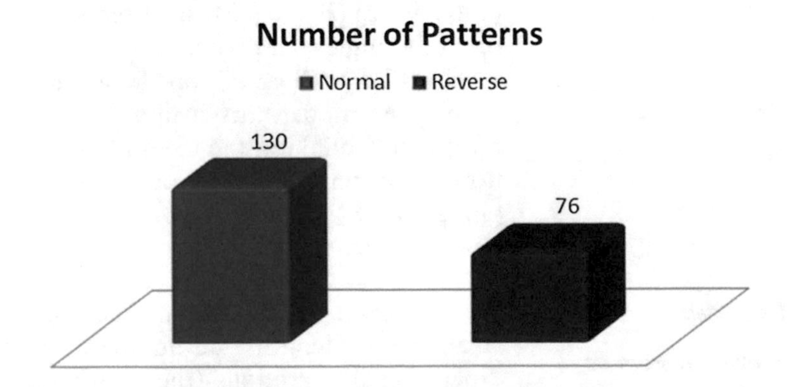

Fig. 7 Second analysis: number of patterns in different conditions

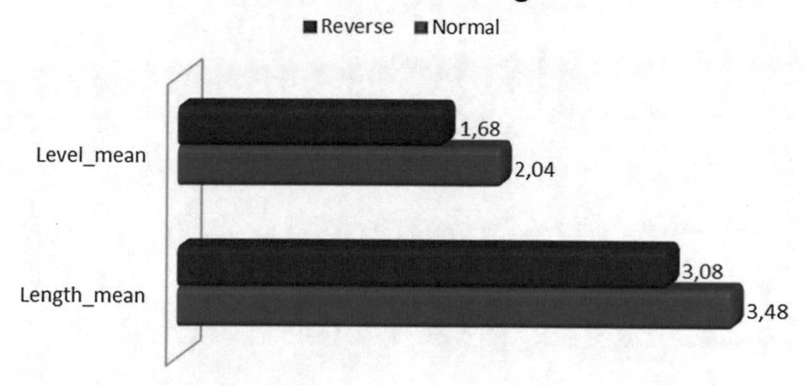

Fig. 8 Second analysis: patterns' level and length in different conditions

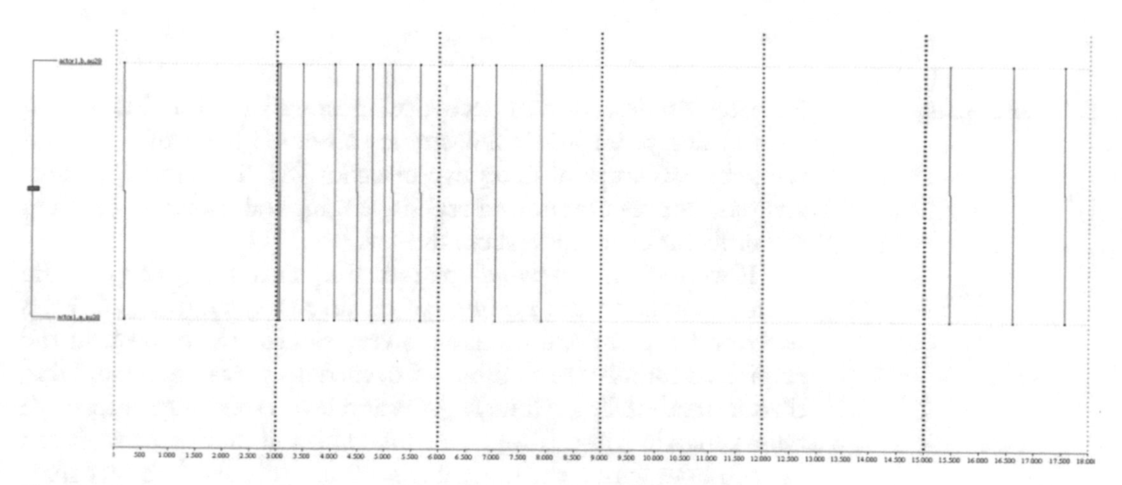

Fig. 9 A significantly distinctive T-pattern of truth in normal condition: au28

Behavioral Patterns

In the condition where the cognitive load was normal, the truth had only one significantly distinctive pattern compared to the entire dataset. It was composed only by the repetition of a single action unit (au28) (Fig. 9).

The lie instead had five significantly distinctive patterns compared to the entire dataset, most were composed by event types belonging to different eyes movements (gaze up, gaze down, and gaze right).

In the condition in which cognitive load was induced (reverse), the truth had four significantly distinctive patterns compared to the entire dataset. Patterns more relevant contained the spontaneous smile, present in three different patterns.

The lie had three significantly distinctive patterns compared to the entire dataset. They concerned the simultaneous presence of gaze up and gaze down (e.g., Fig. 10).

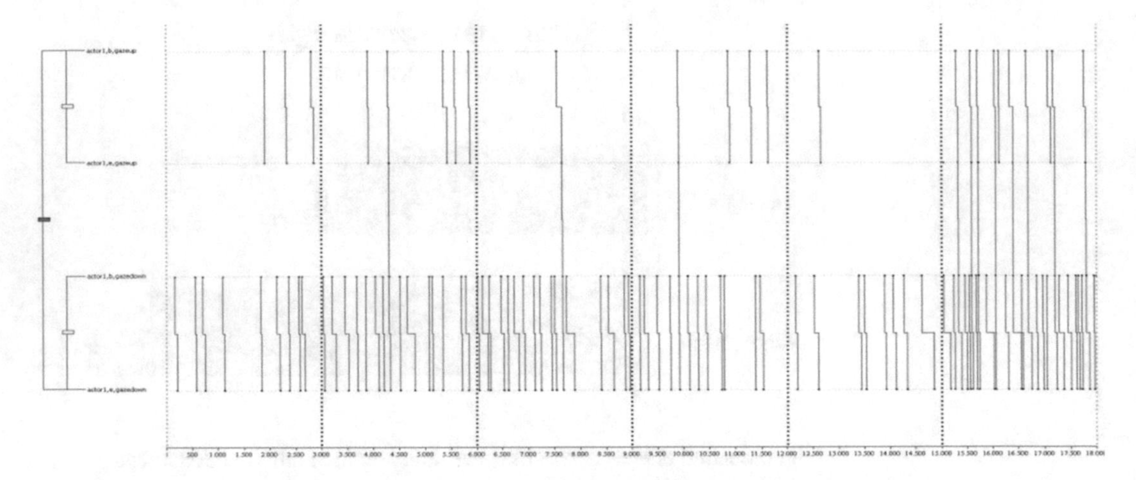

Fig. 10 A significantly distinctive T-pattern of lie in reverse condition: gaze up and gaze down

5.4 Discussion

Research on detection of deception mostly focused on identifying cues to deception, while few studies observed the sequential and temporal structure of deceptive behavior [8]. Through T-pattern analysis, our results showed repetitive temporal patterning among the different experimental conditions.

If we look at the overall pattern frequencies by group, in the normal condition the number of occurrences, lengths and levels between lying and truth telling is very similar. In contrast, in the reverse condition the number of occurrences was higher in lying than in truth telling, while lengths and levels were higher in truth telling than in lying. If we take a look at significant patterns, it can be seen how, within the normal condition, patterns were very similar among lying and truth telling. Moreover, they were difficult to interpret due to a combination of different eyes movements and different action units. Conversely, very regular patterns are more significantly present in the reverse condition, both in the lie and truth conditions (most patterns contain sub-patterns with repetition of event types). Since patterns in the reverse condition are more recurrent, regular and simpler than the ones seen in the normal condition, it is easier to assign a meaning to them. Besides, being the patterns in the reverse—truth condition quite different from the ones in the reverse—lie condition, it could be easier to discriminate nonverbal behaviors between liars and truth tellers in the reverse condition, than doing it in the normal one.

As Vrij stated [8], certain behavioral patterns are associated with honesty and likeability, such as directed gaze to a conversation partner, smiling, head nodding, leaning forward, direct body orientation, posture mirroring, uncrossed arms, articulate gesturing, moderate speaking rates, a lack of –ums and –ers, and vocal variety

[46–49]. Some people show such demeanor naturally even when they are lying (e.g., *natural performers*) [50]. These natural performers are likely to be good liars because their natural behavior is likely to allay suspicion. DePaulo and colleagues [6] examined around 100 different nonverbal behaviors. Significant findings emerged for 21 behaviors. The cues were ranked in terms of their effect sizes. The highest effect sizes were found in the cues that have not often been investigated, such as changes in foot movements, pupil changes, smiling.

In all conditions, except for the reverse-lie one, there are patterns containing event types ascribable to the spontaneous smile [47] (au6 + au12). This behavior is associated to truthful communication in literature [47] and it could be helpful to discriminate lying and truth telling in the reverse condition.

Patterns with a combination of different gaze movements were disclosed in all conditions. Even blinking (au45) does not seem to be significantly discriminative for the different conditions. However, when we compared truth and lies in the reverse condition, gaze movements (e.g., gaze up and gaze down) were significantly more present in lying than in truth telling. Moreover, there is a higher presence of closed eyes (au43) in both normal truth and reverse lie conditions than in other conditions. A number of researchers have linked excessive gaze aversion with increased cognitive load [51, 52]. During difficult cognitive activities, we often close our eyes, look up at the sky, or look away from the person we're talking to [53]. The fact that this behavior is present in the reverse lie condition and not in the normal lie or in the reverse truth could confirm this hypothesis, emphasizing the effectiveness of the reverse order as a method to induce cognitive load and to emphasize the differences between lying and truth telling.

6 Conclusion

Our data seems to support our objectives and proposals, even though it is clearly subject to some limitations. The most important limit, related to the use of self-reports and autobiographical episodes, is a *ground truth bias.* Can we really trust truth tellers? Did the liars lie for real? There is no legal way to assess the veracity status of most of our participants' statements; for example, lie tellers may also give socially desirable versions of their stories, violating the instructions we gave them [52].

In addition, we cannot be sure that participants are perfectly able to fully recall target events, even during the truthful conditions. Moreover, all participants who had to report (truthfully or not) their story in reverse order, experienced serious difficulties in recalling (and even telling) the events in that modality.

There are some factors that had to be excluded from this preliminary study, that we plan to explore in our next studies; we confined our studies to young females participants only, but it is a certain fact that this method should be applied to a wider sample, variable in gender and age range.

Future developments of this research will also take into account interactions between lie/truth tellers and their interlocutor, due to the importance of interpersonal processes involved in deceptive communication, which is created and ruled by a reciprocal game between communicators [27].

References

1. Bond CF, Omar A, Pitre U, Lashley BR, Skaggs LM, Kirk CT (1992) Fishy-looking liars: deception judgment from expectancy violation. J Pers Soc Psychol 63:969–977
2. Buller DB, Burgoon JK (1996) Interpersonal deception theory. Commun Theory 6:203–242
3. Kaplar ME, Gordon AK (2004) The enigma of altruistic lying: perspective differences in what motivates and justifies lie telling within romantic relationships. Pers Relationships 11:489–507
4. Burgoon J, Nunamaker J (2004) Toward computer-aided support for the detection of deception. Group Decis Negot 13:1–4
5. Bond CF Jr, DePaulo BM (2006) Accuracy of deception judgments. Pers Soc Psychol Rev 10:214–234
6. DePaulo BM, Lindsay JJ, Malone BE, Muhlenbruck L, Charlton K, Cooper H (2003) Cues to deception. Psychol Bull 129:74–118
7. Hartwig M, Bond CF Jr (2011) Why do lie-catchers fail? A lens model meta-analysis of human lie judgments. Psychol Bull 137:643–659
8. Vrij A (2008) Detecting lies and deceit: pitfalls and opportunities. Wiley-Interscience, New York, NY
9. Vrij A, Winkel FW, Akehurst L (1997) Police officers' incorrect beliefs about nonverbal indicators of deception and its consequences. In: Nijboer JF, Reijntjes JM (eds) Proceedings of the first world conference on new trends in criminal investigation and evidence. Koninklijke Vermande, Lelystad, pp 221–238
10. Breuer MM, Sporer SL, Reinhard MA (2005) Subjektive Indikatoren von Täuschung. Zeitschr Sozialpsychol 36:189–201
11. Granhag PA, Strömwall LA (2004) The detection of deception in forensic contexts. Cambridge University Press, Cambridge
12. Vrij A, Granhag PA, Mann S, Leal S (2011) Outsmarting the liars: toward a cognitive lie detection approach. Curr Dir Psychol Sci 20:28–32
13. Zuckerman M, DePaulo BM, Rosenthal R (1981) Verbal and nonverbal communication of deception. Adv Exp Soc Psychol 14:1–57
14. Vrij A, Semin GR, Bull R (1996) Insight into behavior displayed during deception. Hum Commun Res 22:544–562
15. Caso L, Gnisci A, Vrij A, Mann S (2005) Processes underlying deception: an empirical analysis of truth and lies when manipulating the stakes. J Invest Psychol Off 2:195–202
16. Vrij A, Mann SA, Fisher RP, Leal S, Milne R, Bull R (2008) Increasing cognitive load to facilitate lie detection: the benefit of recalling an event in reverse order. Law Hum Behav 32:253–265
17. Inbau FE, Reid JE, Buckley JP (2011) Criminal interrogation and confessions. Jones & Bartlett Learning, Gaithensburg, MD
18. Mann S, Vrij A, Bull R (2002) Suspects, lies, and videotape: an analysis of authentic high-stake liars. Law Hum Behav 26:365–376
19. Vrij A, Leal S, Granhag PA, Mann S, Fisher RP, Hillman J, Sperry K (2009) Outsmarting the liars: the benefit of asking unanticipated questions. Law Hum Behav 33:159–166
20. Anolli L, Balconi M, Ciceri R (2002) Deceptive miscommunication theory (DeMiT): a new model for the analysis of deceptive communication. In: Anolli L, Ciceri R, Riva G (eds) Say not to say: new perspectives on miscommunication. Ios Press, Amsterdam, pp 75–104
21. Searle JR (1979) Expression and meaning: structures and theory of speech acts. Cambridge University Press, London
22. McCornack SA (1997) The generation of deceptive messages: laying the groundwork for a viable theory of interpersonal deception.

In: Greene JO (ed) Message production: advances in communication theory. Routledge, New York, NY, pp 91–126

23. Gilbert JAE, Fisher RP (2006) The effects of varied retrieval cues on reminiscence in eyewitness memory. Appl Cogn Psychol 20:723–739
24. Kahana MJ (1996) Associative retrieval processes in free recall. Mem Cognit 24:103–109
25. Geiselman RE, Callot R (1990) Reverse versus forward recall of script-based texts. Appl Cogn Psychol 4:141–144
26. Briggs GE, Peters GL, Fisher RP (1972) On the locus of the divided-attention effects. Atten Percept Psychophys 11:315–320
27. Anolli L (2002) MaCHT-miscommunication as chance theory: toward a unitary theory of communication and miscommunication. In: Anolli L, Ciceri R, Riva G (eds) Say not to say: new perspectives on miscommunication. Ios Press, Amsterdam, pp 3–43
28. Vrij A, Akehurst L, Soukara S, Bull R (2004) Detecting deceit via analyses of verbal and nonverbal behavior in children and adults. Hum Commun Res 30:8–41
29. Vrij A, Edward K, Roberts K, Bull R (2000) Detecting deceit via analysis of verbal and nonverbal behaviour. J Nonverbal Behav 24:239–263
30. Frank MG, Ekman P (1997) The ability to detect deceit generalizes across different types of high-stake lies. J Pers Soc Psychol 72:1429–1439
31. Ekman P, O'Sullivan M, Friesen WV, Scherer KR (1991) Face, voice, and body in detecting deceit. J Nonverbal Behav 15:125–135
32. Davis M, Markus KA, Walters SB, Vorus N, Connors B (2005) Behavioral cues to deception vs. topic incriminating potential in criminal confessions. Law Hum Behav 29:683–704
33. Heilveil I, Muehleman JT (1981) Nonverbal clues to deception in a psychotherapy analogue. Psychother Theory Res Pract 18:329–335
34. Magnusson MS (2006) Structure and communication in interactions. In: Riva G, Anguera MT, Wiederhold BK, Mantovani F (eds) From communication to presence: cognition, emotions and culture towards the ultimate communicative experience. Festschrift in honor of Luigi Anolli. IOS, Amsterdam, pp 127–146
35. Eibl-Eibesfeldt I (1970) Ethology. The Biology of Behavior. Holt, Rinehart and Winston, Inc., New York, NY
36. Magnusson MS (2000) Discovering hidden time patterns in behavior: T-patterns and their detection. Behav Res Methods 32:93–110
37. Magnusson MS (2005) Understanding social interaction: discovering hidden structure with model and algorithms. In: Anolli L, Duncan S Jr, Magnusson MS, Riva G (eds) The hidden structure of interaction. From neurons to culture patterns. IOS, Amsterdam, pp 3–22
38. Magnusson MS (2004) Repeated patterns in behavior and other biological phenomena. Evol Commun Syst 7:111–128
39. Arthur BI, Magnusson MS (2005) Microanalysis of Drosophila courtship behaviour. In: Anolli L, Duncan S Jr, Magnusson MS, Riva G (eds) The hidden structure of interaction. From neurons to culture patterns. IOS, Amsterdam, pp 99–106
40. Kerepesi A, Jonsson GK, Miklosi A, Topál J, Csányi V, Magnusson MS (2005) Detection of temporal patterns in dog–human interaction. Behav Processes 70:69–79
41. Riva G, Zurloni V, Anolli L (2005) Patient-therapist communication in a computer assisted environment. In: Anolli L, Duncan S Jr, Magnusson MS, Riva G (eds) The hidden structure of interaction. From neurons to culture patterns. IOS, Amsterdam, pp 159–177
42. Camerino OF, Chaverri J, Anguera MT, Jonsson GK (2011) Dynamics of the game in soccer: detection of T-patterns. Eur J Sport Sci 12:216–224
43. Burgoon JK, Guerrero LK, Floyd K (2009) Nonverbal communication. Allyn & Bacon, Boston, MA
44. Ekman P, Friesen WV (1978) Facial action coding system: a technique for the measurement of facial movement. Consulting Psychologists Press, Palo Alto
45. Borrie A, Jonsson GK, Magnusson MS (2002) Temporal pattern analysis and its applicability in sport: an explanation and exemplar data. J Sports Sci 20:845–852
46. Buller DB, Aune RK (1987) Nonverbal cues to deception among intimates, friends, and strangers. J Nonverbal Behav 11:269–290
47. Ekman P (2001) Smiling. In: Blakemore C, Jennett S (eds) Oxford companion to the body. Oxford University Press, London
48. Ekman P (2009) Telling lies: clues to deceit in the marketplace, politics, and marriage. WW Norton & Company, New York, NY
49. Tickle-Degnen L, Rosenthal R (1990) The nature of rapport and its nonverbal correlates. Psychol Inquiry 1:285–293
50. Ekman P (1997) Deception, lying, and demeanor states of mind: American and post-Soviet perspectives on contemporary issues in psychology. Oxford University Press, Oxford, pp 93–105

51. Beattie GW (1981) A further investigation of the cognitive interference hypothesis of gaze patterns during conversation. Br J Soc Psychol 20(4):243–248
52. Ellyson SL, Dovidio JF, Corson RL (1981) Visual behavior differences in females as a function of self-perceived expertise. J Nonverbal Behav 5(3):164–171
53. Doherty-Sneddon G, Bruce V, Bonner L, Longbotham S, Doyle C (2002) Development of gaze aversion as disengagement from visual information. Dev Psychol 38(3):438–445

Chapter 4

Paraverbal Communicative Teaching T-Patterns Using SOCIN and SOPROX Observational Systems

Marta Castañer, Oleguer Camerino, M. Teresa Anguera, and Gudberg K. Jonsson

Abstract

This chapter focuses on how to analyze the paraverbal communicative fluency of teaching style. Essential paraverbal criteria related to kinesics and proxemics were studied in lecturers offering courses. Some lessons were analyzed using the Observational Systems of Paraverbal Communication SOCIN and SOPROX, both observational instruments that enables a broad analysis of kinesics and proxemics. The recording instrument used to codify SOCIN and SOPROX was LINCE software and the Theme software was used to detect temporal patterns (T-patterns) in the observational data. The results reveal the power of the teachers' illustrative and regulatory kinesics. The regulatory function makes use of clearly defined kinesic gestures such as emblems and kinetographs, whereas the illustrative function is accompanied by largely undefined kinesic gestures.

Key words Paraverbal communication, Kinesics, Proxemics, Observational instruments, T-Pattern analysis

1 Teaching Communication

In order to understand and improve the scenarios to be managed by teachers it is important to identify the essential aspects of communication, such as gestures, voice quality, and the use of teaching time and space, which are associated with the teaching discourse. In this regard, it is clear that one of the keys in optimizing teaching tasks lies in paying close attention to the communication and teaching style that each teacher may develop and rework over time. More recently, accurate and detailed reviews demonstrated that very little educational research has been concerned with the role of gestures in teaching and learning. "The few existing studies that focus on gesture in an education context, often appearing in journals whose primary focus is not educational research, suggest that such research might be of tremendous importance in helping to understand better the role of gestures in knowing and learning science." ([1]: 365).

Magnus S. Magnusson et al. (eds.), *Discovering Hidden Temporal Patterns in Behavior and Interaction: T-Pattern Detection and Analysis with THEME™*, Neuromethods, vol. 111, DOI 10.1007/978-1-4939-3249-8_4,

An intrinsic part of all teaching activity is a constant communicational flow, in which the spontaneous nature of communication is considered to be a habitual feature. The observation of students' reactions may thus be useful for optimizing this communication [2]. As such, there is good reason why communication is regarded as an indicator of the communicator's emotional, as well as symbolic experiences [3]. Symbolic communication is intentional communication that uses learned, socially shared signal systems of propositional information transmitted via symbols.

Observational methodology was used due to the habitual nature of teachers' behavior and the fact that the context is a naturalistic one. The flexibility and rigor of this methodology makes it fully consistent with the characteristics of the study and it has become a standard approach to observational research [4–6], especially in the field nonverbal communication, motor behavior, and dance [7–14]. The empirical results indicate the power of the teachers' illustrative and regulatory kinesics, along with the proxemics used: the regulatory function is combined with the static posture, whereas the illustrative form accompanies movement. The regulatory function makes use of clearly defined kinesic gestures such as emblems and kinetographs, whereas the illustrative function is accompanied by largely undefined kinesic gestures such as beats.

2 The Singular Nature of Paraverbal Communication

Before proceeding it is important to clarify an aspect related to the concepts *nonverbal* and *paraverbal*. In our view the use of the negative prefix implies that the terms "verbal" and "nonverbal" should be understood as being mutually exclusive, when in fact they refer to two forms of communication that go hand in hand with one another. Therefore, and with the aim of respecting the meaning of the concepts under study, we opt to use the concept of paraverbal communication. The communicative reality in which humans live is understood in terms of the linearity and sequential nature of verbal language, which is produced by a single phonatory organ that is unable to emit simultaneous sounds; in other words, we cannot say *a* and *b* at the same time, and, therefore, verbal discourse can be assimilated to the concept of melody. A further issue is that all discourse which is not strictly verbal is characterized by simultaneity. The diverse—and at the same time, bilateral—structure of our corporeity enables us to generate bodily postures (dynamism), gestures (dynamism), and attitudes (meaning) in a simultaneous way [10] and also "…gestures are often subsequently replaced by an increasing reliance upon the verbal mode of communications" [15]. Paraverbal teaching style refers to the ways in which a teacher conveys his or her educational discourse, and this is why it is sometimes associated with the idea of expressive movement [16]. The paraverbal structure of communication will be addressed here according to four dimensions: kinesics, proxemics, chronemics and paralanguage. All these dimensions of analysis have been considered for many years by key authors in the field [17–19]. In teaching discourse these dimensions can be defined as follows:

(a) *Kinesics*: the study of patterns in gesture and posture that are used by the teacher with or without communicative meaning.
(b) *Proxemics*: the study of how the teacher uses the space in which teaching takes place.
(c) *Chronemics*: the study of how the teacher uses the temporal factors that influence the teaching setting.
(d) *Paralanguage*: the study of all those vocal emissions that are not included in arbitrary verbal language, but which do accompany it.

These dimensions are associated with the study of bodily gesture, the use of space, the use of time, and voice-related paralanguage. Here we outline a theoretical framework for teachers' communicative behavior that delineates kinesic and proxemics dimensions in the Observational System of Kinesic Communication SOCIN and Observational System of Proxemic Communication SOCIN [11], integrating them in an exhaustive and mutually exclusive way. These dimensions can appear simultaneously or concurrently, functioning in an integrated and systemic way. If communication is to be effective, it is necessary to ensure that all the paraverbal dimensions are congruent, i.e., that they seek to transmit the same message, strengthening, confirming, and heightening it in accordance with the educational circumstances [20]. The present study is focused on two of these dimensions, proxemics and kinesics, and the next section provides a more detailed conceptual description of these.

2.1 What Gestures Come From?

It is important to clarify a conceptual aspect that continues to be overlooked in the field of kinesic language based on human motor behavior. Firstly, it is necessary to distinguish between kine, posture, gesture, and attitude with respect to the body.

(a) *Kine*: the basic unit of movement, comparable to the phoneme of verbal language.
(b) *Body posture* refers to the static nature of the body in relation to the position of its various osteoarticular and muscular parts.
(c) *Body gesture* refers to the dynamic nature of the body, without forgetting that each gesture is comprised of multiple micro-postures.
(d) *Body attitude* refers to the meaning that each social group gives to the emotional and expressive ways of using postures and gestures.

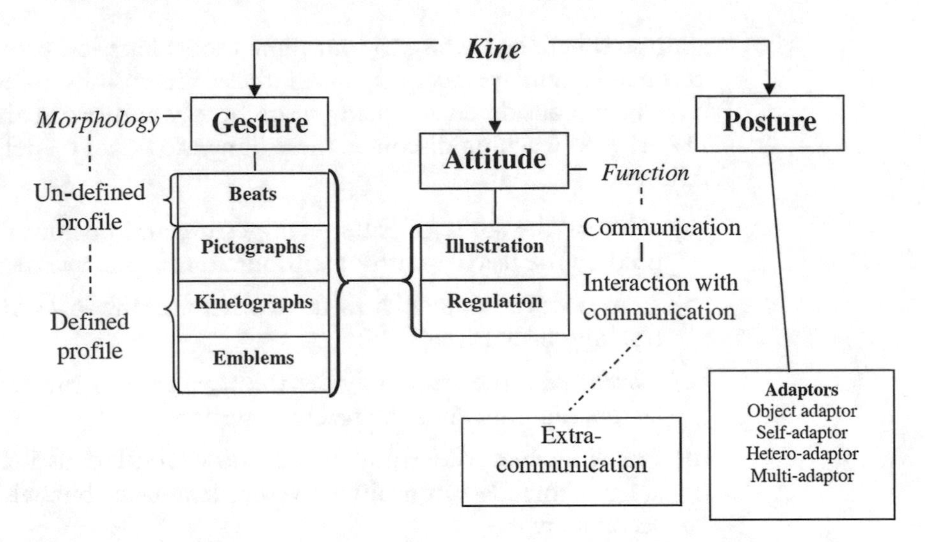

Fig. 1 Relationship between the morphology and function of kinesic gestures

On the basis of this initial clarification, gesture can be regarded as the basic unit of meaning for constructing the paraverbal kinesic observational system. Consider the chart shown in Fig. 1.

Related to the *morphology* of the categories (see Fig. 2) we establish a continuum that encompasses:

(a) Gestures that offer a highly defined profile and which are clearly observable by the receiver.

(b) Gestures with a less well-defined and weaker profile.

A clear example of kines that have their own meaning, and which offer a highly defined gesture profile, is emblems [21].

With respect to their *functionality* we establish a continuum that encompasses:

(a) Gestures with a purely communicative purpose.

(b) Gestures whose purpose is communication with interaction.

(c) Extra-communicative gestures, i.e., those without any explicit interactive or communicative purpose.

2.2 Kinesic and Proxemics Functions

The present paper focuses on how to analyze and optimize the paraverbal communicative fluency of teaching style using SOCIN and SOPROX observational systems [11], an observational instrument that provides a clear analysis of the use of kinesics and proxemics in teaching. Each teacher will have his or her own style of communication and verbal and paraverbal expression, but despite this diversity, gestures are inscribed in the conventions of the society in which a person lives. An individual's

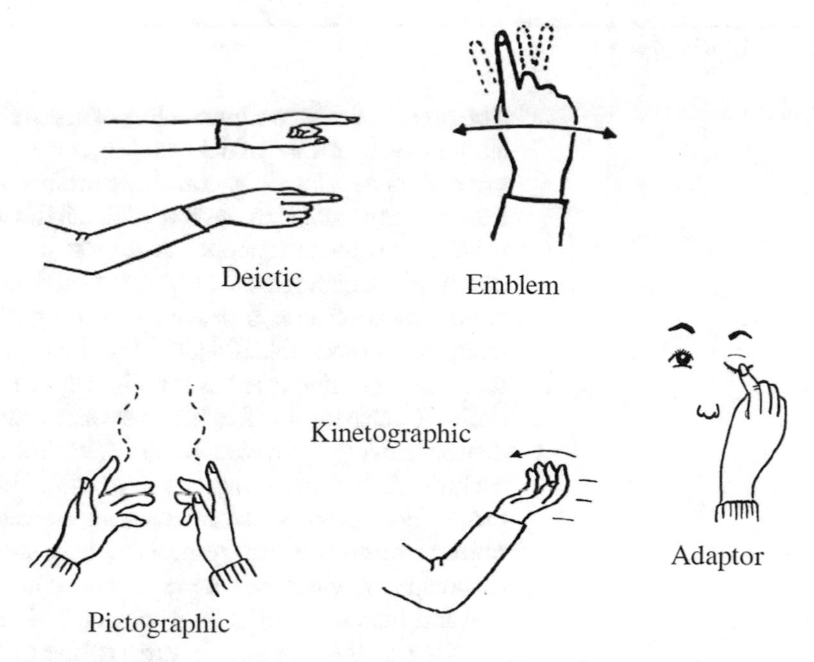

Fig. 2 Examples of gesture morphology: deictic, emblems, pictographic, kinetographic, and adaptor

identity, however unique, is never constructed in isolation as it has to be communicated, and as such it must draw upon social conventions. Thus, paraverbal communication is subject to certain social standards regarding forms of gesture and proxemics that cannot escape the ethnoaesthetics of each historical moment. Therefore, despite the concrete and unique nature of each body [22] it is possible to identify certain kinesic and proxemic functions and morphologies that are sufficiently generalized and which are of considerable interest with respect to teaching. Identifying these features is a central aim of the present study. Given that we use observational methodology [23] in the habitual setting of teachers' behavior, the context is a naturalistic one. In sum, the present study seeks to provide answers to the following questions:

1. Is it possible to codify, in an exhaustive, clear, and manageable way all the possible forms of paraverbal communication used by teachers?
2. How we can obtain behavioral patterns from teachers' communicative competencies that reveal both kinesic and proxemic aspects at the same time?
3. Does teachers' use of proxemics and illustrative and regulative gestures differ between expository situations and interactive ones?

3 Method

The present study continues a line of research being conducted in our laboratory, by means of field studies, since the award of a pedagogical prize. Paraverbal actions constitute an important source of information when the aim is to observe in detail the versatility of human communication. Indeed, people spend less time vocalizing than they do emitting behaviors that are not strictly verbal. In the field of paraverbal communication one is dealing with a type of behavior that despite being very perceivable [24], tends to be largely overlooked due to the sociocultural value that has traditionally been ascribed to it. Within a chain of behavior it is feasible to separate and demarcate behaviors of diverse etiology provided we are referring to discreet and mutually exclusive behaviors. However, in practice this is not always a realistic proposition, since social interaction, as well as interaction between objects, means that many aspects of behavior are interlinked and occur in synchrony, which, of course, is of enormous interest from a conceptual and methodological point of view [25, 26].

One of the inherent features of the optimum form of observational methodology is precisely the absence of standardized instruments. Here, observational methodology was used due to the habitual nature of teachers' behavior and the fact that the context is a naturalistic one. The flexibility and rigor of this methodology makes it fully consistent with the characteristics of the study and it has become a standard approach to observational research [4, 27]. Of particular relevance is its multidimensional nature, which enables it to be adapted to the successive events of paraverbal behavior, as well as to each of its components. In sum, observational methodology can be applied to many different facets of human communication [11, 28–31], and the wide range of possibilities it offers enables us to optimize the demarcation of units or the development of ad hoc instruments such as SOCIN and SOPROX, which combines the field format with category systems. The field format is an open, multidimensional system with multiple and self-regulating codes.

3.1 Design

Here we observed the kinesic and proxemic communication used by teachers in high education courses. Given that the aim of this paper is to present the observation instrument we present a Nomotetic, Punctual, and Multidimensional (N/P/M) [27] design that aims to identify the "intention" of the behavior rather than its extension. Thus, kinesic and proxemic responses are subjected to detailed, in-depth, and specific investigation.

3.2 Participants

Classroom-based lessons on various subjects and taught by four lecturers offering high education courses were recorded. A total of four lessons, each taught by a different teacher, were analyzed. In this study we were not interested in analyzing the individual communicative style of each teacher involved, but rather we sought to identify teachers' communicative style as a whole. Therefore, the sample used comprised four teaching sessions with a mean duration of 50 min, and this entailed analyzing a total of 1120 observation frames ($\overline{x}$ =280 frames/session). The procedure was in line with APA ethics and was approved by the university departments involved. The project did not involve any experiments or manipulation of subjects. The results are based on data obtained from recordings from public TV, and adjusted to the Belmont Report (National Commission for the Protection of Human Subjects of Biomedical and Behavioral Research, 1979) in order to assure that subjects' rights have been protected.

3.3 Instruments

The observation instruments used were SOCIN and SOPROX, which enables the different levels of kinesic and proxemic response to be systematically observed. Kinesic responses were recorded by means of the Observation of Kinesic Gestures (SOCIN; see Table 1), while proxemic gestures were recorded via the Observation of Proxemics (SOPROX; see Table 2). Both systems have been

Table 1
SOCIN: system of observation for kinesic communication [11]

Dimension	Analytical categorization	Code	Description
Function Dimension that refers to the intention of the spoken discourse that the gesture accompanies	Regulatory	RE	Action by the teacher whose objective is to obtain an immediate response from receivers. It comprises imperative, interrogative, and instructive phrases with the aim of exemplifying, giving orders or formulating questions and answers
	Illustrative	IL	Action that does not aim to obtain an immediate response from the receiver (although possibly at some future point). It comprises narrative, descriptive, and expository phrases with the aim of getting receivers to listen
Morphology Dimension that refers to the iconic and biomechanical form of gestures	Emblem	EMB	Gesture with its own preestablished iconic meaning
	Deictic	DEI	Gesture that indicates or points at people, places, or objects
	Pictographic	PIC	Gesture that draws figures or forms in space
	Kinetographic	KIN	Gesture that draws actions or movements in space
	Beats	BEA	Iconically undefined gesture used exclusively by the sender and which usually only accompanies the logic of spoken discourse

(continued)

Table 1 (continued)

Dimension	Analytical categorization	Code	Description
Situational Dimension that refers to a wide range of bodily actions, which usually coincide with parts of the teaching process that cover a certain period	Demonstrate	DE	When the teacher performs in gestures that which he or she wishes the students to do
	Help	HE	When the teacher performs actions with the intention of supporting or improving the contributions of students
	Participate	PA	When the teacher participates alongside students
	Observe	OB	Period of time during which the teacher shows an interest in what is happening in the classroom with the students
	Provide material	PM	When the teacher handles, distributes, or uses teaching material in accordance with the educational setting
Adaptation Dimension that refers to gestures without communicative intentionality in which the teacher makes contact with different parts of his/her body, or with objects or other people	Situational	AF	When the teacher uses an emotionally charged gesture with respect to the students
	Object adaptor	OB	When the teacher maintains contact with objects but without any communicative purpose
	Self-adaptor	SA	When the teacher maintains contact with other parts of his/her body but without any communicative purpose
	Hetero-adaptor	HA	When the teacher maintains bodily contact with other people but without any communicative purpose
	Multi-adaptor	MUL	When several of these adaptor gestures are combined

successfully used in previous research to observe the behavior of expert and novice teachers [11]. These instruments combine the field format (since the investigation is multidimensional) with the category systems (SOCIN and SOPROX), which fulfil the essential criteria of observational methodology as they are exhaustive and mutually exclusive.

SOCIN and SOPROX offer the general communicative structure found in every classroom-based teaching discourse and enables the exhaustive and mutually exclusive observation of the chain of kinesic and proxemic actions that are produced during

Table 2
SOPROX: system of observation for proxemic communication [11]

Dimension	Analytical categorization	Code	Description
Group Dimension that refers to the number of students to whom the teacher speaks	Macro-group	MAC	When the teacher speaks to the whole class/group
	Micro-group	MIC	When the teacher speaks to a specific sub-group of students
	Dyad	DYA	When the teacher speaks to a single student.
Topology Dimension that refers to the spatial location of the teacher in the classroom	Peripheral	P	The teacher is located at one end or side of the classroom
	Central	C	The teacher is situated in the central area of the classroom
Interaction Dimension that refers to the bodily attitude which indicates the teacher's degree of involvement with the students	At a distance	DIS	Bodily attitude that reveals the teacher to be absent from what is happening in the classroom, or which indicates a separation, whether physical or in terms of gaze or attitude, with respect to the students
	Integrated	INT	Bodily attitude that reveals the teacher to be highly involved in what is happening in the classroom, and in a relation of complicity with the students
Orientation Dimension that refers to the spatial location of the teacher with respect to the students	Tactile contact	TC	When the teacher makes bodily contact with a student
	Facing	FAC	The teacher is located facing the students, in line with their field of view
	Behind	BEH	The teacher is located behind the students, outside their field of view
	Among	AMO	The teacher is located inside the space occupied by the students
	To the right	RIG	The teacher is located in an area to the right of the classroom and of the students, with respect to what is considered to be the facing orientation of the teaching space
	To the left	LEF	The teacher is located in an area to the left of the classroom and of the students, with respect to what is considered to be the facing orientation of the teaching space
Transitions: dimension that refers to the body posture adopted by the teacher in space	Fixed bipedal posture	FB	The teacher remains standing without moving
	Fixed seated posture	FS	The teacher remains in a seated position
	Locomotion	LOC	The teacher moves around the classroom
	Support	SU	The teacher maintains a support posture by leaning against or on a structure, material, or person

the teaching process. The instruments described here have a molar structure that is easy to use and readily adaptable to various naturalistic communicative contexts. We believe that the instruments offer greater applicability and flexibility than do other existing tools which, in our view, are hindered by a degree of molecularization that is too complex; for example, the kinesic analyses [32] in the field of nonverbal human communication, or the notation systems [33] provide a considerable amount of information but they are very difficult to use in many natural contexts in which communicative teaching might be observed.

The instrument SOCIN, for kinesic actions, is based on four dimensions (morphology, function, adaptor, and situation). Similarly, the instrument SOPROX, for proxemic actions, is based on five dimensions (group, topology, location, orientation, and transition), each of which gives rise to a system of categories that are exhaustive and mutually exclusive. Observational methodology requires a clear and exhaustive definition of each of the categories included in the observation system or field format. Each of the dimensions, categories, and codes that form part of the SOCIN (Table 1) and SOPROX (Table 2) are defined below.

4 Procedure

The recording instrument used to codify SOCIN and SOPROX was LINCE software [34], an interactive video coding program which allows effective recording processes. It is easy to use and integrates a wide range of necessary functions: coding, recording, calculation of data quality, and the analysis of information in specific formats, thereby enabling it to be directly exported to several applications already used in observational data analysis. LINCE has been designed to facilitate the systematic observation of sport and motor practices in any situation or habitual context in which behavior is spontaneous.

Sessions were digitized to make them available for frame-to-frame analysis and enable them to be coded in LINCE software. The behavior of teachers was observed uninterruptedly across all the sessions, the mean duration being 50 min ($\bar{x} = 280$ frames/session).

Two different observers analyzed all the recordings from observation sessions. In order to control the quality of data [35] the kappa coefficient was obtained (0.94 for all sessions). This coefficient provides a satisfactory guarantee of data quality.

Temporal patterns were detected and analyzed with the *Theme* v.5 software [36]. *Theme* not only detects temporal patterns but also indicates the relevance and configuration of recorded events. The approach is based on a sequential and real-time pattern type, known as T-patterns, which, in conjunction with detection algorithms, can describe and detect behavioral structure in terms of repeated patterns [36]. It has been shown that such patterns, while common in behavior, are typically invisible to observers, even when aided by standard statistical and behavior analysis methods. The T-pattern algorithm has been implemented in the specialized software package, *Theme* (see www.patternvision.com and www.noldus.com). *Theme* also displays event frequency charts based on the occurrences of recorded events and the frequency of each category independently of the other categories. The detection of T-patterns has proven to be extraordinarily productive and fruitful for the study of the multiple facets or fields of body movement as we have pointed before.

5 Results

Obviously, each teacher has his or her own paraverbal communicative style. However, the objective of this paper is not to compare styles but, rather, to reveal the trends in this dimension of communication among teachers working in a similar naturalistic context. The observation of a natural context requires the use of the abovementioned observational instrument, as well as the detection of temporal patterns (T-patterns) in the transcribed actions. The *Theme* program grouped all the recordings of each teacher (nomothetic view) and derived T-patterns that reveal the trends in kinesic and proxemic paraverbal communication from an ideographic perspective.

In the current data sets, *Theme* detected several relevant T-patterns. As an example, let us consider three T-patterns[1] that are of interest with respect to the generation of paraverbal communicative responses. The T-pattern in Fig. 3 demonstrates dyadic interaction between teacher and student, while the T-pattern in Fig. 4 describes and interaction sequence between the teacher and the whole classroom.

[1] *How to read the pattern tree graph*: The upper left box of Figs. 3 and 4 shows the events occurring within the pattern, listed in the order in which they occur within the pattern. The first event in the pattern appears at the top and the last at the bottom. The upper right box shows the frequency of events within the pattern, each dot means that an event has been coded. The pattern diagram (the lines connecting the dots) shows the connection between events. The number of pattern diagrams illustrates how often the pattern occurs. Sub-patterns also occur when some of the events within the pattern occur without the whole of the pattern occurring. The lower box illustrates the real-time of the pattern. The lines show the connections between events, when they take place and how much time passes between each event.

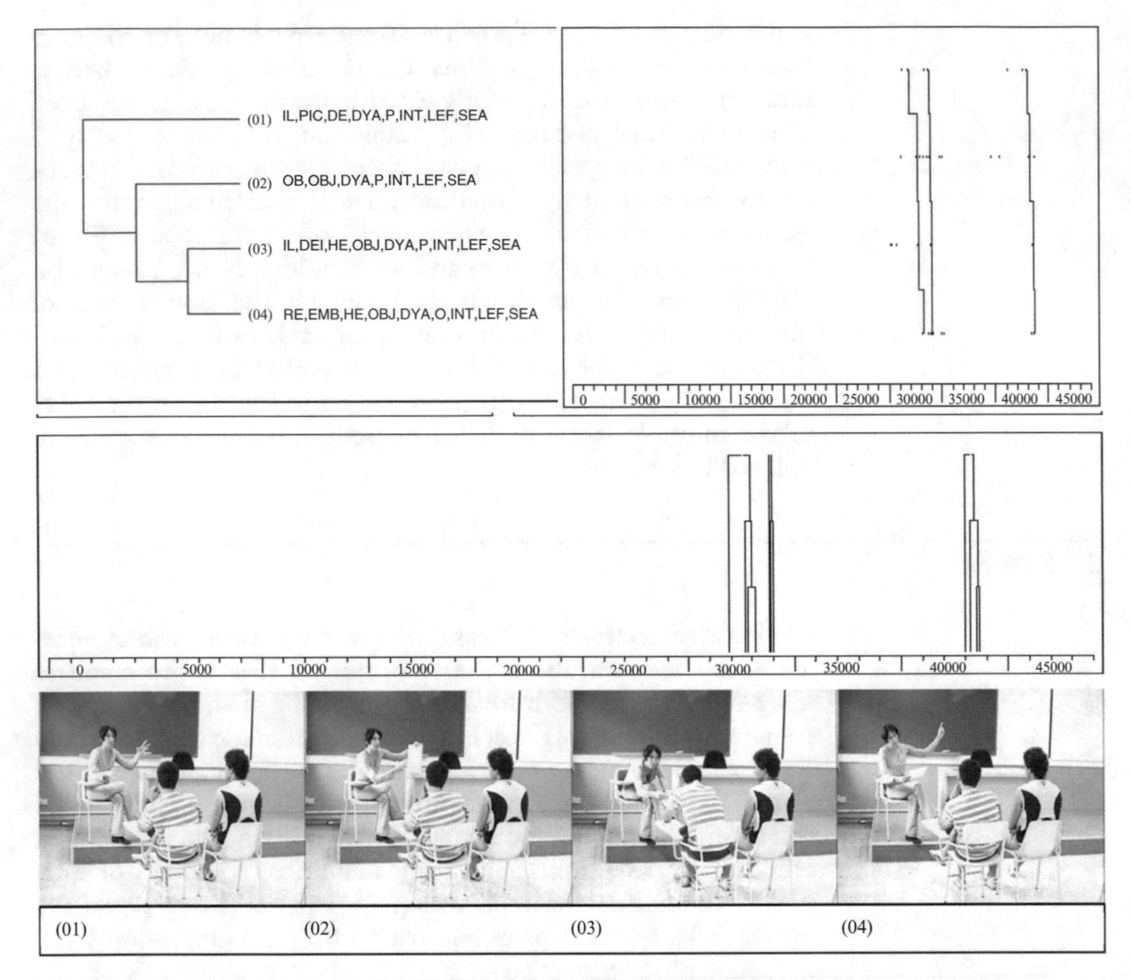

Fig. 3 This relevant T-pattern focuses on interaction with a dyad. It consists of three levels and a sequence of four events, each one of which comprises a complex combination of codes (combinations formed by between five and nine codes), occurring on three occasions during the observation period with the same sequence of events and significantly similar time intervals between each event occurrences. This T-pattern shows how the regulatory and illustrative functions follow one another, and indicates the type of morphology, both in terms of gesture and proxemics, which the teacher uses to accompany these functions. The interpretation that can be derived from the four steps of this T-pattern sequence can be described step by step as follows: (*01*) The teacher begins with an illustrative function (IL), in this case demonstrating (DE) something by using pictographic (PIC) gestures that draw an object or idea in space. He then relates to a partner (DYA) with an integrative (INT) attitude while situated in a peripheral area (P) of the classroom to the left (LEF) of the group and with a fixed bipedal (SEA) posture. (*02*) The teacher begins to observe (OB) what the dyad (DYA) is doing, makes an object-adaptor gesture (OB) and maintains the same proxemic criteria (P) (INT) (LEF) (SEA). (*03*) The teacher begins to offer help (HE) while illustrating (IL) by means of a deictic (DEI) gesture to point; he maintains the object adaptor (OB) and the same proxemic trend. (*04*) The teacher changes to regular (RE) in a dyadic situation (DYA) of help (HE), and is therefore integrated (INT) in a seated position (SEA) to the left of the space (LEF), maintaining an object adaptor (OBJ) and using a well-defined emblem (EMB) gesture

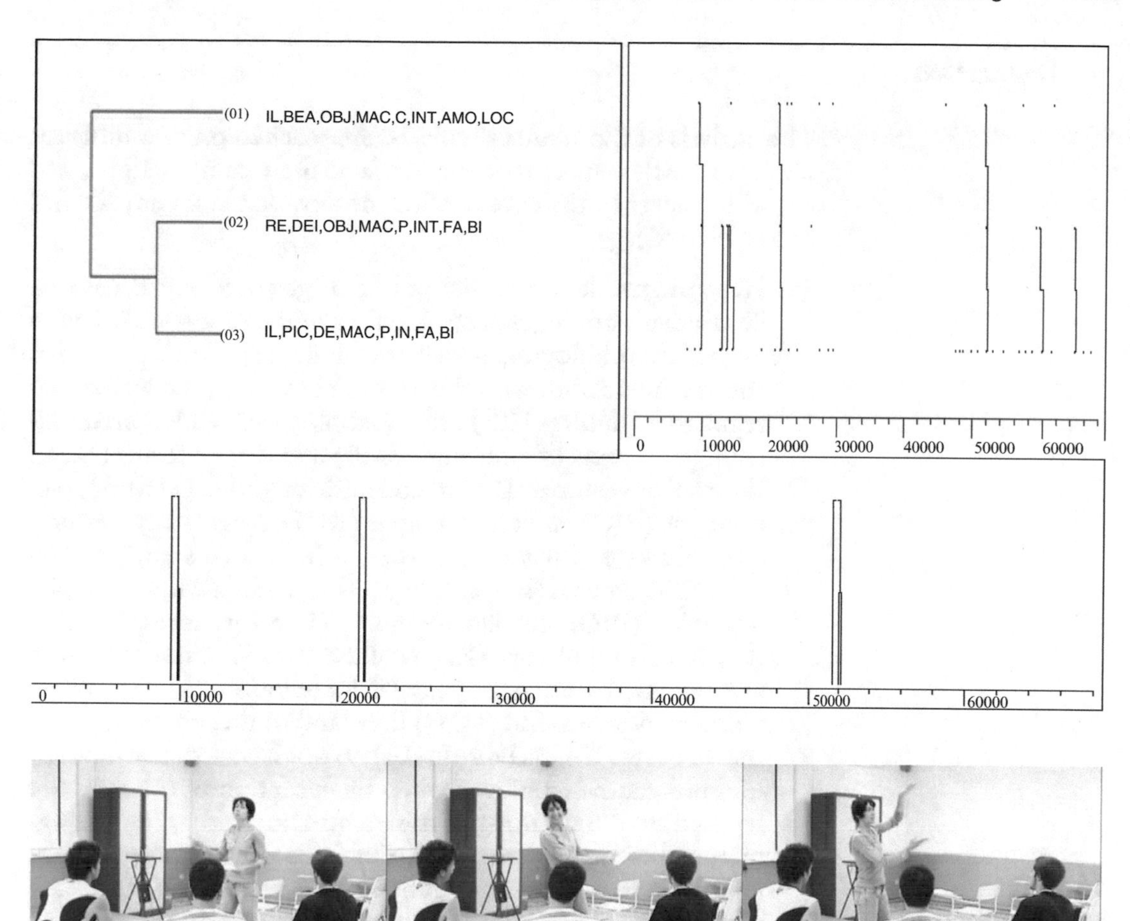

Fig. 4 This relevant T-pattern is related to interaction with the whole class/group of students. It consists of three levels and a sequence of three events, each one of which comprises a complex combination of codes (combinations of eight codes), occurring on three occasions during the observation period with the same sequence of events and significantly similar time intervals between each event occurrences. This T-pattern shows the alternation between the use of illustration (IL) and regulation (RE), as defined previously. As can be seen in (*01*), most illustrative (IL) situations involve expository, narrative, and descriptive phrases that are usually accompanied by gestures whose morphology takes the form of beats (BEA), and also accompanied by locomotion (LOC) or movement by the teacher around the classroom or among the students (AMO). As can be seen in (*02*), situations of regulation (RE), in which the teacher uses imperative, interrogative, or instructive phrases, are usually accompanied by deictic (DEI) gestures and made from a peripheral (P) area of the classroom. In (*03*) one can see another trend in the illustrative function accompanied by more defined gestures, in this case pictographs (PIC) that are usually used when the teacher has a fixed bipedal (BI) posture. It seems that maintaining a fixed posture helps to focus the attention required to make highly defined gestures such as kinetographs or pictographs. The object adaptor (OBJ) appears frequently, except when the teacher begins to demonstrate (DE) (for example, holding a piece of chalk in his hand (OBJ) but then beginning to write with it (DE), or putting it down so as to demonstrate something with his hands more clearly)

6 Discussion

The analysis of the results should be approached on two different levels: (a) with respect to the codes and their combinations; and (b) with respect to the criteria of the observation systems (SOCIN and SOPROX).

(a) Here the results reveal the power of gestures which have an illustrative and regulatory function and are associated with various morphologies, as well as with key aspects of proxemics and the interaction with the group/class. It can be stated that regulatory gestures (RE) are morphologically coded predominantly by means of emblems (EMB) and deictic forms (DEI). Illustrative gestures (IL) are coded through beats (BEA), pictographs (PIC) and kinetographs (KIN). Regulatory gestures (RE) are usually made while the teacher is in a static position (especially bipedal (BI), although also seated (S); however, pictographic (PIC) and kinetographic (KIN) gestures, whether they have a regulatory (RE) or illustrative (IL) function, may also be used during periods of locomotion (LOC). When teachers move around (LOC) they tend to use, above all, illustrative beats (IL). As beats (BEA) do not have their own meaning they can be employed when the attention of others is not focused on the meaning of the gesture but, rather, on the significance of what is being said. Emblems (EMB), deictic forms (DEI) with a regulatory function, pictographs (PIC), and kinetographs (KIN) are usually used from a static position so that recipients are not distracted by any movement (LOC) and, instead, pay attention to the meaning conveyed by the gesture. When a teacher demonstrates (DE) he or she tends to use illustrative gestures (IL), which may be morphologically coded as pictographs (PIC) or kinetographs (KIN), and to a lesser extent as beats (BEA).

(b) As regards the criteria defined by the observation instruments SOCIN and SOPROX the results enable us to highlight a series of trends in both kinesic and proxemic communication, as well as in combinations of the two. The *Function* criterion reveals that most teachers use the regulatory function 30 % of the time, the remaining 70 % corresponding to the illustrative function; in other words, actions that do not require an immediate response such as explaining and providing information account for the largest proportion of time to the detriment of regulatory actions, which do seek an interaction or responses such as asking questions, giving orders, offering help, etc. Concerning the combination of the criteria *Morphology* and *Function* of gestures it can be seen that emblems, deictic forms, pictographs, kinetographs, and beats are used without distinction in order to convey each function, whether it be regulatory or illustrative; however, gestures that are less well-defined in terms of morphology, such as beats, are more likely to accompany the illustrative function, whereas most emblems and deictic forms, both of which are gestures with a well-defined morphology, tend to accompany more the regulatory function.

In our view the criterion *Adaptation* is of less interest as it refers to extra-communicative aspects and is associated with unconscious contact gestures made by the sender; the results show a highly frequent use of these.

With respect to the criterion *Transitions*, fixed bipedal postures are usually alternated with periods of locomotion as the teacher moves from one area of the classroom to another. Occasionally one can observe support postures, generally in conjunction with tables or chairs, whereas when posture is static in the seated position this tends to be maintained for some time.

Concerning the relationship between the criteria *Function* and *Transitions* the results reveal a common association between the regulatory function and static bipedal postures, whereas the illustrative function is combined with locomotion or movement around the classroom. It appears that when giving an illustration, which does not require a gesture of interaction, the teacher feels freer to move around. In contrast, the regulatory function, which does call for gestures that indicate interaction, seems to require greater concentration on the part of teachers and leads them to fix their posture and thus focus their vision on a single point while asking questions, making comments or giving orders, etc.

With respect to the criterion *Orientation* the predominant position tends to be facing the group. Teachers rarely take up a position behind the group.

The criterion *Group* shows that interaction mostly occurs with the whole group, followed by that with micro-groups and, occasionally, with dyads. However, and as is shown in the T-patterns of Figs. 3 and 4, the combinations of codes in each event are more numerous and varied for interactions with student dyads (Fig. 3) than they are for those with micro-groups or the group as a whole (Fig. 4).

7 Conclusion

In addition to observing the particular style of paraverbal communication associated with each teacher the present study also illustrates the ways in which they tend to use such communication from both a kinesic and proxemic perspective. We can provide answers according to the questions proposed for this study: (1) The observation instruments SOCIN and SOPROX have been shown to be effective tools for recording in an exhaustive, clear and manageable way all the possible forms of paraverbal communication used by teachers, whether this be kinesic (by means of SOCIN) or proxemic (via SOPROX). (2) As pointed out in the results section it has been possible to obtain trends in relation to each criterion of both SOCIN and SOPROX, as well as combinations of them; for example, observing the relationship between the criterion *Orientation* and *Topology* or *Transitions* enables us to investigate whether there is a significant relationship between the way in which teachers move around, the spatial orientation chosen and the topology used with respect to the dimensions of the classroom and the space in which teaching takes place. (3) Regulatory gestures (RE) are usually made while the teacher is in a static position (especially bipedal (BI), although also seated (S); however, pictographic (PIC) and kinetographic (KIN) gestures. The regulatory function (RE) always appears in the context of interactions between teachers and students, whereas the illustrative function (IL) appears in expositive situations. Although we have seen how both are associated with specific proxemic behaviors, their power resides in their being effectively combined with verbal expression. Thus, for example, the illustrative function of gestures, regardless of whether these are iconically well defined, is interesting in terms of fostering students' learning, but too much illustration can have a negative effect [37].

A complementary and desirable objective for this type of research would be that teachers, via observation of their classes and a debriefing interview, could become aware of their particular style of paraverbal communication. In light of such information, teachers tend to be highly motivated to improve the efficacy of the paraverbal communication associated with their everyday teaching discourse. In this regard, strategies and techniques based on mixed methods research [12, 38, 39] provide different points of reference and indicators that may help new professionals to understand modulate and adjust the development of their self-perception and behavior.

Given the inevitable limits to the rationality and reflective capacities of educational professionals, we are obliged to consider the intentions of teachers, the tasks of whom can be approached through descriptive and qualitative instruments, as well as with data derived from more quantitative observation of their behavior.

For teachers, having an optimum paraverbal communicative style (both kinesic and proxemic) in combination with effective verbal communication is important in terms of the efficacy of instruction. We firmly believe that the optimisation of these communicative styles would have a direct positive effect on students' learning.

Acknowledgements

We gratefully acknowledge the support of the Generalitat de Catalunya Research Group [*GRUP DE RECERCA I INNOVACIÓ EN DISSENYS (GRID). Tecnología i aplicació multimedia i digital als dissenys observacionals*], Grant number 2014 SGR971.

We gratefully acknowledge the support of the Spanish government project *Observación de la interacción en deporte y actividad física: Avances técnicos y metodológicos en registros automatizados cualitativos-cuantitativos* (Secretaría de Estado de Investigación, Desarrollo e Innovación del Ministerio de Economía y Competitividad) during the period 2012-2015 [Grant DEP2012-32124].

References

1. Roth WM (2001) Gestures: their role in teaching and learning. Rev Educ Res 71(3):365–392
2. Moore A (1996) College teacher immediacy and student ratings of instruction. Commun Educ 45(1):29–39
3. Le Poire BA, Yoshimura SM (1999) The effects of expectancies and actual communication on nonverbal adaptation and communication outcomes: a test of interaction adaptation theory. Commun Monogr 66(1):1–30
4. Anguera MT (2005) Microanalysis of T-patterns. Analysis of symmetry/asymmetry in social interaction. In: Anolli L, Duncan S, Magnusson M, Riva G (eds) The hidden structure of social interaction. From Genomics to Culture Patterns. IOS Press, Amsterdam
5. Sánchez-Algarra P, Anguera MT (2013) Qualitative/quantitative integration in the inductive observational study of interactive behaviour: impact of recording and coding predominating perspectives. Qual Quant 47(2):1237–1257
6. Chacón-Moscoso S, Sanduvete S, Portell M, Anguera MT (2013) Reporting a program evaluation: needs, program plan, intervention, and decisions. Int J Clin Health Psych 13(1):58–60
7. Castañer M, Torrents C, Anguera MT, Dinušová M, Jonsson JK (2009) Identifying and analyzing motor skill responses in body movement and dance. Behav Res Methods 41(3):857–867
8. Castañer M, Andueza J, Sánchez-Algarra P, Anguera MT (2012) Extending the analysis of motor skills in relation to performance and laterality. In: Camerino O, Castañer M, Anguera MT (eds) Mixed methods research in the movement sciences: cases in sport, physical education and dance. Routledge, London
9. Castañer M, Franco S, Rodrigues J, Miguel C (2012) Optimizing verbal and nonverbal communication in PE teachers, instructors and sport coaches. In: Camerino O, Castañer M, Anguera MT (eds) Mixed methods research in the movement sciences: cases in sport, physical education and dance. Routledge, London
10. Castañer M, Torrents C, Morey G, Jofre T (2012) Appraising motor creativity, aesthetics and the complexity of motor responses in dance. In: Camerino O, Castañer M, Anguera MT (eds) Mixed methods research in the movement sciences: cases in sport, physical education and dance. Routledge, London
11. Castañer M, Camerino O, Anguera MT, Jonsson GK (2013) Kinesics and proxemics communication of expert and novice PE teachers. Qual Quant 47:1813–1829
12. Saüch G, Castañer M (2014) Observación de patrones motrices generados por los programas de actividad física para la tercera edad y la percepción de sus usuarios. Rev Psicol Deporte 23(1):181–190
13. Torrents C, Castañer M, Dinušová M, Anguera MT (2010) Discovering new ways of moving: observational analysis of motor creativity while dancing contact improvisation and the influence of the partner. J Creative Behav 44(1):45–62
14. Torrents C, Castañer M, Jofre T, Morey G, Reverter F (2013) Kinematic parameters that influence the aesthetic perception of beauty in contemporary dance. Perception 42(4):447–458

15. Roth WM, Lawless D (2002) Scientific investigations, metaphorical gestures, and the emergence of abstract scientific concepts. Learn Instr 12:285–304
16. Gallaher PE (1992) Individual differences in non-verbal behavior dimensions of style. J Pers Soc Psychol 63(1):133–145
17. Ekman P (1957) A methodological discussion of non-verbal behavior. J Psychol 43(1):141
18. Hall ET (1968) Proxemics. Curr Anthropol 9(2-3):83
19. Poyatos F (1983) Language and nonverbal systems in the structure of face-to-face interaction. Lang Commun 3(2):129–140
20. Jones SE, Le Baron CD (2002) Research on the relationship between verbal and nonverbal communication: emerging integrations. J Commun 52(3):499–521
21. Ekman P (1985) Methods for measuring facial action. In: Scherer K, Ekman P (eds) Handbook methods in nonverbal behavior research. Cambridge University Press, Cambridge
22. McNeill D (2005) Gesture and thought. University of Chicago Press, Chicago, IL
23. Anguera MT, Izquierdo C (2006) Methodological approaches in human communication. From complexity of situation to data analysis. In: Anolli L, Duncan S, Magnusson M, Riva G (eds) The hidden structure of social interaction. From Genomics to Culture Patterns. IOS Press, Amsterdam
24. Patterson ML (1995) A parallel process model of nonverbal communication. J Nonverbal Behav 19(1):3–29
25. Woodall WG, Burgoon JK (1981) The effects of nonverbal synchrony on message comprehension and persuasiveness. J Nonverbal Behav 5(4):207–223
26. Gatewood JB, Rosenwein R (1981) Interactional synchrony. Genuine or spurious? A critique of recent research. J Nonverbal Behav 6(1):12–29
27. Anguera MT, Blanco-Villaseñor A, Hernández-Mendo A, Losada JL (2011) Diseños observacionales, ajustes y aplicación en psicología del deporte. Cuader Psicol Deporte 11(2):63–76
28. Woolfolk AE (1981) The eye of the beholder: methodological considerations when observers assess nonverbal communication. J Nonverbal Behav 5(3):199–204
29. Boice R, Hanley CP, Gansler D, Shaughnessy P, Dudek BC (1984) Generality of observational skill across verbal and nonverbal modes: literature review and experimental test. J Nonverbal Behav 8(3):172–186
30. Baesler EJ, Burgoon JK (1987) Measurement and reliability of nonverbal behavior. J Nonverbal Behav 11(4):205–233
31. Murphy NA (2005) Using thin slices for behavioural coding. J Nonverbal Behav 29(4):235–246
32. Birdwhistell R (1970) Kinesics and context. UPP, Philadelphia
33. Laban RV, Ullman L (1988) The mastery of movement. Northcote House, Plymouth, MA
34. Gabín B, Camerino O, Anguera MT, Castañer M (2012) Lince: multiplatform sport analysis software. Proc Soc Behav Sci 46:4692–4694
35. Jansen RG, Wiertz LF, Meyer ES, Noldus LPJJ (2003) Reliability analysis of observational data: problems, solutions, and software implementation. Behav Res Methods 35(3):391–399
36. Magnusson MS (2005) Understanding social interaction: discovering hidden structure with model and algorithms. In: Anolli L, Duncan S, Magnusson M, Riva G (eds) The hidden structure of social interaction. From genomics to culture patterns. Ios Press, Amsterdam
37. Berends IE, Van Lieshout ECDM (2009) The effect of illustrations in arithmetic problem-solving: effects of increased cognitive load. Learn Instr 19(4):345–353
38. Camerino O, Castañer M, Anguera MT (eds) (2012) Mixed methods research in the movement sciences: cases in sport, physical education and dance. Routledge, London
39. Anguera MT, Camerino O, Castañer M, Sánchez-Algarra P (2014) Mixed methods in research into physical activity and sport. Rev Psicol Deporte 23(1):123–130

Chapter 5

The Self-Organization of Self-Injurious Behavior as Revealed through Temporal Pattern Analyses

Aaron S. Kemp, Mohammed R. Lenjavi, Paul E. Touchette, David Pincus, Magnus S. Magnusson, and Curt A. Sandman

Abstract

Intentional acts of harm to self are among the most dramatic and disturbing behaviors exhibited by human beings and frequently exact a heavy toll in terms of the emotional and economic burden that must be borne by affected individuals, families, caregivers, and society. One major obstacle to understanding and treating self-injurious behavior (SIB) is the absence of adequate tools and methodologies to identify distinctive behavioral phenotypes or to quantify the complex presentation of SIB across varying time scales and environmental settings. Granted, there are increasingly sophisticated analytic techniques available to study behavior, but the vast majority of existing studies on SIB still rely on measures of frequencies or rates of SIB linked to a single environmental condition or other presumed contingencies. In contrast, our recent investigations of SIB among individuals with severe intellectual and developmental disabilities have employed temporal pattern analyses using the Theme™ program to explore the complex organizational dynamics underlying the presentation of SIB as recurrent patterns across time. Comprehensive behavioral and environmental events were recorded in situ, in real time, by trained, unobtrusive observers using The Observer®. The event codes and their associated times were then imported into Theme™, which was used to identify highly significant (i.e., nonrandom), recurrent, temporal patterns that were not constrained by implicit assumptions about the sequential ordering or hypothesized relations among the constituent events. Principal among our findings are that transitions to episodes of SIB are characterized by greater overall behavioral complexity and order within individuals; that self-injuring acts may serve as singular points that increase coherence within self-organizing patterns of behavior; that temporal patterns associated with SIB are highly correlated with basal beta-endorphin and adrenocorticotropic hormone levels across individuals; and that treatment with the opiate antagonist naltrexone may reduce the temporal patterning of SIB. The implication of these findings is that SIB can never be fully understood within a strictly linear conceptualization of "cause-and-effect" sequential dependencies. Instead, we suggest that SIB is dynamically regulated by "internal" processes which contribute to the emergence of complex, self-organizing patterns. If confirmed, these results may portend the development of innovative new behavioral or pharmacologic interventions designed to disrupt self-organizing regulatory processes, rather than simply focusing on putative antecedents or consequences.

Key words Self-injurious behavior, Self-organization, Temporal pattern analyses, Theme

Magnus S. Magnusson et al. (eds.), *Discovering Hidden Temporal Patterns In Behavior and Interaction: T-Pattern Detection and Analysis with THEME™*, Neuromethods, vol. 111, DOI 10.1007/978-1-4939-3249-8_5,

1 Introduction

Abnormality in human behavior is often characterized on the basis of functional criteria pertaining to whether the behavior is generally regarded as "adaptive" or "maladaptive" in a given social context or environmental setting. Behaviors that serve to increase productive or constructive adjustments to functioning in daily life are normally regarded as "adaptive," while behaviors that lead to impaired adjustments to functioning in a given social situation or setting are generally considered "maladaptive" in that they do not appear to serve a productive or constructive adjustment purpose. As such, maladaptive behaviors are thought to be counter-productive to the individual's own social, psychological, or physical well-being. Some of the most commonly observed maladaptive behaviors are repetitive, perseverative, or stereotyped movements, compulsions, tics, addictions, and intentional acts of physical harm to self.

Frequently termed "self-injurious behavior" (SIB), repetitive acts of self-harm are among the most dramatic and disturbing maladaptive behaviors exhibited by human beings. Although SIB has been observed in association with a variety of psychological disorders, neurological conditions, psychiatric diagnoses, and genetic syndromes, the primary focus of the investigations described herein is the frequent occurrence of SIB among individuals diagnosed with severe intellectual and developmental disabilities (IDD), including autistic spectrum disorders. The prevalence of SIB within this population has been estimated to be as high as 30 % and remains one of the primary reasons that individuals with IDD either are retained in institutional (restrictive) environments or administered psychotropic medication [1]. Despite considerable research effort, the underlying causes of SIB in this population remain a mystery and there is still no consensus among researchers and clinicians regarding the most effective interventions or treatments [2].

A major obstacle to understanding the mechanisms of SIB and developing coherent treatment plans has been an absence of reliable and effective methods for quantifying the complex, recurrent patterns of SIB across varying settings and time scales, such that relations with a variety of factors may be empirically assessed [3]. Indeed, the topographies and circumstances surrounding recurrences of SIB vary so considerably from one individual to the next that establishing a metric or analytical technique to reliably assess severity, change, or temporal contingencies is a major methodological impediment. Individuals exhibiting SIB employ an assortment of methods of self-harm that include cutting, hitting, or biting themselves, ingesting foreign objects, hurling themselves to the ground, and banging their head against solid objects often resulting in broken bones, disfigurement, blindness, or even loss of life [4–9]. This broad spectrum of self-harm phenotypes, the range of methods used to commit these acts and the various motives proposed to explain them has militated against a unifying mechanism and the development of a universally effective intervention.

After considerable debate on this topic, a panel of experts convened by the US National Institute for Child Health and Human Development (NICHD) reached a consensus that it should be possible to define a distinct behavioral phenotype for SIB, perhaps with greater precision than most complex human behaviors because it is directly observable and can be reliably counted [10]. However, the NICHD group argued that data collection and analysis of SIB was primitive and that new methods of increasingly sophisticated sequential and temporal analysis of in situ observations of behavior were available [11–14] so that measures and analyses of patterns of SIB should replace or supplement measures of rate and frequency [4, 15]. Nonetheless, the majority of existing studies on this topic still rely on measures of frequencies or rates of SIB, often linked to a single environmental manipulation or experimental condition.

In contrast, our recent investigations of SIB among individuals with severe IDD have employed temporal pattern analyses using the Theme™ program [16, 17] to explore the complex organizational dynamics underlying the presentation of SIB as recurrent patterns across time. In conjunction with the application of temporal pattern analyses, our approach has also explored the heuristic utility of concepts derived from nonlinear dynamical systems theory to provide a unique perspective on the "internal" regulatory processes believed to subserve the persistent recurrence of SIB in this population. In so doing, it was hoped that this novel approach would catalyze the development and empirical testing of new hypotheses regarding the apparent self-organization of self-injurious and other maladaptive behaviors and eventually lead to innovative new treatment modalities for these troubling conditions.

2 Identifying Temporal Patterns of SIB

Increasingly sophisticated analytic techniques have been applied to the investigation of observationally recorded behavioral data over the past 30 years. For example, in 1979, Gene Sackett [13] described the application of lag sequential analyses to directly address the complexity and constraints of existing methods for identifying contingent relations across time in multivariate observational data. The conceptual basis for lag analyses derives from the quantitative methods of auto- and cross-correlation. When applied to qualitative behavioral data, the lag principle examines the conditional (or transitional) probability that a criterion event of interest will be sequentially followed by another event of interest (event lag), or that any observed event will fall within a specified temporal window in relation to the criterion event (time lag).

In the early 1990s, Eric Emerson and colleagues were among the first researchers to apply this analytic approach to the study of SIB among individuals with IDD. In a cross-validational comparison of time-based lag sequential analyses with traditional, experimental (functional) analyses, Emerson et al. [18] found a high degree of consistency between the two approaches (86 % agreement in the identification of behavioral processes underlying SIB). These results were interpreted as lending support to the external validity and overall viability of time-based lag analyses for exploring the mechanisms and contextual contingencies underlying SIB in IDD populations. Since then, several studies have applied this statistical method to quantify the conditional probabilities between recurrent acts of SIB and relations to a variety of other presumed behavioral and/or environmental antecedents or consequents.

Researchers from our project team have also employed time-based lag sequential analyses to examine whether successive episodes of SIB are sequentially dependent. In a study published by Marion et al. [19], sequential dependencies were determined by calculating the conditional probabilities that a match event followed a criterion event within four windows of time: 2, 10, 30, and 60 s. The results indicated that the only, highly significant, sequential predictor of SIB was another, antecedent occurrence of SIB. There was no evidence that SIB was sequentially dependent on environmental events or on other observationally recorded behaviors within these temporal windows. Furthermore, the method of analysis controlled for chance pairings of events and revealed that the sequential patterns of SIB were independent of frequency or rate of occurrence. Additionally, the conclusion that SIBs occur in sequentially related "bouts" was also confirmed using survival analyses to quantify the temporal distribution of SIB patterns [20]. The results reported by Kroeker et al. [20], suggested that, within some individuals with severe IDD, SIB followed "contagious" temporal distribution patterns, which could represent a unique behavioral phenotype that is maintained by biological rather than social or environmental factors.

Despite the meaningful application of lag sequential analyses in studies such as these, the method does have some inherent shortcomings that must be taken into consideration. For example, the temporal windows or variables of interest must be specified a priori. Implicit assumptions regarding the sequential or temporal proximity of contingent events, however, present serious limitations that may preclude the identification of "noncontiguous" or "long-range" temporal relations among events of interest. Furthermore, Sackett [13] cautions that lag sequential methods are highly vulnerable to "capitalization on chance," meaning that as the number of observations collected increases sufficiently, so too will the probabilities of finding significant sequential dependencies. While Bakeman and Gottman [21] provide detailed methods for controlling such

Type 1 errors, they caution that this is an issue of concern whenever lag sequential analytic methods are employed.

In the interest of overcoming such limitations in the study of SIB, our recent investigations [1, 3, 22] have utilized a unique, probabilistic, temporal pattern analysis program known as Theme™ (PatternVision Ltd and Noldus Information Technology BV). As developed by Magnusson [16, 17], Theme™ provides a statistical method of detecting temporal patterns (T-patterns) of related behavioral events that may not be obvious to a trained observer or identifiable by traditional or sequential analysis methods. The T-pattern detection algorithm first identifies significant (nonrandom) recurrences of any two events within a statistically similar temporal configuration (critical interval) in a real-time behavioral record and then proceeds to identify hierarchical relations with any other antecedent or subsequent events. Thus, the search algorithm detects highly significant, hierarchically arranged T-patterns that are composed of statistically related behavioral events that repeatedly appear in the same, relatively invariant, temporal configuration regardless of whether they are contiguous or noncontiguous in sequential distribution across time.

T-Patterns "grow" in complexity as simple patterns are incorporated into larger patterns, and are retained for further analysis according to whether they meet the search parameters specified by the user. Among these parameters is the probability that a given T-pattern will occur in a randomized distribution of the current record, the transitional probability that component patterns must possess to be included in a larger pattern, and the minimum number of instances that detected patterns recur across the record. Hence, the major advantages of this method are that it is not constrained by implicit assumptions about the sequential distribution of the behaviors of interest and allows the user to select the relevant probability levels to be tested against a randomized distribution of the actual behavioral record, thereby providing programmatic control over vulnerability to chance findings.

The following sections will provide a brief overview of the key findings that have been identified through the use of Theme™ in our analyses of observational data that were collected unobtrusively via handheld, mobile computers using The Observer® (Noldus Information Technology BV). For each of the studies detailed below, individual participants were observed by research staff throughout their regular daily routines, in situ, with minimal intrusion. The observation of individuals with varying self-injurious behavioral topographies required a coding strategy with a broad selection of the most salient features observed in the field and informed by previous research conducted over the past 30 years by Curt Sandman and colleagues at the University of California, Irvine. Though not described herein, details regarding the specific coding schemes employed, the reliability of the observational procedures utilized, and the neuropeptide assay methodologies mentioned below can all be reviewed in previous publications from Sandman and colleagues [1, 3, 5, 6, 8, 19, 20, 22–28]. Furthermore, it should be noted that the following studies were all conducted in compliance with the Declaration of Helsinki. Informed consent was obtained from conservators and/or guardians of all study participants. The methods of consent and data collection as well as the specific study protocols were all reviewed and approved by the ethics oversight committees of the University of California, Irvine (UCI Institutional Review Board) and the State of California (Committee for the Protection of Human Subjects).

3 Comparing Temporal Patterns in Records *With* and *Without* SIB

A primary question when evaluating the utility of temporal pattern analyses in our studies was whether the occurrence of SIB would have a discernible impact on the overall temporal organization of the observed events. To answer this question we took 10 days of observational records from 32 individuals (18 male, 14 female; mean age = 40 ± 13 years) known to display SIB and separated them into records with and without an observed SIB. These records were analyzed with Theme™ (Version 5) and the quantitative results (mean number of distinct T-patterns, mean number of T-pattern occurrences, mean length, and mean level) were then compared using Paired-Samples *t*-tests. As can be seen in Fig. 1, records that included SIB produced significantly more distinct T-patterns ($t_{31} = 2.33$, $p < 0.03$; Fig. 1a), more T-pattern occurrences ($t_{31} = 2.14$, $p < 0.04$; Fig. 1b), longer T-patterns ($t_{31} = 2.19$, $p < 0.04$; Fig. 1c), and more complex T-patterns ($t_{31} = 2.37$, $p < 0.03$; Fig. 1d) when compared with records from the same subjects without an observed SIB. Furthermore, these comparisons remained significant even after controlling for an increased opportunity of detecting T-patterns (i.e., more recorded events) in records that included SIB by removing all SIB and staff-interaction codes from the records prior to rerunning the analyses.

As reported by Sandman et al. [22], these results indicate that SIB may function as a "singularity" around which a complex temporal configuration of behavioral patterns becomes increasingly self-organized. For example, it is possible that the temporal patterning of behaviors associated with SIB reflects the dynamical influence of an internal regulatory mechanism that drives the overall system toward greater behavioral coherence and complex structural integrity. Indeed, the dynamical processes underlying transitions between periods of relative calm and the occurrence on an episode of SIB may reflect a system that is in a critical state

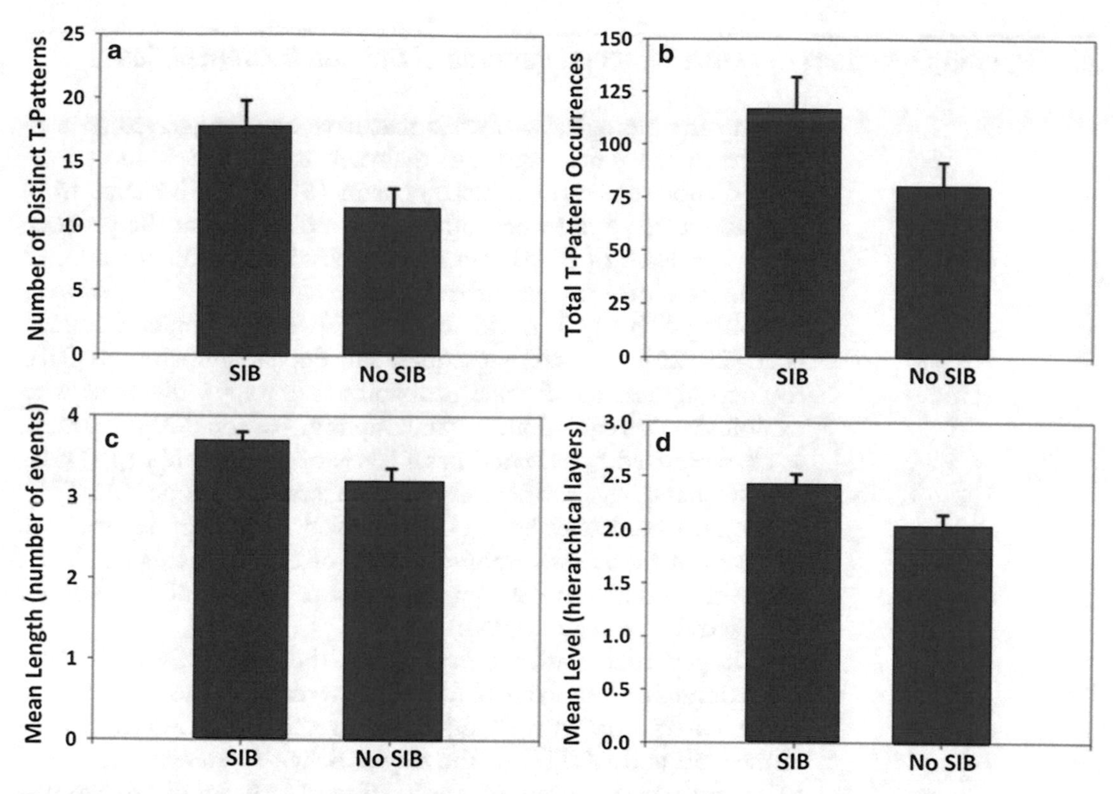

Fig. 1 Quantitative results of the temporal pattern (T-pattern) analyses on files with and without an observed episode of SIB. All of these mean values were found to be significantly different between files with and without an observed SIB episode on Paired *t*-test comparisons

at the "edge of chaos" [29], or some other process of self-organization [30]. Future research could serve to extend these results by investigating the specific mechanisms underlying the emergence of order that appears in the proximity of SIB (e.g., fractal patterns) [31].

Regardless of the specific mechanisms involved, however, transitions were observed between relatively disorganized behavioral states with few temporal patterns to a relatively organized state characterized by significantly more patterns, of greater complexity. It was surprising that the transition to a more organized state was observed when SIB was exhibited because increased complexity and system integrity typically are associated with "adaptive" states [32] and biopsychosocial resilience [33]. It is possible that the movement from behavioral calm to periods of self-inflicted injury confers an adaptive advantage or gain for the individual. For example, SIB could actually produce a movement toward equilibrium in biologically mediated processes associated with the regulation of arousal [34], reward, or pleasure [25], particularly if those processes are already in a state of dysregulation [26].

4 Exploring Relations between Temporal Patterns of SIB and Neuropeptides

Among the biological processes that have been suggested to contribute to the emergence and maintenance of SIB is the stress-related molecule proopiomelanocortin (POMC). Evidence from our laboratory, and several others, has indicated that the processing and release of POMC in the hypothalamic–pituitary–adrenal (HPA) axis may be perturbed among subgroups of individuals exhibiting SIB [1, 5, 6, 25–28, 35–55]. For example, Sandman et al. [25, 26] reported that sequentially dependent patterns of SIB reached highest conditional probabilities among individuals who exhibited a dysregulation, or "uncoupling," of the POMC system, as characterized by elevated basal levels of β-endorphin (βE) relative to basal levels of adrenocorticotropic hormone (ACTH). These two hormones are POMC-derived neuropeptides that are involved in the stress response as part of the HPA axis (ACTH), and in the modulation of pain and pleasure because of their affinity for the opiate receptors (βE).

Despite such evidence implicating the role of POMC, measures of rate or frequency of SIB have never been shown to have a direct correlation with blood levels of the POMC-derived hormones, βE and ACTH. As the sequentially dependent patterns of SIB revealed a significant (though indirect) association, we hypothesized that a more robust pattern detection method such as Theme™ could further elucidate the hypothesized relations between disturbed basal levels of βE and ACTH and a unique behavioral presentation of temporally patterned recurrences of SIB. Accordingly, we used Theme™ (Version 5) to identify T-patterns that included SIB within a dataset of in-situ observational recordings spanning 8 days (~40 h) in 25 individuals (13 male, 12 female; mean age = 40.5 years) with IDD and a history of persistent SIB for whom we also had basal (morning) blood levels of βE and ACTH.

As reported by Kemp et al. [3], and summarized in Table 1, the results of this investigation indicated that the within-subject percentages of detected T-patterns containing SIB were highly correlated with basal levels of βE ($r = 0.79$, $P < 0.001$) and ACTH ($r = 0.79$, $P < 0.001$). These correlations were even higher for the proportion of detected T-patterns that included both SIB and agitated behaviors (AB), but were not significant for the proportion of T-patterns containing AB without SIB or for any of the other "control" behaviors such as non-injurious motor stereotypies or staff interactions. Furthermore, such high correlations were not found, and have not been previously reported, between "raw" frequency counts or rates of SIB and these hormone levels. This indicates that the detection of temporally distributed patterns of SIB may yield measures more directly relevant to the underlying

Table 1
Pearson's *r* correlations (sig.) between hormone levels, behavior counts and rate (rows) and Theme results (columns), including total T-patterns and the proportion of T-patterns containing SIB, agitated behavior (AB), both SIB and AB, motor stereotypy (STER), and staff behaviors (STAFF)

	Total T-patterns	% SIB T-patterns	% AB T-patterns	% SIB and AB T-Patterns	% STER T-patterns	% STAFF T-patterns
βE level	0.29 (0.147)	**0.79 (<0.001)**	0.17 (0.403)	**0.83 (<0.001)**	−0.09 (0.668)	−0.06 (0.759)
ACTH level	0.23 (0.265)	**0.79 (<0.001)**	0.18 (0.371)	**0.85 (<0.001)**	−0.14 (0.494)	−0.09 (0.660)
Total SIB	0.08 (0.690)	0.20 (0.337)	0.06 (0.755)	0.15 (0.481)	0.22 (0.228)	0.04 (0.839)
SIB per hour	0.12 (0.585)	0.22 (0.274)	0.06 (0.748)	0.17 (0.401)	0.21 (0.243)	0.03 (0.892)
Total behaviors	**0.57 (0.003)**	0.37 (0.073)	−0.23 (0.270)	0.29 (0.163)	**0.44 (0.011)**	−0.26 (0.213)

biological mechanisms than traditional methods of quantifying occurrences. Furthermore, these results suggest the potential utility of temporal pattern analyses to identify a sub-type of subjects that may respond most beneficially to certain treatment approaches (e.g., opiate antagonists).

5 Examining the Effects of Naltrexone on Temporal Patterns of SIB

In the early 1980s, Sandman and colleagues were among the first to present evidence that opiate antagonists could be used to reduce the occurrences or attenuate the severity of SIB among individuals with IDD [46]. Since then numerous studies have explored the putative efficacy of opiate antagonists (e.g., naltrexone or naloxone) for the treatment of SIB. Although there have been some reports that have not shown clear support for the efficacy of this intervention [56], several reviews of this approach [1, 57–59] have reported improvements ranging from 57 to 80 % across numerous studies of individuals treated with naltrexone, with "unequivocal responders" comprising from 25 to 47 % of cases reviewed. One of the outcomes of research on this topic has been the suggestion that there may indeed be subgroups of individuals that respond most favorably to treatment with opiate antagonists. Furthermore, as noted above, one of the obstacles in identifying possible biological indicators of treatment responsiveness has been the reliance on relatively simple methods of analysis (e.g., rate or frequency of occurrence) that do not capture the complex expression of SIB as patterns across time.

In order to evaluate whether temporal patterns of SIB may provide a more sensitive measure of the effects of opiate antagonists, we conducted a small study in which six individuals (four male, two female; mean age = 38 years) with IDD and a persistent history of SIB were administered three dose levels of naltrexone (0.5, 1, and 2 mg/kg) for 1 week each, with intervening weeks receiving placebo, and the ordering of doses counterbalanced across subjects. Observational recordings were analyzed separately for each week using Theme™ (Version 5) and the results were then aggregated to compare the proportion of T-patterns containing SIB with the proportion of SIB events in the overall behavior frequency counts observed during the baseline week, and the weeks during which the subjects received either a placebo or one of the three doses of naltrexone. The results of these comparisons are shown in Fig. 2.

These results clearly indicate that SIB T-patterns provide a more sensitive measure of the effects of naltrexone, despite the fact that the small sample size precluded statistical significance for these comparisons. With regard to "clinical significance," one could argue that a reduction of SIB T-patterns would be relatively meaningless if the frequency of SIB shows no changes. However, an alternative interpretation is that the effects of naltrexone may serve to disrupt the underlying biological processes that subserve the organization of SIB into patterns, as expressed across time, thereby creating a condition which could make SIB more amenable to behavioral interventions.

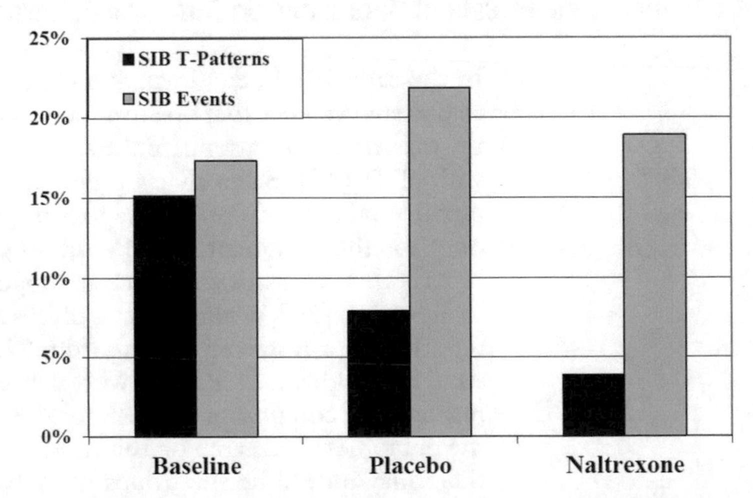

Fig. 2 Mean percentages of SIB T-patterns (out of total detected T-patterns) and SIB Events (out of total recorded behavioral events) detected during Baseline week and during weeks that the subjects were receiving either placebo or naltrexone treatments

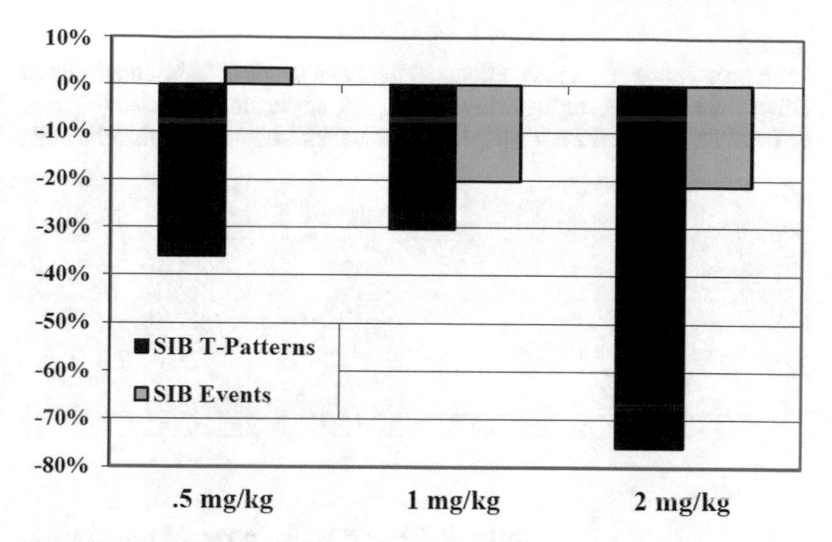

Fig. 3 Mean percentage decreases from placebo to respective naltrexone dose levels for the percent of T-patterns containing SIB (out of total detected T-patterns) and the percent of SIB events (out of total recorded behavioral events). These changes are matched for each subject for the weeks immediately preceding each dose of naltrexone

With regard to the apparent "placebo effect" seen in the change from baseline for the SIB T-patterns in Fig. 2, it should be noted that the dosing of placebo and naltrexone was counterbalanced across subjects on alternating weeks. Accordingly, the aggregation of respective placebo and naltrexone weeks includes potential "carryover" effects from a previous dose of naltrexone. This is more likely to be the reason for the decrease from baseline levels than the "expectation" bias commonly associated with an actual placebo response. Alternatively, Fig. 3 provides a comparison of SIB T-pattern and SIB event percentages that are matched for each subject to show the percent change from the placebo week immediately preceding each of the doses of naltrexone. Again, it appears clear that SIB T-patterns revealed a more pronounced effect in the percent change from weeks during which the subjects received placebo to the following weeks during which the subjects received the respective doses of naltrexone.

In addition to the observational data collected during this small trial, we also had blood levels of the N-terminal fragment of beta-endorphin (βEN) that were collected from each of these subjects prior to the initiation of the drug trial. Samples were collected within 2–5 min after an episode of SIB (defined as the observation of at least one self-injurious act). Control samples were collected on separate days, following a period of at least 30 min without an episode of SIB, and were matched for time of day with the post-SIB samples. To further control for the possible influence of physical exertion, samples were also collected on a separate day following

Table 2
Pearson's *r* coefficients (significance) between levels of beta-endorphin (N-terminal) collected in either the A.M., P.M., or following SIB, no SIB, or physical exercise (PE) and the change in the percent of T-patterns (by behavior type) between weeks treated with placebo and naltrexone

Change in T-pattern % by behavior type	Beta-endorphin (N-terminal) levels				
	A.M.	P.M.	Post-SIB	No-SIB	Post-PE
SIB	0.37 (0.47)	0.25 (0.64)	**0.82 (0.045)**	−0.45 (0.44)	0.66 (0.22)
Stereotypy	−0.76 (0.07)	−0.26 (0.96)	0.52 (0.29)	−0.29 (0.64)	−0.10 (0.87)
Staff interactions	0.42 (0.41)	0.54 (0.27)	0.35 (0.49)	−0.63 (0.25)	0.26 (0.68)

a brisk 10-min walk (physical exercise). Finally, on a separate day from the other samples, morning (8 A.M.) and afternoon (4 P.M.) samples were collected (on the same day) to estimate the peak and nadir, respectively, of the peptide levels under investigation.

As shown in Table 2, these βEN levels were compared with the percentage change in T-patterns (by behavior type) from the weeks the subjects were treated with placebo to the following weeks in which they received naltrexone. Despite the small sample, a highly significant correlation was detected between post-SIB levels of βEN and the change from placebo in the percentage of T-patterns containing SIB to the weeks subjects were receiving naltrexone. No such correlations were found for changes from placebo on any "control" T-patterns (i.e., those containing non-injurious, stereotyped behaviors or staff interactions). These findings suggest post-SIB βEN levels may have direct relevance to the effects of naltrexone on the temporal organization of SIB.

Though limited by a relatively small sample size, these findings may be the first to demonstrate a relation between response to a pharmacological intervention for SIB and endogenous levels of the N-terminal fragment of βE. At least one study [55] has suggested a relatively unrecognized role of the N-terminal in the processes thought to subserve SIB; however, there are no previous studies that have suggested that post-SIB levels may have direct relevance to whether an individual may respond favorably to treatment with naltrexone. Such relations may portend a more complete understanding of the biological mechanisms underlying the persistent and disturbing recurrence of SIB among certain individuals with developmental disorders and clearly warrant further investigations. Without the utilization of temporal pattern analyses in these investigations, however, such relations might continue to lie unrecognized, as traditional linear analyses may not possess adequate sensitivity to reveal the complex emergent structuring of these behaviors across time.

6 Comparing Temporal Pattern Analysis with Lag Sequential Analysis

The studies described above have undoubtedly demonstrated the utility of temporal pattern analyses in the investigation of SIB. Nonetheless, we have also sought to evaluate the relative merits of this method in a direct comparison with the other method of analysis that has been extensively applied in the study of observational data: lag sequential analysis. As noted earlier in this chapter, researchers from our laboratory have applied lag sequential analyses on a number of studies, but were uncertain as to how those findings might have differed if temporal pattern analyses had been applied instead. Accordingly, we conducted a direct comparison of these two complementary methods and examined how they differ with regards to the type of information each may yield when applied to the study of SIB. Furthermore, this study also provided an opportunity to compare the sensitivity of each method to diurnal levels of a stress-related hormone (cortisol) and their respective utility for detecting changes associated with acute and long-term treatment with naltrexone in a case study of a single subject.

This case study included 74, 2-h, observational records collected via unobtrusive, in situ observation of an 11-year-old boy diagnosed with autism and a persistent history of severe SIB. These observational records were collected during three separate time periods, each spanning approximately 1 month, with approximately 6 months between each period. The subject was not receiving any medications during the first observational period and began taking a daily dose of 2 mg/kg of naltrexone 2 weeks into the second period and continued taking this dose throughout the remaining time periods. As such, the "acute" effects of naltrexone described below were evaluated within the second time period, whereas the "long-term" effects were evaluated using data from the third time period. During the third time period, salivary cortisol samples were also collected for 21 consecutive days at four daily time-points: 6:30, 7:00, and 7:30 A.M. to detect the cortisol awakening response, and at 4:00 P.M. to measure the approximate nadir.

All 74 observational records were analyzed using both lag sequential and temporal pattern analyses. Lag sequential analyses were conducted using The Observer® XT (Version 9) to calculate event-based conditional probabilities across all recorded categories. The comparisons described below included the conditional probability of an SIB event contiguously following another SIB event, as this measure previously had been found to be highly sensitive to the treatment effects of naltrexone and to biological variables of interest [25, 26]. Temporal pattern analyses were conducted with Theme™ (Version 5) using the following search parameters: Minimum Occurrence of 3, Significance Level less than 0.05, and a Lumping Factor of 0.99, with all other parameters set to default values.

The conditional probabilities that an SIB event would be followed by a subsequent SIB event (SIB-SIB CPs), as calculated using lag sequential analyses, was found to be significantly correlated with the total number of T-patterns detected by Theme™ ($r=0.44$; $p<0.001$), the number of T-patterns containing SIB ($r=0.32$; $p<0.005$), and the proportion of T-patterns that contained an SIB event ($r=0.63$; $p<0.001$). The raw frequency counts of SIB were also found to be highly correlated with SIB-SIB CPs ($r=0.70$; $p<0.001$), the total number of T-patterns ($r=0.64$; $p<0.001$), the number of T-patterns containing SIB ($r=0.56$; $p<0.001$), and the proportion of T-patterns that contained an SIB event ($r=0.59$; $p<0.001$). This indicates that these measures are highly related and share ~9 to ~40 % common variance. What a direct correlational comparison does not address, however, is whether there are differences in the types of information that can be derived from each method. To address this issue, we also examined the sensitivity of each method to diurnal cortisol levels and their relative utility for detecting changes associated with acute and long-term effects of naltrexone.

Table 3 provides a comparison of how well SIB frequency counts, SIB-SIB CPs, and SIB T-patterns correlated with salivary cortisol levels. All three measures showed modest, though nonsignificant, trends for inverse correlations with the A.M. cortisol levels. For the 4:00 P.M. samples, however, elevated levels of cortisol were significantly associated with the number of T-patterns that included SIB ($r=0.61$; $p<0.005$). Correlations between afternoon cortisol levels and SIB frequency counts were also positive but only marginally significant, while SIB-SIB CPs showed no relations at all with this measure. Finally, the largest correlations with cortisol were for the percent decrease from A.M. to P.M. levels. Significant negative correlations with this measure were found for both SIB frequencies ($r=-0.57$; $p<0.01$) and SIB T-patterns ($r=-0.81$; $p<0.001$), but not for SIB-SIB CPs, indicating that days with flatter diurnal variation were associated with a greater probability of

Table 3
Pearson's correlation coefficients (significance) for comparisons between salivary cortisol levels and various measures of SIB

	SIB frequency	SIB-SIB CPs	SIB T-patterns
Cortisol—6:30 A.M.	−0.27 (0.29)	−0.19 (0.44)	−0.28 (0.26)
Cortisol—7:00 A.M.	−0.26 (0.29)	−0.22 (0.38)	−0.35 (0.15)
Cortisol—7:30 A.M.	−0.28 (0.26)	−0.26 (0.29)	−0.43 (0.08)
Cortisol—4:00 P.M.	0.43 (0.06)	0.00 (0.99)	**0.61 (0.005)**
% Decrease A.M.–P.M.	**−0.57 (0.01)**	−0.02 (0.92)	**−0.81 (0.0005)**

Table 4
ANOVA results comparing the changes in three measures of self-injurious behavior (SIB) following treatment with naltrexone for 3 weeks (acute) and 6 months (long-term)

	Baseline	Acute effects		Long-term effects	
	Mean (SD)	Mean (SD)	*F*-value (sig.)	Mean (SD)	*F*-value (sig.)
SIB T-patterns (%)	55.8 (16.2)	30.4 (23.5)	**5.61 (0.03)**	13.6 (25.2)	**15.59 (0.0005)**
SIB-SIB CPs	0.04 (0.01)	0.04 (0.02)	0.25 (0.62)	0.02 (0.02)	**12.54 (0.001)**
SIB frequency	55.0 (51.9)	26.4 (18.4)	3.05 (0.10)	8.3 (12.9)	**24.13 (0.0005)**

SIB and SIB T-patterns. This would appear to indicate that this subject's cortisol levels remained unusually high well into the afternoon on days when SIB occurred at higher levels and a high proportion of his temporally organized behavioral patterns included SIB among their constituent events, regardless of SIB-SIB sequential dependencies. Overall, these findings indicate that temporal pattern analyses offer superior sensitivity to this biological measure and merit further application in future investigations into the possible contributory role of disturbed diurnal patterns of cortisol in relation to the temporal patterning of SIB.

Table 4 provides a summary of the differential sensitivity of these measures to changes associated with the administration of naltrexone. Repeated-measures ANOVA comparisons revealed that the proportion of T-patterns containing SIB decreased significantly ($F = 5.61$; $p < 0.03$) in the weeks immediately following the initiation of naltrexone. By contrast, no significant acute changes were seen in SIB-SIB CPs or in the raw frequency counts of SIB, though the latter did show a sizable decrease. The repeated-measures ANOVA comparisons of the "long-term" effects, following 6 months of treatment with naltrexone, however, revealed highly significant decreases in all three of these measures: the proportion of T-patterns containing SIB ($F = 15.59$; $p < 0.001$), SIB-SIB CPs ($F = 12.54$; $p < 0.001$), as well as the raw frequency count of SIB events ($F = 24.13$; $p < 0.001$).

Of note, the proportion of T-patterns containing SIB was the only measure to show a statistically significant change during the first few weeks of treatment. Although SIB frequency counts also decreased by more than 50 % this change was not found to be statistically significant. This may mean that SIB T-pattern percentages could yield important predictive information with regards to intervention effectiveness, even when clinically observable effects could be masked by the higher variability in raw frequency counts. The clinical implications of this finding are that physicians may prematurely decide to discontinue treatment with naltrexone on the basis that no notable effects on SIB are observed by the caregivers within the first few weeks. However, these findings would suggest that such a decision, made on the basis of subjective impressions or SIB frequency counts alone, may discount the possibility that naltrexone may produce "subclinical" acute effects on processes subserving SIB, which may only be observable with more sophisticated analytic methods. As suggested earlier in the chapter, naltrexone may serve to disrupt the underlying temporal organization of emergent SIB patterns by blocking the contributory role of the endogenous opiate β-endorphin, thereby creating a condition which could potentiate the success of other behavioral interventions.

7 Examining the Self-Organization of SIB Dynamics

As described by Pincus et al. [60], self-organizing systems are essentially systems in nature that exhibit emergent order through the interactions of sufficiently complex coupling relationships among interacting components. Once such a system emerges, it is maintained over time through reciprocal feedback from the global level back down to the parts, leading to subsequent emergence over time. Self-organization is considered adaptive within "living" systems (e.g., biological, behavioral, or social systems) because it allows such systems to adjust their levels of structural organization towards rigidity or flexibility depending upon environmental demands. Numerous examples of self-organization exist in nature, such as flocking behaviors (e.g., migratory birds, humans in a crosswalk, or driving in traffic), collective survival behaviors (e.g., swarming insects or ant colonies), and also across the various domains of psychological science [61] including: small group dynamics [62], psychotherapy processes [63], emotional dynamics [64], and symptom covariations underlying psychopathology [65].

In the interest of examining whether temporal patterns of SIB displayed the hallmark characteristics of self-organization, Pincus et al. [60] utilized orbital decomposition (OD) [66, 67] to analyze the observational data collected by Sandman et al. [22] and described earlier in this chapter. OD is a technique based on symbolic dynamics designed to identify patterns and measure complexity within categorical time-series data. The use of OD provided a means to test the role of SIB within self-organizing behavior–environment pattern dynamics by producing several theoretically grounded measures of systemic complexity (i.e., Shannon entropy, topological entropy, Lyapunov dimension, and fractal dimension), essentially measures of order and disorder. These entropy measures allow one to empirically examine self-organization in behavioral flows and allowed us to compare the results of this method of analysis with those obtained using Theme™.

The results of this analysis revealed that the dynamics observed in these data were generally what would be described as low dimensional chaos or "edge of chaos" dynamics as described by Kauffman [68] and others as hallmarks of self-organization. This conclusion is based on the Lyapunov dimension value between 1 and 2 (mean $D_l = 1.2$), the fractal dimension between 2 and 3 (mean $D_f = 2.542$), and strong fits to inverse power-law distributions (mean IPL $R^2 = .93$). These results strongly suggest that the behavior-environment dynamics described in the Sandman et al. [22] study would be accurately characterized as self-organizing behavioral flows.

Furthermore, a comparison of records with and without SIB also revealed that the series with SIB contained significantly longer deterministic patterns (mean SIB = 12.48, mean No-SIB = 8.47; $p = 0.008$; $t = 2.658$), a higher number of behaviors within the same time-period (mean SIB = 173, mean No-SIB = 134; $p < 0.001$; $t = 3.702$), and a greater observed variety of behavioral codes (mean SIB = 10.70, mean No-SIB = 9.47; $p = 0.001$; $t = 3.404$). Altogether, longer patterns, higher activity, and higher variety culminated in higher levels of Shannon entropy in series containing SIB (mean SIB = 4.61, mean non-SIB = 4.36; $p < 0.001$; $t = 3.642$). It appears that SIB promotes more behavioral shifting, variety, and unpredictability of patterns. These results suggest that series containing SIB are more coherent (i.e., pattern length), yet also more complex (i.e., higher Shannon entropy), which is consistent with the results reported by Sandman et al. [22] using Theme™ to quantify the temporal patterning of SIB.

8 Theoretical Implications and Future Directions

Theories regarding the persistent recurrence of SIB among individuals with IDD have generally embraced either a strict behavioral or biological perspective. The behavioral account proposes that SIB is reinforced through operant conditioning [69], for example as a means of escaping demands [70, 71], relieving anxiety or seeking attention [34]. Indeed, a number of studies have shown positive outcomes for behavioral interventions which lend support to this conditioning perspective [72, 73]. The biological perspective focuses on physiological and neurological processes associated with SIB, and considers its recurrence in relation to the endogenous release of opiates or other disturbances in the pain, pleasure, or arousal centers of the brain [1, 59]. Treatments aimed at blocking pleasure receptors, lowering pain thresholds, and decreasing arousal have also proven effective in a number of studies [57–59]. Nevertheless, no clear consensus has emerged from either line of investigation and there are no universally effective treatments.

In our recent investigations, described above, we have utilized a sophisticated method of time-pattern analysis (Theme™) and explored the utility of concepts derived from nonlinear dynamical systems theory. It was hoped that this novel approach would provide new insights into the mechanisms subserving the persistent recurrence of SIB, instigate the formulation of new hypotheses for future empirical studies, and eventually lead to the development of innovative new treatments. Indeed, the findings reported herein do support the perspective that the temporal patterning of SIB is an overt expression of a self-organizing process that defies classical interpretation from either a prevailing behavioral or biological viewpoint. Perhaps it is time for researchers of SIB to begin exploring new nosological concepts that are not constrained by traditional theoretical frameworks.

Commonly referred to as "Chaos Theory," the dynamical systems perspective has been growing in popularity over the past 25 years thanks in large part to best-selling books such as *Chaos: Making a New Science* by James Gleick [74] and *At Home in the Universe* by Stuart Kauffman [68]. Among the more important recent developments, is the work of Per Bak, Chao Tang, and Kurt Wiesenfeld, who first introduced the concept of "self-organized criticality" [75]. As noted earlier, a self-organizing dynamical system is one in which complex order can emerge through the interactions of lower-order components. The concept of "self-organized criticality" describes the property of some dynamical systems that are balanced at their critical point (i.e., at the "edge of chaos"), such that slight perturbations of the system can trigger large transitions, or the emergence of a complex rearrangement in the overall state of the system. The classic examples of this are a sand pile shifting under the weight of a few added grains of sand, or a massive avalanche being triggered by the added weight of a solitary skier. These systems can be described as having self-organizing, critical-state dynamics as they are poised at the edge of large transitions, whereby a small change in the system can lead to large-scale changes and greater emergent complexity.

Recent studies [76–79] have described "self-organized criticality" in the human brain, particularly in the critical balancing of inhibitory/excitatory projections within cortical–subcortical loop circuits through activity-dependent, dynamic modulation of synaptic receptor densities or other mechanisms of neural plasticity (e.g., long-term potentiation or depression). It has been suggested that these dynamic modulation processes serve to keep the summation of inhibitory and excitatory projections within these neural loop circuits poised at a critical state to maximize information processing capacity and maintain the flexibility of the system to rapidly respond to environmental demands or other changes in the condition of the overall system. The critical-state dynamics of this system include "long-range temporal correlations" that fluctuate with the amplitude envelopes of neuronal oscillations (e.g., electroencephalography), and "neuronal avalanches" which are spatiotemporal cascades of activity that emerge from the aggregation of local field potentials within parallel neural loop circuits.

Parallel "cortico-striatal-thalamo-cortical" loop circuits are believed to regulate the patterning, storage, and elicitation of complex behavioral repertoires, as well as various motoric, affective, and cognitive "control" processes, and have been shown to be disturbed in individuals with schizophrenia and autism [78–84]. These same cortico-striatal loop circuits have also been suggested as a key neurobiological mechanism involved in the generation and maintenance of repetitive "maladaptive" behaviors, such as motor stereotypies, compulsions, addictions, and even SIB [83, 84]. Furthermore, in addition to the well-known dopaminergic and GABAergic projections within these cortico-striatal loop circuits, recent evidence suggests opioid neuropeptides mediate communication between medium spiny striatal projection neurons which may "provide a new cellular substrate for competitive dynamics in the striatum" [85]. In fact, recent findings also suggest that "an intact endogenous opioid system is necessary for normal goal-directed learning and more importantly, reveal that a compromised endogenous opioid system during learning enhances the habitual control of actions" [86]. In short, these studies have implicated a relatively unrecognized role of opioidergic striatal projections in the dynamic modulation of neural loop circuits that are widely believed to subserve the learning and elicitation of complex behavioral sequences which may either be goal-directed ("adaptive") or habitualized ("maladaptive").

If these neural loop circuits do display "self-organized criticality," as has been suggested, then it would be reasonable to expect that the dynamics of this underlying system should be observable at a behavioral level, particularly since the critical-state dynamics of a self-organized system should be multilevel, self-similar, and scale-independent, by definition. As reported by Pincus et al. [60], and described above in this chapter, the temporal patterning of SIB within the observational records collected by Sandman et al. [22] do display the hallmark characteristics of a self-organized dynamical system poised in a critical state at the "edge of chaos." Accordingly, it is quite possible that the self-organization of self-injurious behavior, as revealed through temporal pattern analyses, could be regarded as an endophenotypic expression of systemic perturbation within the critical-state dynamics of the underlying cortico-striatal loop circuits of the human brain.

Granted, such suggestions are rather speculative; however, it is precisely this type of approach that will be required to overcome the limitations of traditional, linear, cause-and-effect, conceptualizations of complex behavioral phenomena like SIB, and begin to define new behavioral phenotypes with direct relevance to understanding the underlying neurobiological mechanisms. In 2009, the US National Institute of Mental Health (NIMH) launched its Research Domain Criteria (RDoC) project "to develop a research classification system for mental disorders based upon dimensions of neurobiology and observable behavior" [87–89]. The basic premise of this approach is that existing diagnostic categories do not provide an adequate foundation for research into the possible neurobiological mechanisms underlying "abnormal" behavior, as expressed across a dimension of "normal" behaviors. From this perspective, research into the neurobiological mechanisms believed to subserve SIB must seek to quantify a dimension of behavioral processes that may be linked with discrete neural circuits which, when perturbed, result in the perpetuation of maladaptive patterns of behavior, with self-injury occurring at the extreme. As such, our results described herein are directly aligned with the RDoC approach recommended by the NIMH.

In addition to the possibility of providing a new theoretical framework for future investigations of SIB, the implications of the findings presented herein also raise many questions that additional studies might address: Are there common neurobiological mechanisms underlying the learning, expression, and temporal patterning of complex human behaviors across a dimension from normal to abnormal? Do these mechanisms display self-organizing, critical-state dynamics that could be quantified across multiple levels of investigation? What are the "tuning" parameters that could modulate the dynamics of this system? Could a "dysregulation" of the opioidergic system (or other biological process) perturb the critical-state dynamics within the neural loop circuits that are thought to regulate the storage and elicitation of behavioral patterns? Are there other processes that could be used to modulate the dynamics of these circuits?

Finally, we sincerely hope that the current findings and somewhat speculative discussion presented in this chapter will serve to catalyze the development of innovative new treatments for SIB, as well as other maladaptive behaviors. For example, there are several neuromodulatory techniques that offer the potential of directly "tuning" the processes implicated in the dysfunctional perpetuation of maladaptive behaviors. In our previous studies exploring the use of repetitive transcranial magnetic stimulation (rTMS) in the treatment of schizophrenia [90, 91], we demonstrated that the "resonant tuning" of intrinsic alpha-frequency stimulation could be used to increase the amplitude and selectivity of frontal alpha oscillations and produce clinically significant improvements in symptom severity. There are several investigators that are currently exploring the efficacy of rTMS in the treatment of individuals with autistic spectrum disorders. At present, there is at least one published report that offers a preliminary indication that this technique could be beneficially applied in the treatment of SIB, as evidenced by a significant reduction in repetitive behaviors following low-frequency rTMS [92]. Future studies should seek to provide a theoretical basis for the mechanism of action of such neuromodulatory techniques, and we believe that the current chapter provides a viable framework for exploring the putative efficacy of such innovative new interventions for the treatment of SIB. Without the use of Theme™ to identify the dynamics underlying the temporal patterning of this complex and horribly debilitating behavior, however, the foundational tenets of our approach would likely remain unseen and undetected.

References

1. Sandman CA, Kemp AS (2011) Opioid antagonists may reverse endogenous opiate "dependence" in the treatment of self-injurious behavior. Pharmaceuticals 4:366–381
2. Schroeder SR, Oster-Granite ML, Thompson T (2002) Self-injurious behavior: gene-brain-behavior relationships. American Psychological Association, Washington, DC, USA
3. Kemp AS, Fillmore P, Lenjavi M, Lyon M, Chicz-DeMet A, Touchette PE, Sandman CA (2008) Temporal patterns of self-injurious behavior correlate with stress hormone levels in the developmentally disabled. Psychiatry Res 157:181–189
4. Bodfish JW, Lewis MH (2002) Self-injury and comorbid behaviors in developmental, neurological, psychiatric, and genetic disorders. In: Schroeder SR, Oster-Granite ML, Thompson T (eds) Self-injurious behavior: gene-brain-behavior relationships. American Psychological Association, Washington, DC, pp 23–39
5. Sandman CA, Touchette P (2002) Opioids and the maintenance of self-injurious behavior. In: Schroeder SR, Oster-Granite ML, Thompson T (eds) Self-injurious behavior: gene-brain-behavior relationships. American Psychological Association, Washington, DC, pp 191–204
6. Sandman CA et al (2003) β-Endorphin and ACTH are dissociated after self-injury in adults with developmental disabilities. Am J Ment Retard 108:414–424
7. Thompson T et al (1994) Opioid antagonist effects on self-injury in adults with mental retardation: response form and location as determinants of medication effects. Am J Ment Retard 99:85–102
8. Sandman CA et al (1993) Naltrexone reduces self-injury and improves learning. Exp Clin Psychopharmacol 1:242–258
9. Claes L, Vandereycken W (2007) Self-injurious behavior: differential diagnosis and functional differentiation. Compr Psychiatry 48(2):137–144
10. Schroeder SR et al (2001) Self-injurious behavior: gene–brain–behavior relationships. Ment Retard Dev Disabil Res Rev 7:3–12
11. Emerson E et al (1996) Time-based lag sequential analysis and the functional assessment of challenging behaviour. J Intellect Disabil Res 40(Pt 3):260–274
12. Hall S, Oliver C, Murphy G (2001) Early development of self-injurious behavior: an empirical study. Am J Ment Retard 106(2): 189–199
13. Sackett GP (1979) The lag sequential analysis of contingency and cyclicity in behavioral interaction research. In: Osofsky JD (ed) Handbook of infant development. Wiley and Sons, New York, NY, pp 623–649
14. Symons FJ et al (2001) Sequential analysis of the effects of naltrexone on the environmental mediation of self-injurious behavior. Exp Clin Psychopharmacol 9(3):269–276
15. Thompson T, Caruso M (2002) Self-injury: knowing what we're looking for. In: Schroeder SR, Oster-Granite ML, Thompson T (eds) Self-injurious behavior: gene-brain-behavior relationships. American Psychological Association, Washington, DC, pp 3–21
16. Magnusson MS (1996) Hidden real-time patterns in intra- and inter-individual behavior: description and detection. Eur J Psychol Assess 12:112–123
17. Magnusson MS (2000) Discovering hidden time patterns in behavior: T-patterns and their detection. Behav Res Methods Instrum Comput 32(1):93–110
18. Emerson E, Thompson S, Reeves D, Henderson D, Robertson J (1995) Descriptive analysis of multiple response topographies of challenging behavior across two settings. Res Dev Disabil 16:301–329
19. Marion SD, Touchette PE, Sandman CA (2003) Sequential analysis reveals a unique structure for self-injurious behavior. Am J Ment Retard 108:301–313
20. Kroeker R, Touchette PE, Engleman L, Sandman CA (2004) Quantifying temporal distributions of self-injurious behavior: defining bouts versus discrete events. Am J Ment Retard 109:1–8
21. Bakeman R, Gottman JM (1997) Observing interaction: an introduction to sequential analysis, 2nd edn. Cambridge University Press, Cambridge

22. Sandman CA, Kemp AS, Mabini C, Pincus D, Magnusson M (2012) The role of self-injury in the organization of behaviour. J Intellect Disabil Res 56:516–526
23. Sandman CA et al (2000) Computerized-assessment of treatment effects among individuals with developmental disabilities. In: Thompson T, Felces D, Symons F (eds) Behavioral observations: technology and applications in developmental disabilities. Brookes Publishing Co, Baltimore, MA, pp 271–293
24. Hetrick WP et al (1991) ODAP: a stand-alone program for observational data acquisition. Behav Res Methods Instrum Comput 13:453–454
25. Sandman CA et al (2008) The role of proopiomelanocortin (POMC) in sequentially dependent self-injurious behavior. Dev Psychobiol 50:680–689
26. Sandman CA et al (2002) Disregulation of proopiomelanocortin and contagious maladaptive behavior. Regul Pept 108:179–185
27. Sandman CA, Spence MA, Smith M (1999) Proopiomelanocortin (POMC) disregulation and response to opiate blockers. Ment Retard Dev Disabil Res Rev 5:314–321
28. Sandman CA et al (1997) Dissociation of POMC peptides after self-injury predicts responses to centrally acting opiate blockers. Am J Ment Retard 102:182–199
29. Kitzbichler MG, Smith ML, Christensen SR, Bullmore E (2009) Broadband criticality of human brain network synchronization. PLoS Comput Biol 5, e1000314
30. Guastello SJ, Liebovitch LS (2009) Introduction to nonlinear dynamics and complexity. In: Guastello SJ, Koopmans M, Pincus D (eds) Chaos and complexity in psychology: the theory of nonlinear dynamical systems. Cambridge University Press, Cambridge, MA
31. Pincus D, Ortega D, Metten A (2010) Orbital decomposition for the comparison of multiple categorical time-series. In: Guastello SJ, Gregson R (eds) Nonlinear dynamical systems analysis for the behavioral sciences: real data. CRC Press, Boca Raton, FL
32. Pezard L, Nandrino JL (2001) Dynamic paradigm in psychopathology: "chaos theory", from physics to psychiatry. Encéphale 27: 260–268
33. Pincus D, Metten A (2010) Nonlinear dynamics in biopsychosocial resilience. Nonlinear Dynamics Psychol Life Sci 14:253–280
34. Nixon MK, Cloutier PF, Aggarwal S (2002) Affect regulation and addictive aspects of repetitive self-injury in hospitalized adolescents. J Am Acad Child Adolesc Psychiatry 41:1333–1341
35. Sandman CA, Hetrick WP (1995) Opiate mechanisms in self-injury. Ment Retard Dev Disabil Res Rev 1:130–136
36. Sandman CA et al (1991) Brief report: plasma beta-endorphin and cortisol levels in autistic patients. J Autism Dev Disord 21:83–87
37. Sandman CA, Kastin AJ (1990) Neuropeptide modulation of development and behavior: implications for psychopathology. In: Deutsch SI, Weizman A, Weizman R (eds) Application of basic neuroscience to child psychiatry. Plenum Press, New York, NY, pp 101–124
38. Sandman CA, Barron JL, DeMet E, Chicz-DeMet A, Rothenburg S (1990) Opioid peptides and development: clinical implications. In: Koob GF, Strand FL (eds) A decade of neuropeptides, past, present and future. Annals of the New York Academy of Sciences, New York, NY, pp 91–107
39. Sandman CA et al (1990) Plasma B-endorphin levels in patients with self-injurious behavior and stereotypy. Am J Ment Retard 95:84–92
40. Sandman CA (1988) Beta-endorphin disregulation in autistic and self-injurious behavior: a neurodevelopmental hypothesis. Synapse 2:193–199
41. Sandman CA et al (2000) Long-term effects of naltrexone on self-injurious behavior. Am J Ment Retard 105:103–117
42. Sandman CA, Barron JL, Colman H (1990) An orally administered opiate blocker, naltrexone, attenuates self-injurious behavior. Am J Ment Retard 95:93–102
43. Sandman CA et al (1987) Influence of naloxone on brain and behavior of a self-injurious woman. Biol Psychiatry 22:899–906
44. Barron JL, Sandman CA (1985) Paradoxical excitement to sedative-hypnotics in mentally retarded clients. Am J Ment Defic 90:124–129
45. Barron JL, Sandman CA (1984) Self-injurious behavior and stereotypy in an institutionalized mentally retarded population. Appl Res Ment Retard 5:499–511
46. Sandman CA et al (1983) Naloxone attenuates self-abusive behavior in developmentally disabled clients. Appl Res Ment Retard 4:5–11
47. Barron J, Sandman CA (1983) Relationship of sedative-hypnotic response to self-injurious behavior and stereotypy by mentally retarded clients. Am J Ment Defic 88:177–186
48. Ernst M et al (1993) Plasma beta-endorphin levels, naltrexone, and haloperidol in autistic children. Psychopharmacol Bull 29(2): 221–227
49. Leboyer M et al (1994) Difference between plasma N- and C-terminally directed beta-endorphin immunoreactivity in infantile autism. Am J Psychiatry 151(12):1797–1801

50. Bouvard MP et al (1995) Low-dose naltrexone effects on plasma chemistries and clinical symptoms in autism: a double-blind, placebo-controlled study. Elsevier, Kidlington
51. Gillberg C (1995) Endogenous opioids and opiate antagonists in autism: brief review of empirical findings and implications for clinicians. Dev Med Child Neurol 37:239–245
52. Leboyer M et al (1999) Whole blood serotonin and plasma beta-endorphin in autistic probands and their first-degree relatives. Biol Psychiatry 45(2):158–163
53. Tiefenbacher S et al (2000) Physiological correlates of self-injurious behavior in captive, socially-reared rhesus monkeys. Psychoneuroendocrinology 25(8):799–817
54. Novak MA (2003) Self-injurious behavior in rhesus monkeys: new insights into its etiology, physiology, and treatment. Am J Primatol 59(1):3–19
55. Crockett CM et al (2007) Beta-endorphin levels in longtailed and pigtailed macaques vary by abnormal behavior rating and sex. Peptides 28(10):1987–1997
56. Willemsen-Swinkels SHN, Buitelaar JK, Nijhof GJ, Van England H (1995) Failure of naltrexone hydrochloride to reduce self-injurious and autistic behavior in mentally retarded adults. Arch Gen Psychiatry 52:766–773
57. Casner JA, Weinheimer B, Gualtieri CT (1996) Naltrexone and self-injurious behavior: a retrospective population study. J Clin Psychopharmacol 16:389–394
58. Sandman CA (2009) Psychopharmacologic treatment of non-suicidal self-injury. In: Nock MK (ed) Understanding non-suicidal self-injury: current science and practice. American Psychological Association Press, Washington, DC, pp 291–322
59. Symons FJ, Thompson A, Rodriguez MC (2004) Self-injurious behavior and the efficacy of naltrexone treatment: a quantitative synthesis. Ment Retard Dev Disabil Res Rev 10:193–200
60. Pincus D, Eberle K, Walder CS, Kemp AS, Lenjavi M, Sandman CA (2014) The role of self-injury in behavioral flexibility and resilience. Nonlinear Dynamics Psychol Life Sci 18:277–296
61. Guastello SJ, Koopmans M, Pincus D (eds) (2009) Chaos and complexity in psychology: theory of nonlinear dynamics. Cambridge University Press, New York, NY
62. Pincus D, Guastello SJ (2005) Nonlinear dynamics and interpersonal correlates of verbal turn-taking patterns in group therapy. Small Group Res 36:635–677
63. Pincus D (2009) Self-organization in psychotherapy. In: Guastello SJ, Koopmans M, Pincus D (eds) Chaos and complexity in psychology: the theory of nonlinear dynamical systems. Cambridge University Press, Cambridge, MA
64. Kuppens P, Allen NB, Sheeber LB (2010) Emotional inertia and psychological maladjustment. Psychol Sci 21:984–991
65. Katerndahl D, Wang CP (2007) Dynamic covariation of symptoms of anxiety and depression among newly-diagnosed patients with major depressive episode, panic disorder, and controls. Nonlinear Dynamics Psychol Life Sci 11:349–365
66. Guastello SJ (2010) Orbital decomposition: identification of dynamical patterns in categorical data. In: Guastello SJ, Gregson R (eds) Nonlinear dynamical systems analysis for the behavioral sciences: real data. CRC Press, Boca Raton, FL
67. Pincus D, Ortega D, Metten A (2010) Orbital decomposition for the comparison of multiple categorical time-series. In: Guastello SJ, Gregson R (eds) Nonlinear dynamical systems analysis for the behavioral sciences: real data. CRC Press, Boca Raton, FL
68. Kauffman SA (1995) At home in the universe. Oxford University Press, New York, NY
69. Smith RG, Lerman DC, Iwata BA (1996) Self-restraint as positive reinforcement for self-injurious behavior. J Appl Behav Anal 29:99–102
70. Blindert HD, Hartridge CL, Gwadry FG (1995) Case study: controlling self-injurious escape behaviors. Behav Interv 10:173–179
71. Durand VM, Carr EG (1987) Social influences on 'self-stimulatory' behaviour: analysis and treatment application. J Appl Behav Anal 20:119–132
72. Iwata BA, Roscoe EM, Zarcone JR, Richman DM (2002) Environmental determinants of self-injurious behaviour. American Psychological Association, Washington, DC
73. Hanley GP, Iwata BA, McCord BE (2003) Functional analysis of problem behaviour: a review. J Appl Behav Anal 36:147–185
74. Gleick J (1988) Chaos: making a new science. Penguin, New York, NY
75. Bak P, Tang C, Wiesenfeld K (1987) Self-organized criticality: an explanation of $1/f$ noise. Phys Rev Lett 59:381–384
76. Poil SS, Hardstone R, Mansvelder HD, Linkenkaer-Hansen K (2012) Critical-state dynamics of avalanches and oscillations jointly emerge from balanced excitation/inhibition in neuronal networks. J Neurosci 32:9817–9823

77. Palva JM, Zhigalov A, Hirvonen J, Korhonen O, Linkenkaer-Hansen K, Palva S (2013) Neuronal long-range temporal correlations and avalanche dynamics are correlated with behavioral scaling laws. Proc Natl Acad Sci 110:3585–3590
78. Yizhar O, Fenno LE, Prigge M, Schneider F, Davidson TJ, O'Shea DJ, Sohal VS, Goshen I, Finkelstein J, Paz JT, Stehfest K, Fudim R, Ramakrishnan C, Huguenard JR, Hegemann P, Deisseroth K (2011) Neocortical excitation/inhibition balance in information processing and social dysfunction. Nature 477:171–178
79. Thorn CA, Atallah H, Howe M, Graybiel AM (2010) Differential dynamics of activity changes in dorsolateral and dorsomedial striatal loops during learning. Neuron 66: 781–795
80. Graybiel AM (2008) Habits, rituals, and the evaluative brain. Annu Rev Neurosci 31:359–387
81. Graybiel AM, Mink JW (2009) The basal ganglia and cognition. In: Gazzaniga M (ed) The cognitive neurosciences IV. MIT Press, Cambridge, MA
82. Barnes TD, Kubota Y, Hu D, Jin DZ, Graybiel AM (2005) Activity of striatal neurons reflects dynamic encoding and recoding of procedural memories. Nature 437:1158–1161
83. Muehlmann AM, Lewis MH (2012) Abnormal repetitive behaviours: shared phenomenology and pathophysiology. J Intellect Disabil Res 56:427–440
84. Lewis MH, Kim SJ (2009) The pathophysiology of restricted repetitive behavior. J Neurodev Dis 1:114–132
85. Blomeley CP, Bracci E (2011) Opioidergic interactions between striatal projection neurons. J Neurosci 31:13346–13356
86. Wassum KM, Cely IC, Maidment NT, Balleine BW (2009) Disruption of endogenous opioid activity during instrumental learning enhances habit acquisition. Neuroscience 163:770–780
87. Sanislow CA, Pine DS, Quinn KJ, Kozak MJ, Garvey MA, Heinssen RK, Wang PS, Cuthbert BN (2010) Developing constructs for psychopathology research: research domain criteria. J Abnorm Psychol 4:631–639
88. Morris SE, Cuthbert BN (2012) Research domain criteria: cognitive systems, neural circuits, and dimensions of behavior. Dialogues Clin Neurosci 14:29–37
89. Cuthbert BN, Insel TR (2013) Toward the future of psychiatric diagnosis: the seven pillars of RDoC. BMC Med 11:126
90. Jin Y, Potkin SG, Kemp AS, Huerta S, Alva G, Thai T, Carreon D, Bunney WE (2006) Therapeutic effects of individualized alpha-frequency transcranial magnetic stimulation (αTMS) on the negative symptoms of schizophrenia. Schizophr Bull 32:556–561
91. Jin Y, Kemp AS, Huang Y, Thai TM, Zhaorui L, Xu W, He H, Potkin SG (2012) Alpha EEG-guided transcranial magnetic stimulation (TMS) in schizophrenia. Brain Stimul 5:560–568
92. Sokhadze E, Baruth J, Tasman A, Mansoor M, Ramaswamy R, Sears L, Mathai G, El-Baz A, Casanova MF (2010) Low-frequency repetitive transcranial magnetic stimulation (rTMS) affects event-related potential measures of novelty processing in autism. Appl Psychophysiol Biofeedback 35:147–161

Chapter 6

Detecting and Characterizing Patterns of Behavioral Symptoms of Dementia

Diana Lynn Woods, Maria Yefimova, Haesook Kim, and Linda R. Phillips

Abstract

The high prevalence of dementia among older adults over the age of 85 years, the fastest growing sector of the population in the USA, constitutes an impending health care crisis. One major contributor to the costs and distress experiences is the management of behavioral symptoms in persons with dementia (BSD), such as yelling, restlessness and wandering. There is a critical need for effective interventions that either prevent or reduce the occurrence of BSD. Part of the reason that we have failed to create satisfactory treatments that produce safe and consistent results is that currently we have very crude measures for outcomes. Outcomes are usually assessed with a single dichotomous indicator, focusing on either the presence or absence of a particularly troublesome behavior, rather than viewing behavior as occurring in a cluster. The current conceptualization of BSD leads to our inability to quantify and detect subtle behavior changes that might signify intervention responses or to appropriately detect and characterize these behaviors such that interventions can be tailored. Temporal pattern analysis (T-pattern) can be used to detect and characterize subtle behavior changes that might signify intervention responses. In this chapter we present work completed by our research team at the University of California, Los Angeles, School of Nursing using pattern recognition software, THEME 5.0. The aim of the following sections is first, to review and critique the current measurement strategies for BSD and their limitations, and second, to offer an alternative method of conceptualizing and analyzing behavioral data. The purpose of using THEME™ is to identify and quantify behavior patterns with regard to intensity, frequency and complexity. These temporal associations may reveal characteristics of behavior escalation, which in turn will enhance the tailoring and timing of interventions to reduce BSD burden. The T-pattern algorithm was used to detect and characterize BSD using videotaped and direct observation data. Three exemplars are presented: first, a case study of interaction leading to the escalation of behavior, second, the relationship between dementia and culturally appropriate interaction and BSD, and third the effect of a non-pharmacological treatment effect on BSD. The analysis of the exemplar data is presented and discussed in the context of a direction for future research.

Key words Behavioral symptoms of dementia, Escalation patterns, Intervention validation

1 Introduction

The burgeoning population of elders over the age of 85 years, the fastest growing sector of the population in the USA [1–3], coupled with the high prevalence of dementia (45 %), constitutes an

Magnus S. Magnusson et al. (eds.), *Discovering Hidden Temporal Patterns in Behavior and Interaction: T-Pattern Detection and Analysis with THEME™*, Neuromethods, vol. 111, DOI 10.1007/978-1-4939-3249-8_6,

impending health care crisis [4]. One major contributor to the costs is the management of behavioral symptoms in persons with dementia (BSD) [5, 6]. BSD, such as yelling, restlessness, and wandering, are some of the most difficult problems faced by caregivers, consuming enormous amounts of time and effort [7]. Although 32 % of behaviors occur alone, 66.4 % co-occur [8] and 18 % co-occur within the same hour [9]. Behaviors manifested by persons with dementia (PWD) are frequently accompanied by alterations in affect and response to a particular interaction. BSD can result in fear and avoidance in caregivers and others in close proximity, embarrassment for affected older adults, and nursing home (NH) placement [10, 11]. These behaviors also contribute to staff turnover [12–14], and accidents involving both elders and care providers [15]. Moreover, high restlessness among PWD is strongly associated with reduced caregiver quality of life [16]. As a consequence, there is a critical need for effective interventions that either prevent or reduce the occurrence of BSD. Finding these interventions has remained elusive despite years of research.

The aim of this chapter is, first, to review and critique the current measurement strategies for BSD and their limitations, and second, to offer an alternative method of conceptualizing and analyzing behavioral data. The purpose of using pattern recognition software, such as THEME™, is to identify and quantify behavior patterns with regard to intensity, frequency, and complexity. These temporal associations may reveal characteristics of behavior escalation, which in turn will enhance the tailoring and timing of interventions to reduce BSD burden. To this end, the analysis of three exemplar data is discussed, highlighting future research directions.

1.1 Problem and Measurement Issues

We define BSD as restlessness, vocalization, pacing, searching and wandering, tapping and banging, and attempts to escape from restraint. Studies indicate that 38 % of individuals with dementia have repetitive purposeless activity (restlessness), [17] and between 29 and 45 % exhibit problematic vocalization [9]. Although non-pharmacological interventions are recommended as a first line of treatment for BSD [18], overall these treatments show only small to moderate effects [19]. Some speculate [20–22] that at the heart of the difficulty of developing safe and effective interventions is a measurement problem [20].

Part of the reason that we have failed to create satisfactory treatments that produce safe and consistent results is that currently we have very crude measures for outcomes. Outcomes are usually assessed with a single dichotomous indicator, focusing on either presence or absence of a particularly troublesome behavior, such as yelling (range 0–1). Rarely are outcomes of 0 considered (meaning that yelling is absent, but so are functional behaviors). The current measurement strategy is based on the belief that BSDs are separate or that they may occur with one other behavior. This view does not include the notion that these behaviors may form complex, nonrandom clusters that occur repeatedly in the same individual [23].

The current conceptualization of BSD leads to our inability to quantify and detect subtle behavior changes that might signify intervention responses. Strongly influenced by the ABC (antecedent–behavior–consequence) model of behavior [24–26] BSDs are usually viewed as single behaviors ("target behaviors") that occur following a stressful event or stimuli. Under the single behavior approach, some researchers [27, 28] conceptualize BSD as consisting of subtypes of troublesome or unsafe behaviors. For example, Cohen Mansfield [27] considers twenty-nine BSD divided into four subscales (physical agitation; physical aggression; verbal agitation; verbal aggression). This approach to studying particular disruptive behaviors is common. For example, wandering [29–31], aggressive behavior [32], and problematic vocalizations [14, 17, 33, 34] have been studied extensively. Most measures of BSD under the single behavior approach rely on frequency of the target behavior. In the Cohen Mansfield Agitation Inventory (CMAI) [35], the most widely used instrument for measuring BSD, the 29 behaviors are rated on a frequency scale from 1 to 7 (never occurs to occurs several times an hour). Scores are summed within and then across subscales to derive the total score. Rarely considered in this type of measurement is that the absence of the target behavior can also mean absence of functional behaviors. This frequently happens with use of psychotropic drugs where the PWD becomes somnambulant.

Since the current measurement strategy is based on the belief that BSD are separate or occur with one other behavior, it is impossible to determine the interrelationships among behaviors and their therapeutic importance. Moreover, it is impossible to link triggers to troublesome behaviors because the measurement begins at the point when the target behavior occurs. A trigger may precipitate a behavioral chain that begins with the appearance of more benign behaviors (e.g., restless tapping) and escalates to the target behavior (yelling). Given the current conceptualization and strategies available for measuring outcomes, it is difficult to know if the problem is with the interventions or with the measurement of outcomes. The conceptual and measurement problems in studies of BSD have been recognized for some time. In 1996, for example, Cohen-Mansfield commented that "we need to acknowledge and clarify the complexity of BSD" [36], a challenge that we still face today.

1.2 Reconceptualization

Consistent with the view of von Gunten and colleagues [22] and based on our preliminary data, we conceptualize BSD as clusters of behavior rather than being single events [23]. Within each individual, behaviors vary by intensity and the number of different clusters displayed in a specific time period. Behavior patterns can escalate over time, becoming more complex (increased in variability), more intense (more troublesome), and more frequent [23].

This suggests that individual behaviors in BSD may be linked together in predictable ways.

Behavioral patterns are key characteristics of new conceptualization of BSD. Two types of patterns are interesting for clinical consideration: temporal patterns and patterns of escalation. Temporal patterns have been examined for the past several years using a variety of methods in an attempt to determine the time of peak behavior, albeit using aggregated data, and instruments designed to be used in cross sectional studies [37]. Patterns of escalation involve an increase in the frequency and intensity of a behavior. Escalation temporality, the period of time within which a behavior changes, can be demonstrated in the case of vocalization, which begins as repetitive mumbling and escalates to louder calling, then yelling within a period of 20–40 min [23]. The "ideal" sequence of escalation follows a clearly demarcated linear continuum from agitation to aggression to violence [38, 39]. The reverse, de-escalation, is assumed but rarely described. However, in contrast to the that model, Woods et al. [23] examined patterns of BSD escalation using videotaped data and found that the "ideal" pattern was rarely observed. Rather than escalating within categories in a linear fashion, patterns moved back and forth between behavioral categories. This suggests that, in addition to containing more frequent and severe behaviors, escalation patterns become more complex, increasing in variability among co-occurring behaviors.

We conceptualize escalation as involving behavioral chains that begin with benign behaviors and culminate in the target behavior that is very disruptive to family and staff caregivers. For example, our data show particularly troublesome behaviors like yelling may be accompanied by benign behaviors (e.g., fidgeting, tapping) that may actually be part of a pattern of escalation. We present a new strategy to measure outcomes validly and reliably, representing characteristics of behavioral patterns, the interrelationship of behaviors and the escalation of behavior over time.

1.3 Temporal Pattern Analysis

Our approach to the detection and characterization of BSD is pattern recognition using THEME™5.0 [PatternVision Ltd. and Noldus Information Technology] [40] for measuring complex behavior patterns that do not necessarily occur in a proscribed sequence within individuals. THEME™ makes no a priori assumptions about the characteristics of the behavior patterns. Rather, it allows the data to inform the resultant patterns, providing a more accurate characterization of temporal behavior patterns, behavior complexity and escalation. By detecting the complexity, frequency, and interrelationships of behavior patterns then quantifying these factors within person, THEME™ provides a method of detecting within-individual sequential and nonsequential temporal patterns (T-patterns) of related behavior clusters that are not obvious to the trained observer or identifiable by traditional sequential methods [40, 41].

This pattern recognition software first identifies significant recurrences of any two events in a realtime behavioral record, such as low-intensity restlessness or vocalization. A simple pattern is defined as two behaviors that occur nonrandomly. This relationship becomes a behavioral pair that is specified as level one. T-patterns increase in complexity as simple patterns are incorporated into more complex patterns. We define a complex pattern as incorporating at least eight behaviors within four levels. Moreover the search algorithm can be set (filtered) such that only high-intensity and/or complex behavior patterns are detected and quantified. Thus, the number of high-intensity and complex patterns within a time period can be examined in addition to escalation from simple patterns (lower intensity and few behaviors) to complex patterns (higher intensity and more varied behaviors). Escalation measured in this way shows behavior clusters that include behaviors of low intensity becoming more complex and including different behaviors of higher intensity.

2 Three Exemplars

Three exemplars of BSD data analyzed with THEME™ are presented: two case studies and one intervention study. For the intervention study, the software was used to confirm and expand the analysis of aggregated behavioral data. Each of the three exemplars highlights a different aspect for the use of THEME™. All behavioral data in the parent studies were collected using the modified Agitated Behavior Rating Scale (mABRS), which codes for the frequency and intensity of six different behaviors: restlessness, escape restraints, searching or wandering, tapping or banging, pacing and walking, and vocalization [42] (see Table 1). Behaviors were scored on a 4-point, Likert-type scale ranging from "not present" (0) to "high-intensity" (3).

To ensure the validity and reliability of behavioral observations research assistants (RAs) completed a 2 h training session, then simultaneously observed and scored a resident every 20 min for 2 h. Interclass correlation was calculated to ascertain inter-rater reliability using a two-way mixed effects model where people effects were random and measures effects were fixed. Cohen's kappa, ranged 0.66–1.00, considered good to excellent agreement [43, 44] For the frequency measures percentage agreement was 85.5–96.8 %, with kappa values in the range of 0.71–0.93. For intensity measures percent agreement was 83–89.5 %, with kappa values ranging from 0.60 to 0.64. The mABRS [42] is a valid and reliable tool that allows the capture of both the frequency and intensity of multiple co-occurring behaviors. This instrument fits the outlined measurement strategy necessary for the pattern recognition and analysis using THEME™.

Table 1
Examples of mABRS coding and descriptions

Behavior	Description
Restlessness	
1 LowIntensity	Rhythmic, purposeless movements of hands in rubbing/picking motion
2 RepetiManipulat	Repetitive manipulation of an object, as if a nervous habit; may display facial distress
3 LargeUrgency	Large amplitude rubbing/picking; may see rocking motions of the torso
Escaping restraints	
1 FidgetRest	Purposeless fidgeting w/restraints; non-stressful, nonurgent behavior
2 TuggingRest	Purposeful tugging at restraints; urgency reflected in motions
3 Violent/exit	Violent rattling of or attempts to slip out underneath restraints
Searching/wandering	
1 Search/table	Undirected/haphazard searching through pockets, purse, lap blanket, or tray
2 SearchDrawer	Searching thru pockets, rooms, drawers, the purpose of which is disoriented to reality
3 WanderRoom	Wandering into various rooms in search of something
Tapping/banging	
1 TapRhythmHands	Purposeless, rhythmic tapping of a few fingers or the feet; denture clacking
2 TapSwingH&F	Irregular tapping or swinging motions which involve the whole hand/arm, feet and legs
3 ViolentBang/Hit	Violent banging or hitting on self or object. Involves hands, arms, and possibly legs
Pacing/walking	
1 WalkingAimless	Walking aimlessly around the halls or unit. Movement is relaxed and steady
2 WalkingPurpose	Purposeful walking toward door or exit; very directed activity—difficult to redirect
3 Pacing/Marching	Pacing or marching with increasing speed around hall or unit
Vocalization	
1 LowMumble	Mumbling to self or others; words are not necessarily distinct. Low volume
2 RepeatModIntens	Continuously asking questions of others; repetitive. Mild intensity groaning, moaning
3 ScreamYell	Complaining, demanding, temper outbursts, yelling, screaming

2.1 Case Study 1: Behavior, Affect, and Interaction

The first exemplar focuses on using THEME™ to detect temporal relationships among behaviors exhibited by a 93-year old Caucasian female nursing home resident with a Mini Mental State Exam (MMSE) score of 13 indicating moderate-to-severe dementia. Of the eighteen 20-min videotapes recorded over a 4 day period, one videotape that demonstrated the majority of BSD was selected for further analysis by THEME™. As well as the six behaviors included in the mABRS [42], persons with dementia may express their emotions through facial expression, body movements, posture, gestures, and non-word vocalizations [45], all of which are considered behavior. Affect is closely related to behavior and can be a sign of unfulfilled needs that may prompt behavior. The quality of interaction with others affects affect and potentially, if positive, alleviate outbursts of behavior or change the escalation trajectory of a behavior. THEME™ was used to find patterns that include all behavior affect and interaction that may shed light on inter-relationships and provide an avenue for intervention.

2.1.1 Methods

A trained research assistant used Noldus The Observer 5.0 (Noldus Information Technology, the Netherlands) to recode BSD, affect and interaction in the videotaped observation. BSD was coded using mABRS. Affect and interaction behaviors were coded with the Observable Displays of Affect Scale (ODAS) [46]. This scale contains 41 behaviors categorized into six subscales of positive and negative facial displays, vocalizations, and body movement/posture. During the observation presence/absence of each behavior was indicated. Inter-rater reliability for the categories ranged 0.68–0.98. Test-retest reliability for the categories ranged from 0.97 to 1.00. Ten experts in gerontological nursing with an average of 17.8 years in the field established content validity [47]. Interaction was coded when staff, family, other resident or observer approached the subject. Inclusion of ODAS behaviors gives more contextual information to the interaction and provides an important avenue toward understanding internal and external interaction factors influencing BSD.

The coded observational record was transferred from The Observer into THEME™. The search settings were set at (1) minimum occurrence= 3, specifying that the pattern must occur at least three times in the record; and (2) significance level of <0.05, specifying the probability of random occurrence. Filters were set to detect all behaviors patterns, those containing BSD and those only containing high-intensity BSD and interaction.

2.1.2 Results

In the selected 20-min videotape, the resident appears agitated, and is trying to get out of a geri-chair (similar to a lounger with raised feet), that has a table for meals locked across the chair. This is considered a form of restraint. She is yelling multiple times during the 20-min time period. Several nursing assistants approach her making an attempt to calm her, but leave after a short interaction, after which there is another outburst of behavior. There is a direct temporal relationship between the interaction and behavior, but the specific relationship is difficult to discern when observing the videotape.

THEME™ analysis provided a more detailed picture of this videotaped behavior, yielding 1094 different behavior patterns with 4477 occurrences. Thirty-seven percent of the total number of patterns (511 patterns with 2084 occurrences) included some sort of BSD. Of those, 60 patterns included BSD, affect, and interaction. A complex pattern that consisted of 16 behaviors and 8 levels is shown in Fig. 1. The pattern occurs three times during the 20-min videotaped observation period. Figure 1 illustrates the sequence of interaction, with a behavioral outburst when the person was left alone following an interaction with staff. When the staff is nearby, the person exhibits the positive affect of attending to message, however, when the staff leaves and the person is left alone, she begins to bang on the table tray.

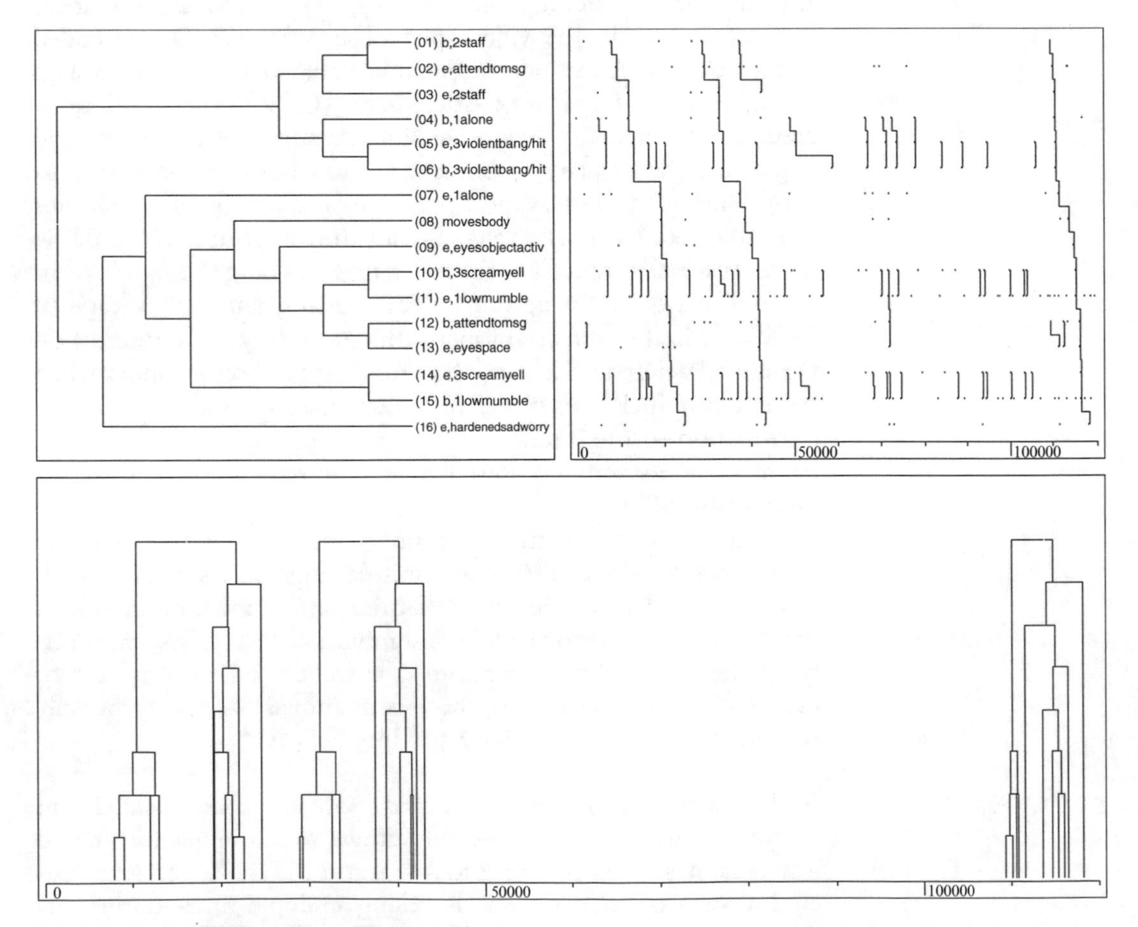

Fig. 1 Example of BSD pattern with interaction

Using traditional analysis methods, a mean behavior score was correlated with staff interaction. There was no identification of the temporal order of the relationships. THEME™ detected a pattern that suggests that once the person was left alone, she started "acting up." An occurrence of three times in a 20-min period indicates that this pattern may be a fairly frequent pattern that may be predicted. The clinical implication of detecting such a pattern is significant. If the type of behavior as well as the clustering of behavior shows a predictable pattern, then interventions can be timed to prevent or decrease the manifestation of this pattern.

2.2 Case Study 2: Behavior and Certified Nursing Assistant's Care

This case study is a part of a larger study [48] that focused on describing the association between the culturally competent care delivered by direct care staff to Korean American (KA) elders and BSD. The specific aim was to explore the relationship between the social interaction competence of non-Korean Certified Nursing Assistants (CNAs) and KA elders exhibiting BSD. We examined this association as a first step in the development of culturally appropriate and supportive interventions for ethnic minority older adults. This case study focuses on detecting the social interaction between the resident with dementia and the CNA, and the influence of the CNA's cultural and dementia competent behaviors on BSD. We focus on two research questions: (1) using THEME™ can we validate the original findings from the parent study; (2) can THEME™ provide more discrete information about the specific high-intensity troublesome behaviors.

Dementia-competent social interaction behaviors indicate the CNAs' ability to be aware of and use dementia appropriate verbal and nonverbal communication skills while caring for NH residents. Culturally competent social interaction behaviors indicate a CNAs' ability to be aware of and understand cultural differences that may affect residents' behaviors, and to use culturally appropriate verbal and nonverbal communication skills while caring ethnic minority nursing home residents.

In this case study, one resident was selected from the parent study [49]. The resident was a 74-year-old KA woman with moderate to severe dementia (MMSE = 10) who resided in the nursing home for 6 years. Her primary language was Korean with minimal understanding of English. Of the three CNAs who provided direct care, two were Hispanic and spoke Spanish primarily, and the other was African-American and English speaking. All CNAs were able to speak a few Korean words related to work, such as "sit," "turn," and "wait a minute". We used THEME™ to detect and characterize the individual pattern of high-intensity BSD, then analyze the association between the CNAs' social interaction behaviors and the resident's BSD.

2.2.1 Methods

Resident's BSD and CNA's social interaction behaviors were directly observed while assisting with morning and afternoon care. Trained research assistants input the codes using handheld computers. The scales used were mABRS [42], Dementia Social Interaction (DSI) and Culture Social Interaction (CSI) coding schemas [49]. DSI consists of 27 behavioral codes, including 12 dementia-competent and 15 dementia-incompetent behaviors. The dementia appropriate social interaction behaviors include making eye contact, speaking slowly, keeping a gentle pace, providing continuous assistance, speaking in a calm tone of voice, using gestures, and informing the resident before providing care. Inter-rater reliability of the DSI coding scheme varied from 0.73 to 0.83 for Cohen's Kappa and 75–84.62 % for percentage agreement [49]. CSI is based on cultural norms among Koreans, respecting older adults, which is embedded in verbal and nonverbal communication. The CSI schema consists of 17 behavioral codes, including eight culturally competent and nine incompetent behaviors. The culturally competent social interaction behaviors for KA older adults include greeting, using polite expression speaking in Korean, keeping proximity around arm length, not staring at older adults, and not calling the resident's first name The inter-rater reliability of the CSI coding schema varied from 0.69 to 0.82 using Cohen's Kappa, indicating good to very good agreement, and 76.92–84.21 % for percentage agreement [49].

Total BSDs per encounter were computed by multiplying the frequency of each behavioral occurrence by the intensity score and then summing the six types of BSDs for each of the six encounters. For each CNA, we computed the proportion of dementia-competent and culturally competent behaviors by summing the competent behaviors and then dividing by the total number of both competent and incompetent behaviors (competent behaviors/competent + incompetent behaviors).

One resident's data was selected for THEME™ analysis. Six observations of morning and afternoon care over a 3-day period were used for analysis. We exported the resident's behavioral data and CNA social interaction behavioral data into THEME™. The goal was to identify specific behavior patterns that contained high-intensity behavior for vocalization and restlessness. The search settings were (1) minimum occurrence= 6, specifying that the pattern must occur at least six times, such that the pattern occurs once in each observation period; (2) significance level of <0.05, specifying the probability of random occurrence.

2.2.2 Results

Using the original coding, the primary behavioral symptoms detected were vocalization and pacing during the observation period. While the CNA assisted the resident with morning or afternoon care, the resident refused care, cursed, screamed, or hit the CNA. When the resident was left alone after care was provided, the

Table 2
Resident's BSD and CNAs' social interaction behaviors

Day	Time of day	Total number of BSD	Culturally appropriate behaviors (%)	Dementia appropriate behaviors (%)
1	a.m.	50	24	87
	p.m.	244	41	81
2	a.m.	176	13	90
	p.m.	260	53	70
3	a.m.	45	15	89
	p.m.	236	55	50

resident searched other resident's drawers/rooms or paced until she napped. In the parent study there was no significant association between the total number of BSD and the CNAs proportion of dementia or culturally appropriate social interaction behaviors for either morning or afternoon care; however, there was a trend toward a decrease in BSD as dementia competent care increased especially in the afternoon.

The total number of BSD and the proportion of competent social interaction behaviors per social interaction encounter are shown in Table 2. The average total number of BSD was 169. The average CNAs' percentage of culturally appropriate behaviors was 33.5 % and dementia appropriate behaviors were 77.83 %. The total number of BSD was much higher in the late afternoon than in the morning. The average count of BSD in the morning was 90.33 rising to 246.67 in the afternoon, a difference of 156.34.

THEME™ analysis results provided more detailed information, by detecting specific types of BSD followed by specific dementia or culturally appropriate CNA behaviors. This relationship and these specific behaviors were not captured previously. Pattern recognition using THEME™ yielded 711 total patterns, 269 of which included BSD and interaction (37.84 % of total patterns). Figure 2a shows a within-person pattern containing both interaction and BSD. The analysis shows specific patterns with the following interaction. The CNA called the resident by her first name, which is culturally inappropriate and spoke a command in English, while assisting with ADLs. The resident then yelled at the CNA. The CNA provided the care step-by-step and used simple familiar words, which are dementia appropriate social interaction behaviors. The intensity of the resident's vocalization decreased to intensity 2 (mild intensity groaning) from a previous intensity of 3 (see Table 1).

THEME™ results validated findings from the original study [48]. The original findings showed that for the group of NH residents with high-intensity behavior, both dementia and culturally

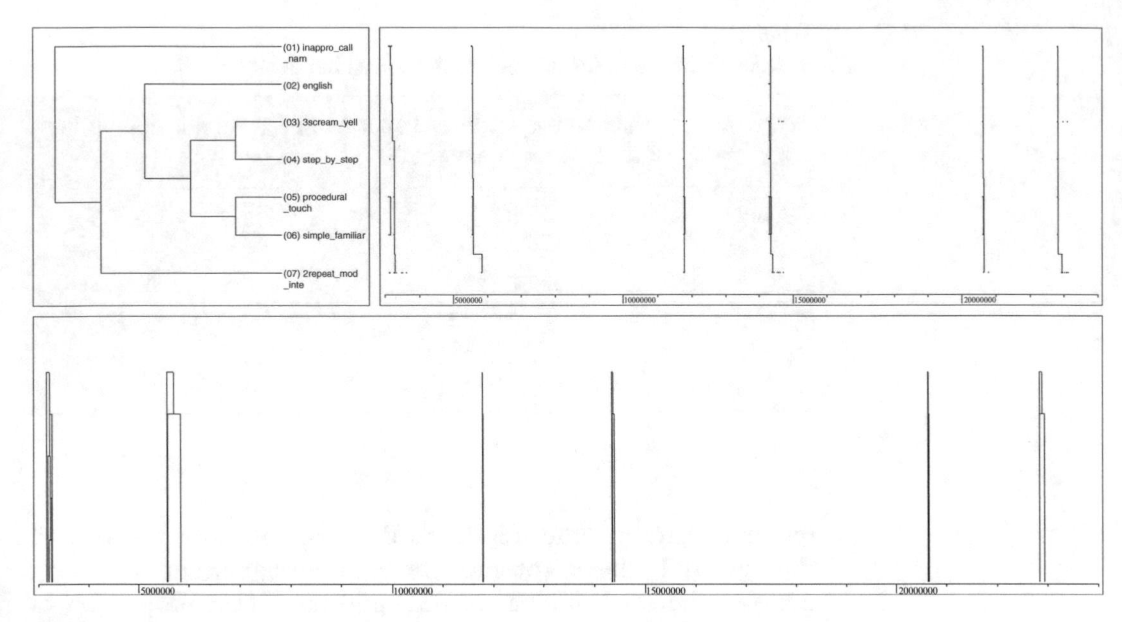

Fig. 2 Example of BSD pattern with CNAs' social interaction behaviors

competent social interaction behaviors are interrelated and influenced BSD with morning and afternoon differences. This association was most prominent among those with high-intensity BSD ($N=10$, $b=-4.46$, $SE=3.59$, $p=0.22$). This specific resident was one of the ten residents who exhibited high-intensity behavior, with more behaviors in the late afternoon than in the morning. However, since the original behavioral data was aggregated within an encounter across several time points, it was hard to capture discrete behavioral characteristics.

On the other hand, the THEME™ analysis expanded the results, elucidating the characteristics of specific behavior patterns, by identifying which behaviors are included in these patterns and their relationship to the time of day (morning or afternoon). This analysis shows that the CNAs' social interaction behavior may contribute to the escalation/de-escalation of BSD. In addition, in the late afternoon there was a delayed response between when the CNA used simple familiar language when providing care and vocalization of intensity of 2 (see Table 1) indicated by the little step between behaviors (06) and (07) in Fig. 2. One possible explanation is a delay in neurological connections. This information has important clinical implications and emphasizes that persons providing care may have to wait for a response. The response will likely not be immediate.

2.3 Case Study 3: Confirmation of Intervention Results

In the third exemplar study we wanted to confirmthe THEME results with known results of an intervention study. The study described elsewhere [50] was a double blind randomized controlled trial (RTC) to evaluate the effect of therapeutic touch (TT) to decrease BSD in NH residents with dementia. The sample of 64 predominantly Caucasian participants aged 67–93 years, with female–male ratio of 1:4 was randomized into three groups. The experimental group received TT with contact on the neck and shoulders; the placebo group received a mimic treatment that looked identical (to the naïve observer), and the control group received routine care. The purpose of using THEME™ to analyze the data was to detect the number of patterns at each of the five phases of the study over a 21-day period (baseline, treatment 1, post-treatment 1, treatment 2, post-treatment 2) and confirm a previously identified intervention response.

2.3.1 Methods

For the original data analysis behavioral data were aggregated into five time periods: 1 = baseline (days 1–4); 2 = treatment 1 (day 5–7); 3 = post-treatment 1 (days 8–12); 4 = treatment 2 (days 13–15), and 5 = post-treatment 2 (days 16–20). The percent of each behavior was calculated in relation to the total behavioral scores. Behavior scores were calculated by frequency × intensity of behavior. The mean score was calculated per time period by summing the frequency-intensity score per hour, per day, per person, computing the mean score per person per time period, and then pooling the scores across persons, for each of the five time periods. A mixed model approach was used to detect significance between the three groups.

Using THEME™ analysis behavioral data were imported and analyzed. The search settings were (1) minimum occurrence= 3, specifying that the pattern must occur at least three times (2) significance level of <0.05, specifying the probability of random occurrence. Then a filter was applied to patterns containing high-intensity behaviors. Percent of behavioral clusters containing high-intensity behaviors was calculated by dividing the number of patterns of high-intensity behaviors by the total number of found patterns. A mixed model was used to look at change in the percent of high-intensity patterns over the course of the study.

2.3.2 Results

For the original study, a mixed methods approach for longitudinal data was used as an omnibus test for group differences. Post hoc analysis using ANOVA methods was used to examine any group differences across time periods. Restlessness decreased significantly by group ($F2, 61 = 3.03$, $p = 0.05$) and time period $F2, 61 = 11.04$, $p < .0001$). [50]. While the original study examined only single behaviors, aggregating data, THEME™ results identified complex repeated patterns of behavior that included high-intensity restlessness and the other five behaviors (Table 1). Therefore a decrease in number of patterns indicates a decrease in several behaviors. As shown in Figure 3 and consistent with the original study, group variability was not equal at baseline. THEME™ showed that high-intensity patterns occurred in 7/22 of the experimental group; 11/21 of the placebo group and in 12/21 of the

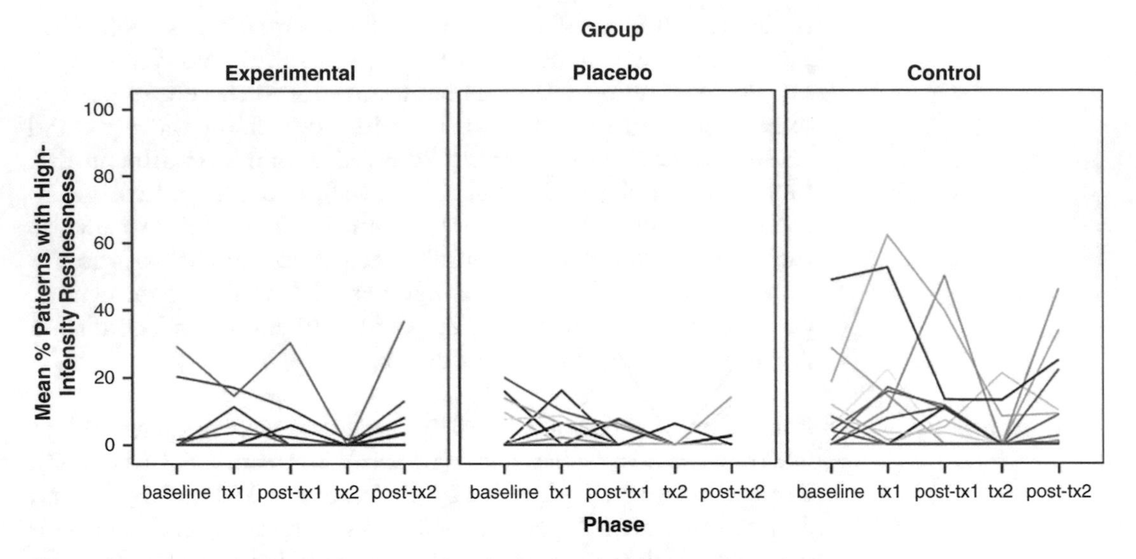

Fig. 3 Percent of patterns with high-intensity restlessness. Each *line* is an individual

control group. THEME™ results showed the same trend of decreasing high-intensity patterns of BSD including restlessness at time period 4 (treatment time 2) in the experimental group compared to the placebo and control groups. There was no significant change over time in the percent of patterns over time when the three groups were compared.

Using THEME™ we were able to detect patterns that contained high-intensity restlessness providing more discrete and specific information about which behaviors are included in these high-intensity patterns, thus providing insight into the specific behaviors that may be linked to an escalation pattern. Consistent with previous findings, the number of patterns decreased at treatment time period 2, although results were not significant, likely related to the small sample size and to the high variability. Moreover, it is important to remember that a decrease in the number of patterns containing high-intensity restlessness means that several behaviors decrease, not only restlessness. For example, a single behavior may decrease in intensity from a 3, maximum intensity to a 2 or 1, much lower intensity. While this may be clinically significant for this one behavior, decreasing the number of high-intensity patterns (those occurring at least 3 times) means that a cluster or constellation of BSD has decreased together. This results in an overall decrease in agitation in several persons and thus an overall decrease of agitation in the environment, potentially preventing others from escalating [53].

3 Discussion and Conclusion

Data show that rather than being single events, BSD are complex, nonsequential, nonrandom, patterned clusters of behavior recurring repeatedly in the same individual [23]. These behavior patterns vary by the number of clusters in a specific time period, the level of intensity of behaviors contained in the clusters, and the number of different clusters displayed in a specific time period. Behavior patterns can escalate over time and become more complex (increased in variability), more intense (more troublesome), and more frequent. Ultimately, to develop and provide safe, effective, and timely treatments we must use measurement strategies that can be used to measure outcomes in a way that validly and reliably represents behavioral patterns, the interrelationship of behaviors and the escalation of behavior over time. The use of T-patterns to detect stable structures no matter how hidden they are [51], is inextricably linked to the valid and reliable measurement of behavioral patterns and thus the development of effective and timely treatments. These patterns can also be used to further characterize the context in which the behavior takes place, increasing a broad understanding of these behaviors and guiding the timing and development of treatment.

The data presented shows that THEME™ can detect complex stable patterns. The use of T-pattern detection in the study of BSD has tremendous potential. The identification of patterns that cannot be identified through simple observation has great benefits in informing clinical practice. First, by detecting and characterizing these patterns we have the potential to predict when these patterns may most likely occur, then time already developed treatment and utilize new treatments to alleviate or prevent these disturbing behaviors. For example, once the relationship between the context, such as a staff interaction or noise, and an individual's behavior is identified, timely measures, related specifically to time of day, can be taken that modify the environment and reduce the manifestation of the behavior. Better utilization of resources has major cost implications.

We used THEME™ to validate and expand previous findings and to confirm an intervention effect. In Kim's study [48], the use of THEME™ provided not only validation but also more discrete information about specific behaviors and their relationship to time of day. Moreover, the detected patterns provide contextual information to better inform underlying factors associated BSD. Similarly, in the Woods study [50], the use of THEME™ confirmed an intervention effect for a non-pharmacological intervention. Moreover, we detected specific complex patterns that were more responsive to the intervention than other patterns. This has specific clinical implications. Since persons with dementia are frequently in a secured area in a nursing home, the behavior of someone exhibiting high distress and intense behavior results in distress for numerous other residents [52]. Being able to recognize and intervene with specific predicable BSD patterns in a timely manner reduces the escalation and distress of not only one person but of several persons.

Using THEME™ the pattern of temporal behavioral clusters can be quantified and related to environmental complexity at baseline. If an intervention is applied to decrease environmental complexity, the effect could be reflected in a decrease in complex patterns of behavioral clusters. Even if the effect influences only one or two individuals with high-intensity behaviors, research shows that decreasing the BSD of specific individuals frequently results in a generalized decrease of BSD on the entire unit, especially in designated dementia units [52]. Interventions tailored more specifically to individuals would result in a more efficient use of resources. Once we understand who responds and the characteristics of the behavioral clusters exhibited by individuals who respond we may be able to tailor and time effective interventions. For example, a non-pharmacological intervention may alter the complexity of BSD by decreasing the number and intensity of behaviors that occur together, by altering number of behavioral complexes that contain high-intensity behavior, or by altering the frequency of the behavioral complexes that occur over time. All of these results are beneficial. They can decrease not only staff time and frustration but also the associated costs [53].

The ability to quantify patterns of behavior and validate these patterns with other biological measures, such as stress hormones and temperature, is the key to the development of tailored interventions. Our vision for clinical practice and management is that by characterizing the patterns of BSD escalation, person-centered interventions will be modified and timed to ameliorate BSD, and prevent escalation and associated accidents and injury to persons with dementia and to staff.

The measures we currently use to evaluate treatment efficacy are crudely devised and basically focus on the presence or absence of specific high-intensity behavior such as yelling and intense restlessness. When these behaviors occur they require high staff involvement to calm both the person with the intense behavior and other residents who show increased agitation [52]. The use of THEME™ to detect temporal changes in behavioral clusters might be the key to detecting subtle behavior changes that herald an unrecognized treatment response or a response that suggests the need for a different treatment approach. Future studies must focus on behavior patterns and not single behaviors to guide the development of safe, timely, and effective treatments.

References

1. Alzheimer's Association (2012) Alzheimer's disease facts and figures. J Alzheimer Assoc 8:131–168
2. Moore MJ, Zhu CW, Clipp EC (2001) Informal costs of dementia care: estimates from the national longitudinal caregiver study. J Gerontol B Psychol Sci Soc Sci 56:S219–S228
3. Small GW, McDonnell DD, Brooks RL, Papadopoulos G (2002) The impact of symptom severity on the cost of Alzheimer's disease. Am Geriatr Soc 50:321–327
4. von Strauss E, Viitanen M, De Ronchi D, Winblad B, Fratiglioni L (1999) Aging and the occurrence of dementia: findings from a population-based cohort with a large sample of nonagenarians. Arch Neurol 56:587–592
5. Apostolova LG, Cummings JL (2008) Neuropsychiatric manifestations in mild cognitive impairment: a systematic review of the literature. Dement Geriatr Cogn Disord 25:115–126
6. Muangpaisan W, Intalapaporn S, Assantachai P (2008) Neuropsychiatric symptoms in the community-based patients with mild cognitive impairment and the influence of demographic factors. Int J Geriatr Psychiatry 23:699–703
7. Seitz D, Purandare N, Conn D (2010) Prevalence of psychiatric disorders among older adults in long-term care homes: a systematic review. Int Psychogeriatr 22:1025–1039
8. Borroni B, Grassi M, Agosti C, Costanzi C, Archetti S, Franzoni S, Caltagirone C, Di Luca M, Caimi L, Padovani A (2006) Genetic correlates of behavioral endophenotypes in Alzheimer disease: role of COMT, 5-HTTLPR and APOE polymorphisms. Neurobiol Aging 27:1595–1603
9. Souder E, Heithoff K, O'Sullivan PS, Lancaster AE, Beck C (1999) Identifying patterns of disruptive behavior in long-term care residents. J Am Geriatr Soc 47:830–836
10. Donaldson C, Tarrier N, Burns A (1998) Determinants of carer stress in Alzheimer's disease. Int J Geriatr Psychiatry 13:248–256
11. Pollak CP, Perlick D, Linsner J (1994) Sleep and motor activity of community elderly who frequently use bedtime medications. Biol Psychiatry 35:73–75
12. Aronson MK, Post DC, Guastadisegni P (1993) Dementia, agitation, and care in the nursing home. J Am Geriatr Soc 41:507–512
13. Cohen-Mansfield J (1997) Turnover among nursing home staff. A review. Nurs Manage 28:59–62
14. Sloane PD, Davidson S, Knight N, Tangen C, Mitchell CM (1999) Severe disruptive vocalizers. J Am Geriatr Soc 47:439–445
15. Beattie E, Song J, LaGore S (2005) A comparison of wandering behavior in nursing homes and assisted living facilities. Res Theory Nurs Pract 19:181–196
16. Shin MD, Carter M, Masterman D, Fairbanks L, Cummings JL (2005) Neuropsychiatric symptoms and quality of life in Alzheimer disease. Am J Geriatr Psychiatry 13:469–474
17. Beck C, Richards K, Lambert C, Doan R, Landes RD, Whall A, Algase D, Kolanowski A, Feldman Z (2011) Factors associated with problematic vocalizations in nursing home residents with dementia. Gerontologist 51:389–405
18. A.A.f.G.P. American Geriatrics Society (2003) The American Geriatrics Society and American Association for Geriatric Psychiatry recommendations for policies in support of quality mental health care in US nursing homes. J Am Geriatr Soc 51:1299–1304
19. O'Connor DW, Ames D, Gardner B, King M (2009) Psychosocial treatments of psychological symptoms in dementia: a systematic review of reports meeting quality standards. Int Psychogeriatr 21:241–251
20. Beck C (2011) Personal Communication
21. Donaldson G (2012) Randomized trials for comparative effectiveness: the bronze standard again? In: Western Institute of Nursing Communicating Nursing Research conference: advancing scientific innovations in nursing 2012, Western Institute of Nursing: Portland MArriott Downtown Waterfront Hotel, Portland, OR, pp 1–17
22. von Gunten A, Alnawaqil AM, Abderhalden C, Needham I, Schupbach B (2008) Vocally disruptive behavior in the elderly: a systematic review. Int Psychogeriatr 20:653–672
23. Woods DL, Rapp CG, Beck C (2004) Escalation/de-escalation patterns of behavioral symptoms of persons with dementia. Aging Mental Health 8:126–132
24. Mather N, Goldstein S (2001) Learning disabilities and challenging behaviors: a guide to intervention and classroom management. Paul H. Brookes Publishing Company, Baltimore, MD
25. Skinner BF (1974) Behavior modification. Science 185:813
26. Smith M, Buckwalter K (1993) Acting up and acting out: assessment and management of aggressive and acting out behaviors. University of Iowa, Iowa City, IA

27. Cohen-Mansfield J, Billig N (1986) Agitated behaviors in the elderly: I. A conceptual review. J Am Geriatr Soc 34:711–721
28. Cohen-Mansfield J (1986) Agitated behaviors in the elderly: II. Preliminary results in the cognitively deteriorated. J Am Geriatr Soc 34:722–727
29. Algase DL, Antonakos C, Yao L, Beattie E, Hong G, Beel-Bates C (2008) Are wandering and physically nonaggressive agitation equivalent? [Article]. Am J Geriatr Psychiatry 16:293–299
30. Algase DL, Beattie ERA, Therrien B (2001) Impact of cognitive impairment on wandering behavior. West J Nurs Res 23:283–295
31. Algase DL, Moore DH, Vandeweerd C, Gavin-Dreschnack DJ (2007) Mapping the maze of terms and definitions in dementia-related wandering. Aging Mental Health 11:686–698
32. Whall AL, Colling KB, Kolanowski A, Kim H, Son Hong G, DeCicco B, Ronis DL, Richards KC, Algase DL, Beck C (2008) Factors associated with aggressive behavior among nursing home residents with dementia. Gerontologist 48:721–731
33. Beck C, Frank L, Chumbler NR, O'Sullivan P, Vogelpohl TS, Rasin J, Walls R, Baldwin B (1998) Correlates of disruptive behavior in severely cognitively impaired nursing home residents. Gerontologist 38:189–198
34. Boustani M, Zimmerman S, Williams CS, Gruber-Baldini AL, Watson L, Reed PS, Sloane PD (2005) Characteristics associated with behavioral symptoms related to dementia in long-term care residents. Gerontologist 45:56–61
35. Cohen-Mansfield J, Marx MS, Rosenthal AS (1989) A description of agitation in a nursing home. J Gerontol 44:77–84
36. Cohen-Mansfield J (1996) Conceptualization of agitation: results based on the Cohen-mansfield agitation inventory and the agitation behavior mapping instrument. Int Psychogeriatr 8:309–315, discussion 351-304
37. Cummings J (1996) Theories behind existing scales for rating behavior in dementia. Int Psychogeriatr 8:293–300
38. Beck C, Heithoff K, Baldwin B, Cuffel B, O'Sullivan P, Chumbler NR (1997) Assessing disruptive behavior in older adults: the disruptive behavior scale. Aging Ment Health 1:71–79
39. Bliwise DL, Carroll JS, Lee KA, Nekich JC, Dement WC (1993) Sleep and "Sundowning" in nursing home patients with dementia. Psychiatry Res 48:277–292
40. Magnusson MS (2000) Discovering hidden time patterns in behavior: T-patterns and their detection. Behav Res Methods Instrum Comput 32:93–110
41. Magnusson MS (1996) Hidden and real-time patterns in intra and inter-individual behavior: description and detection. Eur J Psychol Assess 12:112–123
42. Woods DL, Dimond M (2002) The effect of therapeutic touch on agitated behavior and cortisol in persons with Alzheimer's disease. Biol Res Nurs 4:104–114
43. Bakeman R, Gottman JM (1997) Observing interaction. Cambridge University Press, Cambridge
44. Fleiss JK (1981) Statistical methods for rates and proportions. Wiley, New York, NY
45. Lawton MP, Van Haitsma K, Klapper J (1996) Observed affect in nursing home residents with Alzheimer's disease. J Gerontol B Psychol Sci Soc Sci 51:3–14
46. Vogelpohl TS, Beck CK (1997) Affective responses to behavioral interventions. Semin Clin Neuropsychiatry 2:102–112
47. Beck CK, Vogelpohl TS, Rasin JH, Uriri JT, O'Sullivan P, Walls R, Phillips R, Baldwin B (2002) Effects of behavioral interventions on disruptive behavior and affect in demented nursing home residents. Nurs Res 51:219–228
48. Kim H, Woods DL, Phillips LR, Mentes J, Martin J, Moon A (2012) The relationship between direct-care staff's competence in social interaction and behavioral symptoms in Korean-American nursing home residents with dementia. University of California, Los Angeles, CA
49. Kim H, Woods DL (2012) The development of direct-care staff social interaction coding schemas for nursing home residents with dementia. Geriatr Nurs 33:113–117
50. Woods DL, Beck C, Sinha K (2009) The effect of therapeutic touch on behavioral symptoms and cortisol in persons with dementia. Forsch Komplementmed 16:181–189
51. Anguera MT, Blanco-Villasenor A, Losada JL, Arda T, Camerino O, Castellano J (2003) Match and player analysis in soccer: computer coding and analytic possibilities. Int J Comput Sci Sport 2:118–121
52. Woods DL, Mentes J (2006) Agitated behavior as a prodromal symptom of physical illness: a case of the flu. J Am Geriatr Soc 54:1953–1954
53. Murman DL, Chen Q, Powell MC, Kuo SB, Bradley CJ, Colenda CC (2002) The incremental direct costs associated with behavioral symptoms in Alzheimer's disease. Neurology 59:1721–1729

Chapter 7

Typical Errors and Behavioral Sequences in Judo Techniques: Knowledge of Performance and the Analysis of T-Patterns in Relation to Teaching and Learning the Ouchi-Gari Throw

Ivan Prieto, Alfonso Gutiérrez, Oleguer Camerino, and M. Teresa Anguera

Abstract

The aim of the study was to detect the most frequent errors and their associated behavioral sequences in relation to the judo technique Ouchi-gari, the ultimate objective being to propose improvements to the way in which judo is taught. The novice participants ($n = 31$; 15 men and 16 women) were all students from the Faculty of Educational and Sports Science at the University of Vigo (Spain) and they were filmed while performing the technique in the context of a systematic observational study. The results, based on descriptive statistics and the sequential analysis of T-patterns obtained via the THEME v.5 software, revealed that students committed a series of typical technical errors when learning the Ouchi-gari technique that affected the whole throw sequence. These errors were primarily related to an initial failure to put the adversary off balance, the foot and trunk position, the reaping action, and the final action of the arms. As regards the teaching of judo these findings can be used to propose motor tasks and movement sequences for novices that would ensure successful learning of the technique, this process being based on a range of tasks and the use of corrective feedback.

Key words Observation instrument, Observing judo, Sequence learning, T-Patterns

1 Introduction

The first studies of combat sports focused on the physiological demands associated with the effort required by the combat [1–3]. More recently, research, particularly in judo, has considered ways of improving the training process [4–6] and the psychological response of judokas in competitive settings [7, 8]. However, these studies do not examine in detail the factors that affect the teaching and learning process with novices, and knowledge is still lacking regarding the factors that influence the learning and the overall or nonlinear acquisition of a given combat technique [9].

Magnus S. Magnusson et al. (eds.), *Discovering Hidden Temporal Patterns in Behavior and Interaction: T-Pattern Detection and Analysis with THEME™*, Neuromethods, vol. 111, DOI 10.1007/978-1-4939-3249-8_7,

The model based on knowledge of performance in relation to sports technique [10, 11], and in particular the errors committed during performance [12, 13], is a valuable and novel tool. Correcting a technical movement or gesture from the perspective of errors and knowledge of performance is, when information about the nature of those errors is available, more useful than simply pointing out the outcome of performance [14].

The aim of the present study was to apply the knowledge of performance model to the judo throw Ouchi-gari in order to reveal error sequences which are normally difficult to perceive. This was done by means of a systematic observational analysis, the results of which are used to propose ways of improving the teaching of judo techniques. Such proposals have already been made in relation to other sports [15].

2 Method

The study was based on observational methodology [16, 17], which has the rigor and flexibility required to study the episodes of behavior that emerge naturally during the process of teaching and learning judo. Based on the work of Borrie, Jonsson, and Magnusson [18, 19] the type of observation carried out was systematic, open, and nonparticipant.

2.1 Design

The observational design [20] was nomothetic (i.e. various participants performing the same technique, in this case Ouchi-gari), based on monitoring (a throw technique across five academic years), and multidimensional (different dimensions of the observation instrument). The use of this design implies a series of decisions that are made in relation to the participants, the observational instrument and register, and the procedure of data analysis.

2.2 Participants

Participants were students on a degree course in Physical Activity and Sports Science ($n = 31$; 15 men and 16 women), covering five academic years (from 2003/2004 to 2007/2008) and with an age range of 21–30 years ($M = 24.56$; SD = 2.73). They were all novices in judo and gave their informed written consent prior to being filmed on video. The recordings made were distributed equally across the five academic years (six subjects per year, except for the final year, in which there were seven).

2.3 Observation Instrument

The observation instrument developed for this study was the SOBJUDO-OU (see Table 1), a tool that combines the robustness and flexibility required to observe motor behavior [21, 22]. The criteria of which the SOBJUDO-OU is comprised include the object of the present study: technical errors in performance. The technical model used for both the process of teaching and

Table 1
SOBJUDO-OU observation instrument

Criterion	Code	Description
Grip	BGRIP	*Tori* uses his left hand to grip *Uke's judogi* midway up the forearm. The correct position would be at the elbow.
Off-balance	NOB	*Tori* does not put *Uke* off balance in the first part of the technique. His arms maintain the initial grip and only serve to accompany the action.
	DOB	The frontal off-balancing action and the subsequent initial displacement are performed in a discontinuous way.
Left-foot position	ILFP	*Tori* incorrectly positions his left foot after the initial movement.
Right-arm position	ARMP	At the end of the technique, *Tori's* right hand is in the supine position when gripping the left lapel of his opponent, leaving the dorsal part of his body in contact with the mid-level of the opponent's trunk (similar to the grip of Morote-seoi-nage).
Face position	FAP	The position of *Tori's* face is incorrect while performing the technique.
Trunk position	TRP	*Tori* incorrectly positions his trunk before initiating the leg reap.
	CGA	In the *kake* phase of this throw, *Tori's* center of gravity is too high.
Right-arm action	IRAA	In the *kake* of Ouchi-gari, the right-arm action of *Tori* is insufficient.
Left-arm action	ILAA	In the *kake* of Ouchi-gari, the left-arm action of *Tori* is insufficient.
	RAAR	*Tori* raises his left arm during the final phase of the movement, at the moment of pushing off the opponent's body (elbow at the same level as the wrist).
Reaping	SLWR	Instead of performing a circular reap the leg which should perform this action is used for support, thereby failing to carry out the reap.
	KBR	*Tori* bends his right leg while performing the reap, such that his ankle ends up above the imaginary line that would be described by the initial position of his knee.
	BLO	Instead of reaping his opponent's left leg with his own right leg, *Tori* performs a blocking action on it.
	INCR	*Tori* interrupts the reap, which is therefore incomplete.
	INSR	During the final phase of this technique, the position of *Tori's* leg that is performing the reap is insufficient as regards his opponent's leg.
Throw stage	NR	While performing Ouchi-gari, *Tori* fails to reap *Uke's* leg with his own.
Control stage	FACC	During the *kake* stage *Tori* uses his right arm to accompany *Uke's* fall to the floor.
	FNC	During the final stage of the technique *Tori* performs no action with his left hand and therefore fails to control the fall of his adversary's body.

(continued)

Table 1 (continued)

Criterion	Code	Description
Rebalancing	RRF	After performing the throw *Tori* loses his balance. In order to regain it he steadies himself with his right foot.
	RLF	Upon completion of the technique *Tori* loses his balance, which he regains by steadying himself with his left foot.
	RHR	Upon throwing *Uke*, *Tori* loses his balance and uses his right hand in an attempt to maintain him.
	RFR	After the reap, *Tori* uses the leg that had performed the action to regain his balance.
Globality	SLEX	The throw is executed slowly and without any continuity.

learning the throw and for its observation was based on the approach of the Kodokan school [23].

As the SOBJUDO-OU is a multidimensional instrument, it is compatible with the proposed observational design. Each one of its dimensions gives rise to a system of categories that fulfills the conditions of exhaustiveness and mutual exclusivity (E/ME).

2.4 Recording Instrument

The performance of the technique under study (the Ouchi-gari) was filmed after a training period lasting approximately 4 months. Data were gathered by means of two digital video cameras (JVC GZ-MG21E), and the recordings of the throws were subsequently edited using the video editing suite Pinnacle Studio v.12.

The observational register was created using the software Match Vision Studio Premium v.1.0 [24]. This is an interactive multimedia program that enables the user to visualize and register digitalized video recordings on the same computer screen. The program is highly flexible and allowed us to introduce all the codes corresponding to each of the changing criteria of the SOBJUDO-OU observation instrument, thereby producing a register of their appearance in succession.

2.5 Procedure

The performance of the technique under study (the Ouchi-gari) was filmed after a training period lasting approximately 4 months, involving 3 h of practice per week. Overall, a total of 17 throws were learnt. During the video recording each participant performed five of all the techniques that had been learnt, all without opposition from the other judoka and starting from a static position (i.e. it was technical work). Stratified random sampling was used to assign participants and techniques. The quality of the data registered by two observers was assessed by means of Cohen's kappa [25], with values of this coefficient above 0.8 being regarded as indicative of reliability (inter-observer agreement). This test was conducted using the software GSEQ v.5 for Windows [26, 27] and yielded a kappa value of 0.82. Having ensured the quality of the data, an initial descriptive analysis of the frequency and percentage of occurrence of technical errors was then conducted.

After recording all the throws performed the Match Vision Studio software produces a series of Excel files containing the successive configurations formed by the lines of codes that have changed, with their temporality and duration expressed in frames (25 frames is equivalent to 1 s). These Excel (.xls) files, which provide frequencies for all the registered occurrences of codes, were then transformed successively in order to enable various analyses to be carried out.

The codes of the SOBJUDO-OU observation instrument were then exported to the THEME software [28–30] with the aim of detecting temporal patterns (T-patterns). These T-patterns, which were obtained by means of the algorithm incorporated within THEME v.5 [29], can help to reveal hidden structures and unobservable aspects in sports techniques. The application of the THEME software has proved to be highly effective for studying both team and individual sports [21, 31, 32].

2.6 Data Analysis

The frequency of occurrence of the different errors made when performing the Ouchi-gari throw was determined by means of a descriptive analysis using SPSS 15, the results of which are shown in Table 2. An analysis of temporal patterns among the observed errors was also conducted using THEME, the aim here being to identify the most significant error sequences. The Mann–Whitney *U* test (in SPSS 15, with significance set at $p < 0.05$) was used to analyze the data in relation to the chosen independent variable, in this case the gender of participants.

3 Results

3.1 Statistical Analysis

Following on from the above description of the errors observed during performance of the Ouchi-gari, (Table 1) this section describes the frequency and percentage of occurrence of errors in the study group ($n = 31$).

The most common errors detected were related to an initial failure to put the adversary off balance (NOB), an inadequate position of the right arm (ARMP), an incorrect positioning of the face and trunk (FAP and TRP), the height of the center of gravity during the *tsukuri* and *kake* phases of the throw (CGA), insufficient traction effect of both arms in the final phase of the throw (IRAA and ILAA), and an incorrect reaping action (INCR).

Application of the Mann–Whitney *U* test to compare means and detect possible differences between men and women in their performance of the Ouchi-gari revealed no significant differences ($p > 0.05$).

Table 2
Frequency and percentage of occurrence of technical errors made when performing the Ouchi-gari

	Error	Frequency	Percentage
Grip	BGRIP	9	29
Off-balance	NOB	25	80.6
	DOB	8	25.8
Left-foot position	ILFP	13	41.9
Arm position	ARMP	16	51.6
Face position	FAP	23	74.2
Trunk position	TRP	19	61.3
	CGA	17	54.8
Right-arm action	IRAA	28	90.3
Left-arm action	ILAA	29	93.5
	RAAR	2	6.5
Reaping	SLWR	9	29
	KBR	3	9.7
	BLO	2	6.5
	INCR	22	71
	INSR	9	29
Throw stage	NR	7	22.6
Control stage	FACC	4	12.9
	FNC	4	12.9
Rebalancing	RRF	3	9.7
	RLF	6	19.4
	RHR	2	6.5
	RFR	7	22.6
Globality	SLEX	1	3.2

3.2 Detection of Temporal Patterns (T-Patterns)

In order to examine the errors made in greater detail the THEME software [28, 29] was used to analyze temporal patterns (T-patterns) in the observational data. This kind of analysis is able to reveal important links related to the sequences of errors that emerge.

Figure 1 shows the sequence of errors detected. The left-hand box represents the relationship between the different categories (i.e. the technical errors, as listed in the SOBJUDO-OU observation

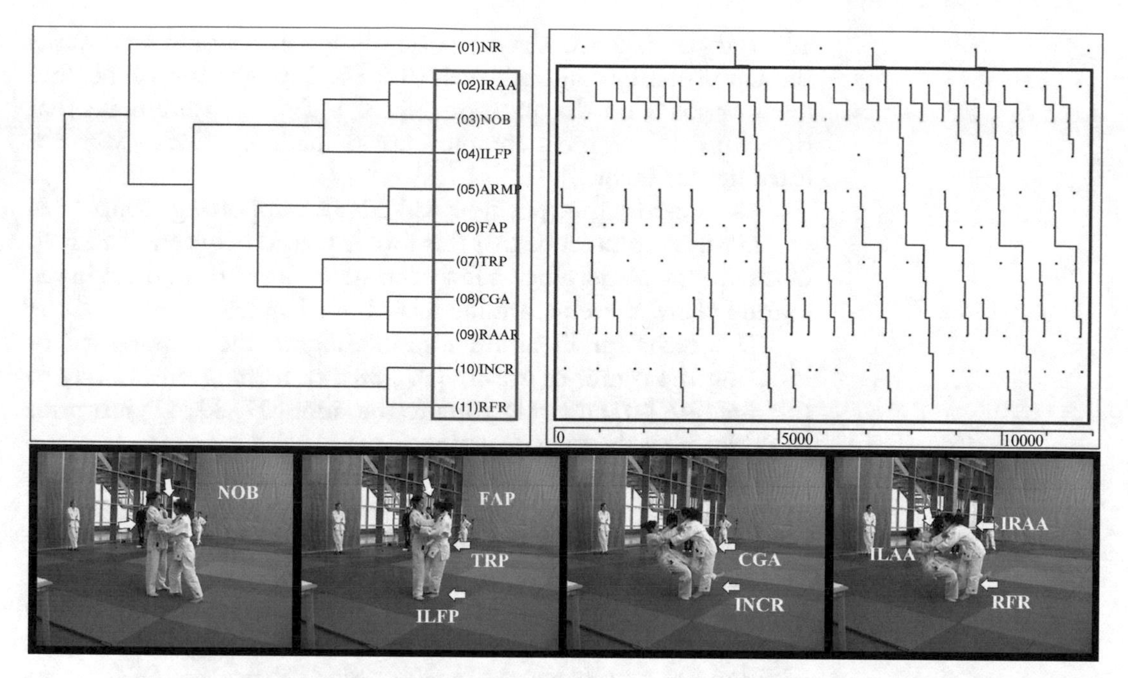

Fig. 1 The most representative tree diagram for Ouchi-gari

instrument). This tree diagram should be read from top to bottom, i.e. the first category to appear is at the top (red rectangle). The right-hand box shows how many times each of these relationships occurs, by means of lines that run from top to bottom (blue rectangle).

These results show that the initial failure to put the adversary off balance (NOB) precedes an incorrect positioning of the left foot (ILFP), which in turn leads to an incorrect trunk and face position (TRP-FAP). As a result of this foot position, *Tori* fails to achieve the optimal lowering of her center of gravity (CGA), which prevents an appropriate reaping action from being performed (INCR) and leads to an insufficient arm action (ILAA-IRAA). Finally, she has to use her right leg in order to regain her balance (RFR).

4 Discussion

Although there is a lack of scientific research on technical errors in judo, a review of the literature reveals that the most prestigious practitioners of the sport do, in their publications, reflect upon key aspects of technique or the most common errors made [33–36]. Interestingly, the points they make, which are no doubt based on their personal and professional experience, often coincide with the typical errors revealed by the present study.

For example, most of the authors consulted recommend that in the first part of the Ouchi-gari, which involves putting the adversary off balance, the direction of the force should be towards the left posterior diagonal, such that *Uke's* weight falls on his/her heels or solely on the left foot [33, 37]. The literature notes that this is one of the most common errors made by novices who are learning this throw [34].

As regards the position of *Tori's* supporting foot [37] specifically recommends that this foot is placed between *Uke's* feet, in the frontal plane, albeit a few centimeters away from them in the sagittal plane, thereby forming a kind of triangle.

The technical literature also highlights the importance of lowering the center of gravity [37, 38], of turning one's body to the side [39, 40] and of looking to the right [37, 39, 41] just prior to performing the reaping action.

Another of the most common errors observed in the present study was related to an insufficient and misdirected arm action. In this regard the experts recommend that *Tori* should aim to shift *Uke* backwards and towards the left posterior diagonal in the final phase of the technique [34, 35, 39], continuing the action that was begun with the initial off-balancing maneuver [38] but using in addition one's body weight to favor this off-balancing [37].

As regards the reaping action the recommendations are clear. For example, Taira [32] states: "The reap is performed with the Achilles tendon in contact with the same part of *Uke's* body, with the tip of the foot remaining in contact with the *tatami* throughout the action and reaping in a circular direction" (p. 267). This point of view is shared by authors such as [33, 36, 40, 41], among others.

The tree diagrams reveal a clear sequence of errors, whereby the initial failure to put the adversary off balance subsequently prevents the left foot from being correctly positioned, which means that the body is not facing the right way during the *tsukuri*. This makes it impossible to execute a sufficient arm action and an effective reaping action during the *kake* phase of the technique.

These sequences are not explicitly referred to in the technical manuals consulted, but the descriptions they contain do refer to aspects of the most representative tree diagram. For example, Uzawa [41] highlights the importance for *Tori* of using his left leg to put *Uke* off balance at the start, so as to be subsequently able to execute a more effective reap (NOB–INCR). Daigo [33] notes that if *Tori* does not bend his knees and lower his center of gravity (CGA) during the *tsukuri* phase he will not be able to execute a sufficient arm action (IRAA–ILAA) or, subsequently, an effective reap. This sequence of errors can be clearly seen in the tree diagram. Finally, Taira [37] stresses that the reaping and arm actions should be performed simultaneously and in a synchronized way, suggesting that the two are complementary as regards the correct execution of each (IRAA/ILAA–INCR).

5 Conclusions

With the aim of avoiding the errors detected here a number of recommendations can be made regarding execution of the Ouchi-gari technique: (1) use your left hand to grip *Uke's* right sleeve at the level of the elbow; (2) put *Uke* off balance in the direction of the left posterior diagonal, such that your body weight falls solely on your left foot; (3) place your left foot between *Uke's* feet (in the frontal plane); (4) turn your hip and head to the right, lowering your center of gravity by bending your knees; (5) reap the adversary's leg with a circular motion, such that he is thrown backwards (specifically, the reap should be centered over the area of *Uke's* Achilles tendon and with the ankle extended); and (6) the reap should be accompanied by a strong arm action.

The teaching and learning of this judo technique could be improved by paying special attention to the following movement sequences, which will ensure that the throw is correctly executed: (1) correctly putting the adversary off balance increases the likelihood that the feet and body will subsequently be in the correct position; and (2) correct positioning of the supporting foot and ensuring that the face and trunk are adequately positioned favors the required arm action (traction) and makes it easier to execute the reap during the throw phase of the technique.

5.1 Practical Implications Based on Knowledge of Performance

The results of this study also enable a number of strategies based on knowledge of performance to be proposed with the aim of improving the teaching and learning of the Ouchi-gari technique:

1. When demonstrating the technique the student's attention should be drawn towards the key points highlighted by this study. In relation to the theoretical aspects of the throw, coaches may wish to incorporate the use of video or other images that illustrate its fundamental features, as well as the most common errors detected here. At all events, teachers or coaches should focus only on the most relevant aspects.
2. Instructors could design tasks or drills that focus the student's attention on the most significant errors and sequences of behavior detected.
3. After a throw is performed in training the subsequent communication between coaches and students could be improved by providing more precise feedback. Coaches should begin by focusing on the most significant errors and sequences identified in the present study, leaving any others for a later stage of training. It is also helpful to focus on a few key aspects so that students do not become overloaded with information. At all events the results of this study can provide a platform for dif-

ferent kinds of feedback (verbal, verbal with a practical demonstration or verbal with hands-on assistance), which should always be positive in nature.

4. Coaches could draw up observation/evaluation sheets based on the category system of the observation instrument used in this study. One model would be for students to work in groups of three, with one of them observing the other two while they perform the throw. The former student would therefore conduct an observational analysis using the evaluation sheet, noting the errors made and providing immediate feedback. The same approach could also be used with video recordings, thereby enabling the observational analysis to be conducted after the throw has been performed.

Acknowledgments

We gratefully acknowledge the support of the Generalitat de Catalunya Research Group [GRUP DE RECERCA I INNOVACIÓ EN DISSENYS (GRID). Tecnología i aplicació multimedia i digital als dissenys observacionals], Grant number 2014 SGR971.

We gratefully acknowledge the support of the Spanish government project Observación de la interacción en deporte y actividad física: Avances técnicos y metodológicos en registros automatizados cualitativos-cuantitativos (Secretaría de Estado de Investigación, Desarrollo e Innovación del Ministerio de Economía y Competitividad) during the period 2012–2015 [Grant DEP2012-32124].

References

1. Borkowski L, Faff J, Starczewska-Czapowska J (2001) Evaluation of the aerobic and anaerobic fitness in judoists from the Polish national team. Biol Sport 18:107–117
2. Degoutte F, Jouanel P, Filaire E (2003) Energy demands during a judo match and recovery. Br J Sports Med 37:245–249
3. Sterkowicz S, Franchini E (2001) Specific fitness of elite and novice judoists. J Hum Kinetics 6:81–98
4. Artioli GG, Iglesias RT, Franchini E et al (2010) Rapid weight loss followed by recovery time does not affect judo-related performance. J Sports Sci 28(1):21–32
5. Calmet M, Ahmaidi S (2004) Survey of advantages obtained by judoka in competition by level of practice. Percept Mot Skills 99:284–290
6. Harrison A, Thompson KG, Cosgrove M et al (2003) Physical characteristics and body mass management of international judo players. J Sports Sci 21(4):275
7. Koral J, Dosseville F (2009) Combination of gradual and rapid weight loss: effects on physical performance and psychological state of elite judo athletes. J Sports Sci 27(2):115–120
8. Sterkowicz S, Blecharz J, Lech G (2000) Differentiation between high class judoists in terms of indices of experience, physical development, psychomotor fitness and their activities during competitions. J Hum Kinetics 4:93–110
9. Chow JY, Davids K, Button C (2006) Nonlinear pedagogy: a constraints-led framework for understanding emergence of game play and movement skills. Nonlinear Dynamics Psychol Life Sci 10:71–103
10. Laguna PL (2008) Task complexity and sources of task-related information during the

observational learning process. J Sports Sci 26(10):1097–1113
11. Mononen K, Viitasalo JT, Konttinen N et al (2003) The effects of augmented kinematic feedback on motor skill learning in rifle shooting. J Sports Sci 21(10):867–876
12. Helsen W, Gilis B, Weston M (2006) Errors in judging "offside" in association football: test of the optical error versus the perceptual flash-lag hypothesis. J Sports Sci 24(5):521–528
13. Tzetzis G, Votsis E, Kourtessis T (2008) The effect of different corrective feedback methods on the outcome and selfconfidence of young athletes. J Sports Sci Med 7:371–378
14. Schmidt R, Lee TD (2005) Motor control and learning. Human Kinetics, Champaign, IL
15. Zivcic K, Breslauer N, Stibilj-Batinic T (2008) Diagnosing and scientifically verifying the methodological process of learning in gymnastics. Odgojne Znanosti-Educational Sciences 10(1):159–180
16. Anguera MT, Jonsson GK (2003) Detection of real-time patterns in sport: interactions in football. Int J Comp Sci Sport 2:118–121
17. Black CB, Wright DL, Magnuson CE et al (2005) Learning to detect error in movement timing using physical and observational practice. Res Q Exerc Sport 76:28–41
18. Borrie A, Jonsson GK, Magnusson MS (2002) Temporal pattern analysis and its applicability in sport: an explanation and exemplar data. J Sports Sci 20:845–852
19. Borrie A, Jonsson GK, Magnusson MS (2001) Application of T-pattern detection and analysis in sports research. Metodología de las Ciencias del Comportamiento 3(2):215–226
20. Anguera MT, Blanco-Villaseñor A, Losada JL (2001) Diseños Observacionales, cuestión clave en el proceso de la metodología observacional. Metodología de las Ciencias del Comportamiento 3:135–161
21. Fernández J, Camerino O, Anguera MT et al (2009) Identifying and analyzing the construction and effectiveness of offensive plays in basketball by using systematic observation. Behav Res Methods 41:719–730
22. Jonsson GK, Anguera MT, Blanco-Villaseñor A et al (2006) Hidden patterns of play interaction in soccer using SOF-CODER. Behav Res Methods Instrum Comput 38(3):372–381
23. Kodokan (no date) Nage Waza: various techniques and their names [Video]. Kodokan Judo Video Series, Tokyo
24. Castellano J, Perea A, Alday L et al (2008) The measuring and observation tool in sports. Behav Res Methods 40:898–905
25. Cohen J (1960) A coefficient of agreement for nominal scales. Educ Psychol Meas 20(1):37–46
26. Bakeman R, Quera V (1992) SDIS: a sequential data interchange standard. Behav Res Methods Instrum Comput 24(4):554–559
27. Bakeman R, Quera V (2001) Using GSEQ with SPSS. Metodología de las Ciencias del Comportamiento 3:195–214
28. Magnusson MS (1996) Hidden real-time patterns in intra- and inter-individual behavior. Eur J Psychol Assess 12:112–123
29. Magnusson MS (2000) Discovering hidden time patterns in behavior: T-patterns and their detection. Behav Res Methods Instrum Comput 32:93–110
30. Magnusson MS (2005) Understanding social interaction: discovering hidden structure with models and algorithms. In: Anolli L, Duncan S, Magnusson MS, Riva G (eds) The hidden structure of interaction: from neurons to culture patterns. Ios Press, Amsterdam, pp 2–21
31. Gutiérrez-Santiago A, Prieto I, Camerino O et al (2011) The temporal structure of judo bouts in visually impaired men and women. J Sports Sci 29(13):1443–1451
32. Louro H, Silva AJ, Anguera MT et al (2010) Stability of patterns of behavior in the butterfly technique of the elite swimmers. J Sports Sci Med 9:36–50
33. Daigo T (2005) Kodokan Judo throwing techniques. Kodansha International, Tokyo
34. FFJDA (1967) La progression française d'eseignement. Tome I. Techniques de projections Nage Waza. FFJDA, Paris
35. Mifune K (2004) The Canon of Judo: classic teachings on principles and techniques. Kodansha International, Tokyo
36. Ohlenkamp N (2006) Black Belt Judo skills and techniques. New Holland, London
37. Taira S (2009) La esencia del judo I. Satori Ediciones, Gijón, Spain
38. Kudo K (1972) Judo in action: throwing techniques. Japan Publications, Tokyo
39. Inogai T, Habersetzer R (2002) Judo pratique. Du débutant à la ceinture noire. Amphora, Paris
40. Kano J (1989) Judo Kodokan. Eyras, Madrid
41. Uzawa T (1970) Tratado de Judo. INEF Madrid, Madrid

Chapter 8

Qualitative Differences in Men's and Women's Facial Movements in an Experimental Situation

Anaïs Racca, Magnus S. Magnusson, César Ades, and Claude Baudoin

Abstract

Studies indicate that men and women show quantitative differences in their production of facial movements. However, less is known regarding the qualitative aspect of facial differences between women and men. The aim of this study was to determine whether men and women could present a gender-specific temporal organization of facial movements. Eighteen Brazilian students (nine men and nine women) were filmed whilst performing a perceptual task involving verbal answers, the aim being to generate various types of facial expressions. The location of 29 facial points involved in facial movements were coded frame by frame and then analyzed with the THEME Software, allowing the detection of temporal patterns (*T-patterns*) of facial movements. Our results highlighted three typically feminine *T-patterns* (expressed only by women) in an experimental context, all involving mouth movements. No typically masculine *T-patterns* were detected in the same situation. This study shows the existence of some qualitative differences between women and men in their facial movements' organization.

Key words Human behavior, Facial movements, Gender, Pattern detection

1 Introduction

Sexual dimorphism is a widespread phenomenon among animal species, and in humans the face constitutes a major cue allowing sex identification. Bruce and her colleagues [1] showed that even without any cultural cues relative to the gender (make up, hair style, etc.) or facial hair, participants are extremely efficient and fast in recognizing someone's gender based on the picture of their face. The differences between men and women allowing this identification relate to the structure of several facial features (notably the brows, eyes, jaw, etc.) [2], to the configuration of these features between them [3] and to the differences related to the skin's texture and pigmentation [4].

This chapter is dedicated to the memory of Professor César Ades who died tragically in an accident in March 2012.

Magnus S. Magnusson et al. (eds.), *Discovering Hidden Temporal Patterns in Behavior and Interaction: T-Pattern Detection and Analysis with THEME™*, Neuromethods, vol. 111, DOI 10.1007/978-1-4939-3249-8_8,

In human social interactions faces are not static entities, studying dynamic aspects of face gender identification is therefore essential. Facial movements are notably involved in recognizing people's emotional expressions [5, 6]. Could facial movements also provide cues about someone's gender? Some studies suggest that it could be the case. The challenge here is to isolate dynamic information from static ones. Using point-light displays or the animation of a three-dimensional androgynous head with male or female facial movements (captured from markers arranged on particular points of actors' face), researchers showed that participants can identify the gender of the stimuli above chance [7–10]. Moreover, internal facial movements appear particularly useful for this gender identification, in opposition to rigid head movements [8]. Studies investigating the nature of sexual dimorphism in human facial movements notably indicate that women seem to express more facial movements than men [10–13] and that men display wider movement than women [14–16]. Other studies also showed that men display more asymmetrical facial movement than women [17, 18].

Human facial movements carry information indicating someone's gender that is perceptible and usable by viewers. To date research focused only on some aspects (i.e. quantitative traits) of gender differences and did not take into account the temporal structure of facial movements. A new methodology, characterized by a bottom-up approach, has been set up in our team in order to consider the temporal organization of facial movements, video recorded in men and women. It relies on manual pointing of facial points involved in facial movements using no a priori organizational assumption and records the movement of these points throughout video sequences, frame by frame. The software THEME [19] (www.noldus.com) permits to search for T-pattern (i.e. Temporal-Pattern) present in one set of data that are totally absent from another set. Applied to our topic, it allows the detection of T-patterns presented only by men and never by women and vice versa. The aim of the present study was to investigate whether humans present such gender-typical temporal organization of facial movements.

2 Method

We video recorded facial movements of young healthy adults in several countries (Bulgaria, Brazil, France, Gabon, Morocco, etc.) following similar experimental conditions for future cross-cultural comparisons. The present study focused on a Brazilian population with a total of 21 volunteer students from the University of Sao Paulo. The experiment was designed to privilege freedom and spontaneity in the expression of facial movements of naïve participants, in opposition to the use of patches on the face of actors in

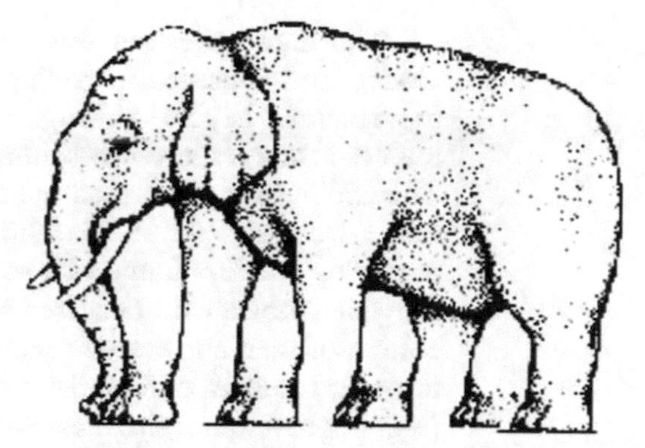

How many legs does this elephant have?

Fig. 1 Example of a picture and question presented to the participants

other studies [8, 9]. To do so, the subjects were in a situation allowing them the expression of various facial movements (including speech) but in which (a) head movements are limited (due to constraints from our coding technique, see below) and (b) a camera could be placed in front of them. Participants were welcomed by an experimenter then left alone in a room and asked to perform a computer-based perceptual task. The total duration of the task was around 5 min per subject. The task consisted in watching series of pictures ("trompe-l'oeil") and to answer questions regarding their ambiguous nature, out loud (see example Fig. 1). The participants were naïve towards the real aim of the study; nevertheless, they were aware of the presence of the camera in order to keep the context indirectly social. All participants gave their consent to be filmed in this study. Some participants had to be excluded from the experiment: one was chewing a gum and two had their hand in front of their mouth during the quasi-total duration of the study. The final sample contained nine men and nine women, aged from 19 to 26 years old.

Three 5-s video sequences were selected from each participant, including 2 s before a verbal answer (easy to identify on the video) and 3 s after its start. The sequences were selected randomly among all "codable" ones for each participant. Indeed, the use of one single video camera provides only two-dimensional information and the coding of facial movements would be distorted in case of rigid head movements. Therefore, only video sequences in which the participants presented facial movements but kept its head stationary were analyzed. In their studies, Hill and Johnston [8] showed that internal facial motions are particularly useful for gender identification of three-dimensional androgynous heads compared to rigid head movements. Consequently we assume that our technique does not neglect important information regarding our purpose.

The data collection was performed using the software Face Coder[1] and consisted in coding 29 facial points involved in facial movements [14, 20] throughout video sequences. Those points involved the eyes, eyebrows, nose, mouth, and chin of the subjects. The coding was performed manually (using the computer cursor), frame by frame (rate: 25 frames/s). All 29 points were coded regarding the first frame of a sequence and then only facial points for which movements occurred were coded for the next frames. The coding of men and women sequences was performed alternatively in order to avoid differential effect of potential changes in coding (e.g. improvement, tiredness etc.) on men and women's data. For each facial point and each frame the Face Coder software attributed coordinates within an orthonormal basis relative to the participants' face (see Fig. 2): the abscissa (X) was built using the two external eye corners (points 1 and 6) and the ordinate (Y) passed by the tip of the nose (point 18) (see illustrations Figs. 2 and 3).

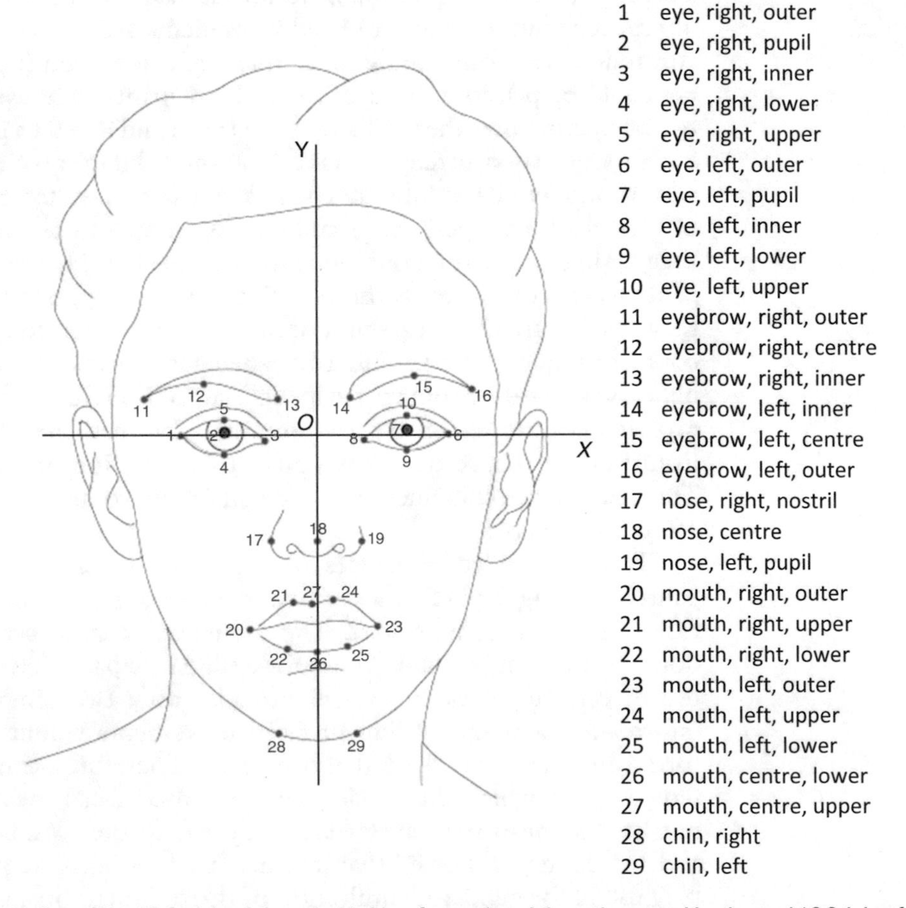

Fig. 2 Illustration of the 29 facial points coded (with reference) and the orthonormal basis used (*O* Origin of the basis)

[1] This software has been developed by Magnus S. Magnusson as Invited Professor at the University of Paris 13.

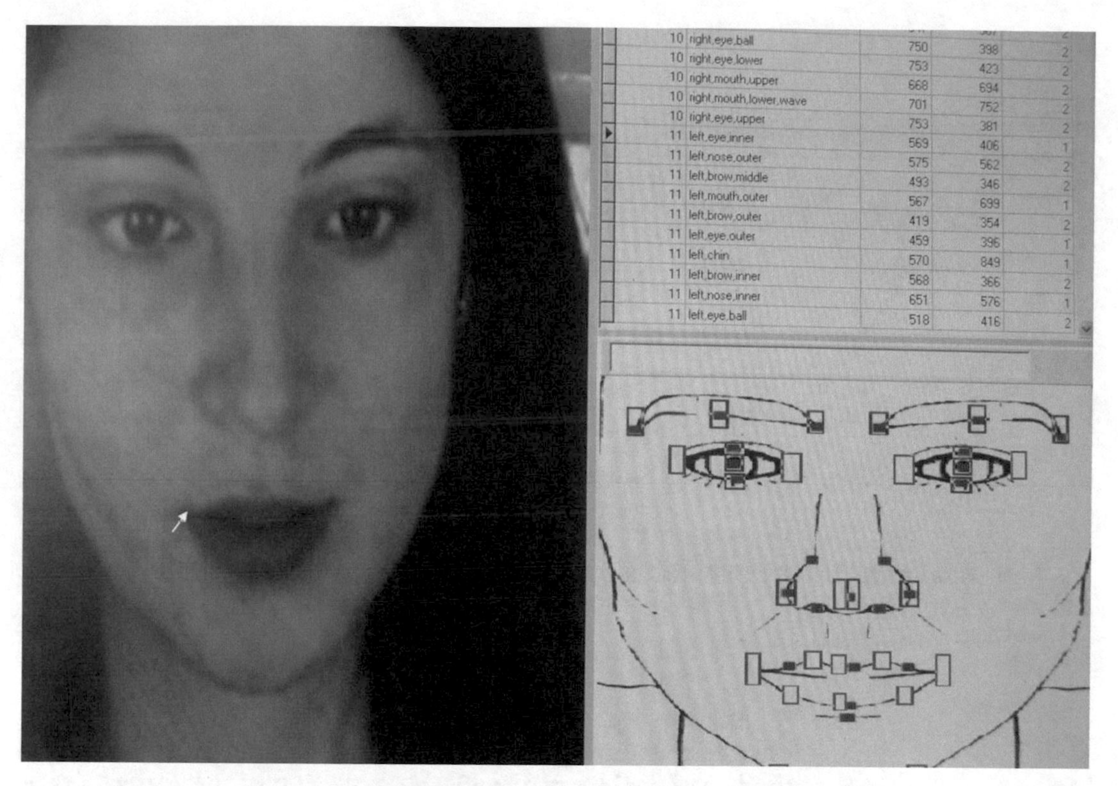

Fig. 3 Coding of a frame under Face Coder (*screen shot*). The points are coded on the videos (*left part of the screen*), a schematic face illustrates the location of these points (*bottom right corner*) and the points' coordinates are listed in a table (*upper right corner*)

For each frame the software calculated the distance of the facial points to the origin of the basis (Point-Origin = PO). A movement was defined as a move of a given point from the origin of the basis, higher than 1.5 standard deviation compared to its averaged distance (calculated through the entire sequence). This criteria was set up in order to avoid taking into account slight inaccuracy in coding (because manual) and appeared the most adequate by looking at our data set. Two types of movements could be detected: a movement towards the origin (decrease of PO) or a movement away from the origin (increase of PO). Both the start and the end of the movements were taken into account (see Fig. 4). The data set then consisted in a list of those movements throughout frames of the sequences.

Data were analyzed with THEME 6.0 BETA. The software's parameters were defined so as to detect all T-patterns present in men sequences exclusively (not present in women ones) and then all T-patterns present in women sequences exclusively (not present in men ones). We used the default parameters for analysis including minimum occurrence number = 3, significance level = 0.005.

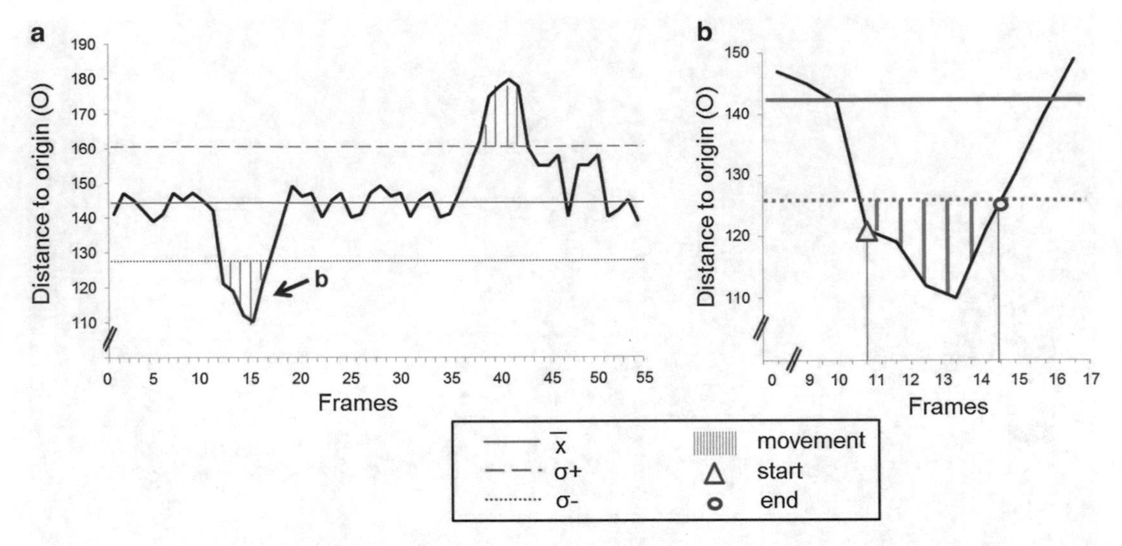

Fig. 4 Evolution through time of the distance of a point to the origin (PO). Illustration of a movement, its start and end

Analyses were conducted regarding quantitative gender differences concerning the size of the pattern repertoire (averaged number of distinct patterns), the repetition of the pattern (averaged frequency of manifestation of the pattern) and the complexity of the pattern (averaged number of movements involved in a pattern).

3 Results

We detected three typically feminine T-patterns, illustrated in Fig. 5. Each of those patterns were displayed by six out of nine women and never displayed by men. Two women presented those three patterns twice within the sequences. These are relatively long patterns (five or six events) within 5 s sequences and all involved the start of a mouth movement going away from the origin. Those patterns seem likely to be related to the way female participants opened their mouth when started to speak. No typically masculine T-patterns were detected. That is, all T-patterns of facial movements produced by men were also produced by women.

Regarding quantitative aspects of gender differences in the expression of T-patterns of facial movements, no significant differences were observed between men and women as exposed in Table 1.

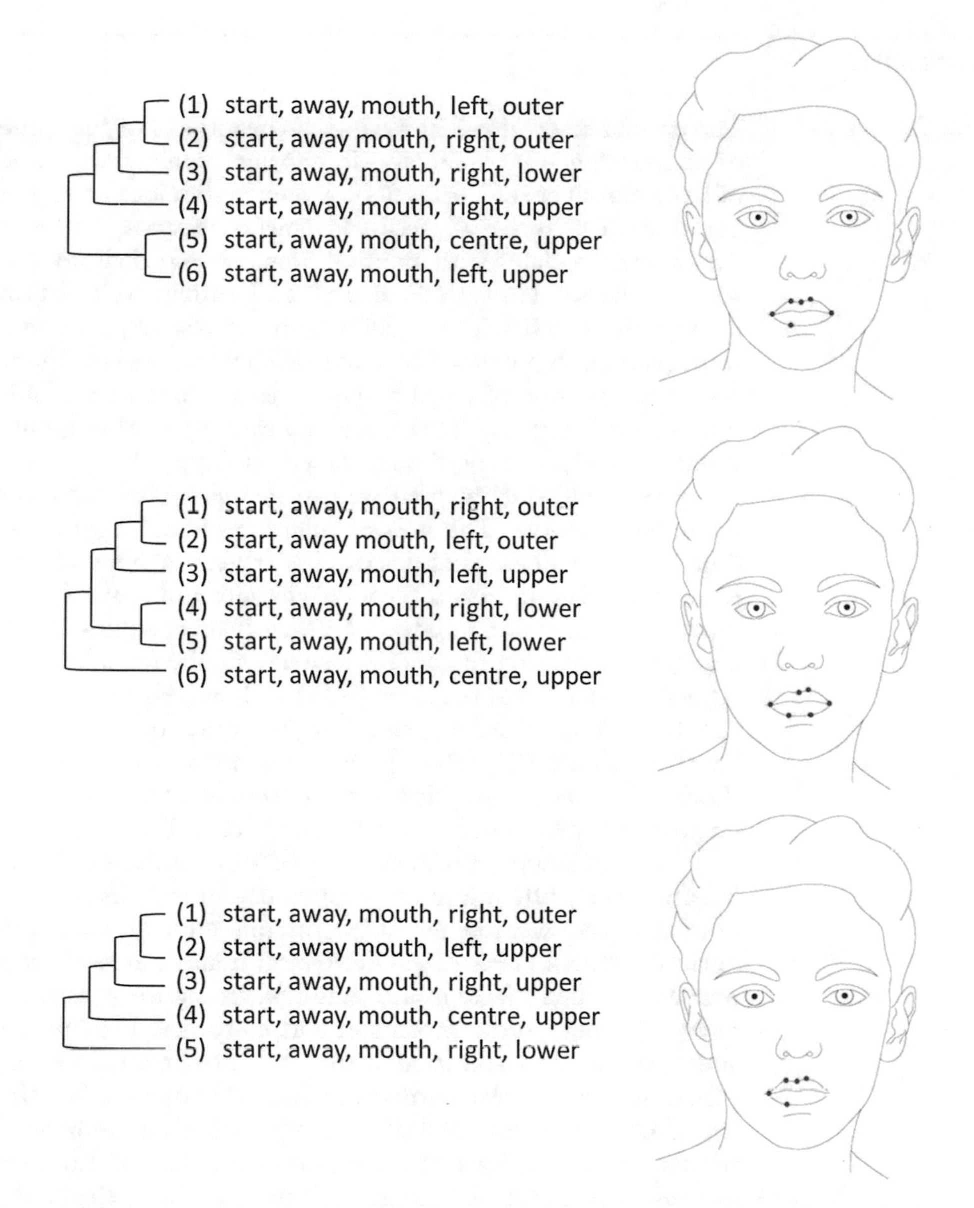

Fig. 5 Typically feminine T-patterns (*numbers* correspond to the order of occurrence of the events within a pattern)

Table 1
Quantitative differences in men and women T-Patterns (Mann–Whitney U tests)

Measures	Men	Women	*Z*	*P*
Size pattern repertoire	110.56	139.78	0.75	0.49
Pattern frequency	5.13	4.71	1.41	0.16
Pattern complexity	2.83	2.74	0.22	0.83

4 Discussion

The present study aimed at testing the hypothesis of the existence of qualitative gender differences in humans' temporal organization of facial movements. Our results corroborate this idea by highlighting three T-patterns of facial movements expressed by women exclusively, associated with mouth movements and likely involved in speech behavior. No typically masculine T-patterns of facial movements were detected. That is, all pattern expressed by men were, at some point, also presented by women. These results constitute the first demonstration of a gender-typical temporal structure of facial movements in humans. They were confirmed in another study with Bulgarian students using the same methodology [21].

Why would only women present gender-typical facial movement organization? The answer might be found regarding the function of gender-typical behavior. Many researchers were interested in the implication of gender-typicality and attractiveness in human's faces. From a static point of view (using pictures) researchers agree concerning the importance of typically feminine traits in attractiveness for women faces [22–26], however, the opposite is not as obvious regarding male face for which discrepant results have been found [25, 27–31]. One possible explanation relates to the fact that menstrual cycle and mating context are known to infer on women's preferences for male faces [32]. Therefore, women may be optimizing reproductive success by preferring feminine-looking males, offering good parental investment, in some situations and masculine looking males, offering good genes, in others. Such differential effect of gender-typical traits in attractiveness in men and women's faces is also found regarding their facial movements. In their study, Morrisson and coworkers [10] presented androgynous head animated with facial movements from either men or women to naïve participants and asked them to identify the sex of the animations as well as rating their attractiveness. They showed that animations whose genders were the easier to identify were also those that were considered the most attractive, but only regarding animations from women's facial movements. Our findings strikingly fit with the idea that gender-typicality is adaptive regarding women faces but not necessarily in men. It remains to be seen whether the women T-patterns found in our study are found attractive or not.

Previous studies reported quantitative differences between the way men and women move their face. Notably, women were found to present more movements than men [10–13] and men presenting larger movements than women [14–16]. Looking at facial movements' organization through time, no such gender distinction could be observed. Thus, T-pattern differences between men and women may be limited to qualitative aspects, at least in the conditions of the present study.

This study demonstrates the presence of typically feminine organization of facial movements in humans. Contrary to other studies in the field, the method we chose here privileged subjects' freedom in the expression of facial movements. However, due to methodological constraints, the experimental situation could not be directly social. Would those typically feminine T-patterns also be found in a directly social context? Similarly, are those T-patterns present in every human culture? Further studies are currently carried out in order to answer those questions, but such research will be greatly facilitated as tools for refined automatic coding of facial movements become generally available.

Acknowledgments

We would like to thank Marion Hacquin as research assistant in a pilot study and Adélaïde Racca (adelaide@superblitz.org) for providing us with the illustration of the face used in Figs. 2 and 5.

References

1. Bruce V, Burton AM, Hanna E, Healey P, Mason O, Coombes A et al (1993) Sex discrimination: how do we tell the difference between male and female faces? Perception 22:131–152
2. Brown E, Perrett DI (1993) What gives a face its gender? Perception 22(7):829–840
3. Baudouin J-Y, Humphrey GW (2006) Configural information in gender recognition. Perception 35(4):531–540
4. Bruce V, Langton S (1994) The use of pigmentation and shading information in recognising the sex and identities of faces. Perception 23:803–822
5. Cunningham DW, Wallraven C (2009) Dynamic information for the recognition of conversational expressions. J Vis 9(13):1–17
6. Nusseck M, Cunningham DW, Wallraven C, Bülthoff HH (2008) The contribution of different facial regions to the recognition of conversational expressions. J Vis 8(8):1–23
7. Berry DS (1991) Child and adult sensitivity to gender information in patterns of facial motion. Ecol Psychol 3(4):349–366
8. Hill H, Johnston A (2001) Categorizing sex and identity from the biological motion of faces. Curr Biol 11:880–885
9. Hill H, Jinno Y, Johnston A (2003) Comparing solid-body with point-light animations. Perception 32(5):561–566
10. Morrison ER, Gralewski L, Campbell N, Penton-Voak I (2007) Facial movement varies by sex and is related to attractiveness. Evol Hum Behav 28:186–192
11. Brody LR, Hall JA (2000) Gender, emotion, and expression. In: Lewis M, Haviland-Jones JM (eds) Handbook of emotions. Guilford Press, New York, pp 338–349
12. Bullis C, Horn C (1995) Get a little closer: further examination of nonverbal comforting strategies. Commun Rep 8(1):10–17
13. Hall JA (1985) Nonverbal sex differences. Johns Hopkins University Press, Baltimore
14. Giovanoli P, Tzou C-HJ, Ploner M, Frey M (2003) Three-dimensional video-analysis of facial movements in healthy volunteers. Br J Plastic Surg 56:644–652
15. Paletz JL, Manktelow RT, Chaban R (1994) The shape of a normal smile: implications for facial paralysis reconstruction. Plast Reconstr Surg 93:784–789
16. Weeden JC, Trotman CA, Faraway JJ (2000) Three dimensional analysis of facial movement in normal adults: influences of sex and facial shape. Angle Orthod 71(2):132–140
17. Borod JC, Koff E, Yecker S, Sanschi C, Schmidt JM (1998) Facial asymmetry during emotional expression: gender, valence, and measurement technique. Neuropsychologia 36(11):1209–1215

18. Hausmann M, Behrendt-Körbitz S, Kautz H, Lamm C, Radelt F, Güntürkün O (1998) Sex differences in oral asymmetries during word repetition. Neuropsychologia 36(12):1397–1402
19. Magnusson MS (2000) Discovering hidden time patterns in behavior: T-patterns and their detection. Behav Res Methods Instrum Comput 32(1):93–110
20. Tzou C-HJ, Giovanoli P, Ploner M, Frey M (2005) Are there ethnic differences of facial movements between Europeans and Asians? Br J Plast Surg 58:183–195
21. Baudoin C, Lopes FA, Magnusson MS, Racca A, Knezevic A, Simeonovska-Nikolova D, Durand J-L (2009) Gender differences in facial movement production: a case study. XXXIst International Ethological Conference, Rennes (France)
22. Cunningham MR (1986) Measuring the physical attractiveness: quasi-experiments on the sociobiology of female facial beauty. J Pers Soc Psychol 50:925–935
23. Fink B, Penton-Voak I (2002) Evolutionary psychology of facial attractiveness. Curr Dir Psychol Sci 11:154–158
24. O'Toole AJ, Deffenbacher KA, Valentin D, Mc-Kee K, Huff D, Abdi H (1998) The perception of face gender: the role of stimulus structure in recognition and classification. Mem Cogn 26:146–160
25. Perrett DI, Lee KJ, Penton-Voak I, Rowland D, Yoshikawa S (1998) Effects of sexual simorphism on facial attractiveness. Nature 394:884–887
26. Perrett DI, May KA, Yoshikawa S (1994) Facial shape and judgements of female attractiveness. Nature 368:239–242
27. Gillen B (1981) Physical attractiveness: a determinant of two types of goodness. Personal Soc Psychol Bull 7:384–387
28. Grammer K, Thornhill R (1994) Human (*Homo sapiens*) facial attractiveness and sexual selection: the role of symmetry and averageness. J Comp Psychol 108:233–242
29. Keating CF (1985) Gender and the physiognomy of dominance and attractiveness. Soc Psychol Q 48:61–70
30. Penton-Voak IS, Chen JY (2004) High salivary testosterone is linked to masculine male facial appearance in humans. Evol Hum Behav 25:229–241
31. Rhodes G, Hickford C, Jefferey L (2000) Sex-typicality and attractiveness: are supermale and superfemale faces super-attractive? Br J Psychol 91:125–140
32. Penton-Voak IS, Perrett DI, Castles DL, Kobayashi T, Burt DM, Murray LK, Minamisawa R (1999) Menstrual cycle alters face preference. Nature 399:741–742

Chapter 9

Understanding Film Art: Moments of Impact and Patterns of Reactions

Monika Suckfüll and Dagmar Unz

Abstract

While there is a wide consensus that in the core of fictional stories is the generation of emotions, little is known about how this happens. The aim of our contribution is to better understand the nature of reception processes that occur while people are watching movies. We analyze patterns of physiological and facial reactions to a movie by using THEME. We refer to data gathered in a study conducted to investigate cognitive and emotional reactions to the animated short film *Father and Daughter* (2000). The narrative structure of the movie and the most important formal features were determined on the basis of dramaturgical models. In the study, heart rate and skin conductance of the participants were measured, and the facial reactions of the consenting participants were videotaped. In summary, the analyses enlighten the dynamic nature of movie reception: Important scenes of a movie are prepared with virtuosity, creating "lines" or repetitions of motives combined with each other; the film maker plays with the expectations and emotions of the viewers.

Key words Film art, Narrative structure, Reception, Emotion, Facial expression, Physiological reaction, Time-pattern

1 Introduction

No one doubts that media has the potential to attract the audience. Nevertheless, little is known about *how* the audience processes media stimuli. Many researchers agree that an essential function of fictional stories is the generation of emotions (e.g., [1]). Tan describes movies as emotion machines [2]; Konijn and ten Holt name emotion as "key construct in processing media messages" ([3], p. 37). Media induces emotions through the presentation of a story by using a huge variety of storytelling techniques and formal features that aim to involve the viewer. In fictional stories, ideas and ideals concerning our social world are presented in a way that is easier to grasp and to comprehend than our complex social environment [4]. Thus, if we aim at understanding film art, it is essential to understand the techniques of emotion elicitation as

Magnus S. Magnusson et al. (eds.), *Discovering Hidden Temporal Patterns in Behavior and Interaction: T-Pattern Detection and Analysis with THEME™*, Neuromethods, vol. 111, DOI 10.1007/978-1-4939-3249-8_9,

well as the emotional processes that de facto occur during the reception of a movie.

In order to understand the nature of such emotional processes, we analyzed patterns of physiological and facial reactions to a movie by using the T-pattern-detection of THEME. The study, we took the data from, was conducted to investigate emotional reactions to the animated short film *Father and Daughter* [5, 6]. The narrative structure of the movie and the most important formal features were determined using different theoretical approaches in film studies. Those structures and features are now—in the context of this contribution—conceptualized as events, i.e. as time intervals within the movie that are hypothetically of relevance for the viewers. We also use THEME to determine reoccurring combinations of events in the movie. In the reception study, heart rate and skin conductance of 30 participants were measured. The faces of the consenting 16 participants were videotaped during the reception and were analyzed using the *Facial Action Coding System* [7]. We will focus on the analysis of reoccurring combinations of physiological and facial reactions associated with particular events in the movie for those 16 participants. The results are interpreted in the light of emotion theories.

2 Media-Induced Emotions

Emotions are key features of our mental architecture. They are activated by means of specific cues in our environment and focus our attention on urgent and relevant information. Furthermore they allow a fast evaluation of environmental stimuli or events for the organism's needs, plans, or preferences. In addition to this they initiate psychological and physiological processes to support rapid reactions for dealing with a given situation [8]. Further, the nonverbal aspects of emotions include "the communication of reactions, states, and intentions by the organism to the social surround" ([9], p. 557). Most theorists support the view that emotions have several components or subsystems (e.g., [10, 11]). Conceptualizing emotion as a process, Scherer defines emotion "as an episode of interrelated, synchronized changes in the states of all or most of the five organismic subsystems" ([12], p. 93). The Component-Process Model by Scherer distinguishes five subsystems of an emotion: (I) the cognitive appraisal, (II) the physiological arousal, (III) the motor system and most notably the facial expression as part of this system, (IV) the subjective feeling, and (V) the motivational system [12, 13] An emotional episode is conceptualized as a component patterning process driven by cognitive appraisals. This means that an emotion is not due to the eliciting event itself, but to the evaluation of the event by an individual.

Thus, emotions are a process of sequential appraisals, which result in a certain pattern of outcomes in the emotion response systems. The process of appraisal itself is a cognitive process, thus it is not accessible to direct observation and also impossible or at least difficult to access via verbalization. But each appraisal evokes an adequate reaction in the subsystems and thus triggers a corresponding reaction, for instance, in facial expression. For several outcomes of stimulus evaluation checks, Scherer and colleagues [14–19] predict related changes in facial expressions. For example, appraising an event as novel is related to raising the eyebrows or frowning; appraising an event as pleasant is related to pulling the lip corners upwards and raising the cheeks (like in smiles). If a certain appraisal does result in a change of facial expression—as proposed—facial expressions are observable indicators of unobservable emotional processes.

Thus, we use knowledge about specific changes in the physiological and the motor system (facial expression) for the interpretation of emotional processes that are operationalized via T-patterns of events in a movie and the reactions of the recipients. Emotions that arise as a result of the reception of a movie may differ from those that are experienced in daily life in many respects due to the specific stimuli and due to the specific (principally safe) reception situation (for more details see [20]). However, we assume that media-induced emotions are *processed* in the same way as naturally occurring emotions ([20, 21], see also [3]).

3 The Movie and Its Structures and Features

Under discussion is the animated short film *Father and Daughter* by the Dutch film maker Michael Dudok de Wit [5]. The 8-min-long movie is centered on a human conflict:

One day a father leaves his little daughter behind by rowing away in a boat for reasons we are not told. The daughter returns again and again with her bicycle to the place on the shore from where her father's boat left. She is obviously longing for his return, but this does not happen, not even by the time the girl has become older. She is shown in different ages and circumstances. Finally, as an old woman, she goes down the hill and into the water; she seems to commit suicide. But then the movie switches to a dream world: The water changes to grass and she finds her father's empty boat in it. She lies down in the boat, and then she stands up and sees her father again; while running to him she becomes a child once more [6, 21].

Movies are compositions of narrative structures; they are a temporal arrangement of events, which serves as a reception-guideline [22]. This arrangement is basically meant to involve the spectator. But, there are also features that allow the spectator to

Table 1
Types of events in *Father and Daughter*

Involvement	Distance
1. Conflict	8. Humor
2. Solution	9. Hole
3. Longing	10. Symbol
4. Futility	11. Counterpoint
5. Encounter	
6. Effort	
7. Change	

create distance by bringing to mind that the movie is an artificial, esthetic product and not reality. Thus, in a first attempt, we distinguish between two opposing modes of reception: involvement and distance. The relation between involvement and distance is not meant to be a dichotomy—involvement or distance. Instead a cognitive and emotional "going into" the movie is possible, because of a distance now and then [23].

Different theoretical approaches in film studies are combined to analyze the movie. The premise hereby is that the effects of isolated cinematic elements, like for instance a certain protagonist or a particular acoustic element, cannot be empirically concretized in an appropriate way without considering the narrative context. The cinematic composition was approximated by determining the basic narrative structures [22]. This analysis was complemented by considering acoustic features and specific features of the movie that allow for distance. The analysis finally results in a list of 11 event types for the short film *Father and Daughter* (see Table 1). In the paragraphs that follow, all 11 types of events are described step by step (for more details see [6], p. 44–48).

The film scientist Wuss [22] describes both, the sequence of central events on screen and the viewers' reception as processes of problem-solving. In movies often problematic or conflicting situations have to be solved. Two different kinds of conflict are connected in *Father and Daughter*. The central conflict that results from the father's leaving and the inner conflict of the daughter, which is intensified as it becomes increasingly clear that she cannot reconcile herself to the loss of her father. The movie contains three scenes which clearly represent the central conflict: (I) The father leaves his daughter at the beginning of the movie; (II) the final resignation of the daughter when she seems to commit suicide, and (III) when she lies down in the boat to just sleep or even to die and rest in peace (see Fig. 1).

Fig. 1 Three conflicting situations

The movie offers three clearly markable solutions to the conflict. At one point in the movie it seems that the daughter may reconcile herself to the loss of her father, when she comes to the shore with a friend, sitting on the back of his bicycle—probably the man she loves. At another point in the movie the director switches from reality to a dream world, in which everything is possible: The daughter lives—a scene shown after the suicide-scene. This also indicates a solution. Finally, she lies down in the boat. The viewer may conclude that now she is dead. But then she stands up and meets her father once more. This is the third solution.

According to Wuss [22], the so-called *Topic Lines* generate sense by the repetition of "small" events, which are difficult to detect for the recipient (as well as for the researcher). The narration of *Father and Daughter* is structured by a main *Topic Line*, which intensifies the inner conflict:

The daughter returns again and again on her bicycle to the place, where her father left. There she looks out over the empty water and then rides away again. Although these attempts seem quite different in their visual expression, they prove to be very similar in terms of content: Each episode invariably leads to the same disappointing result. By the repetition of the invariant situation it becomes clear that the father's leaving represents an essential loss in the life of his daughter, and that her attempts to deal with this loss are futile [6, 21].

Two event types are differentiated: The daughter's looking out over the water that signifies her longing for her father, and her riding away that signifies futility (see Fig. 2).

A third *Topic Line* can be identified: On her way to the place of farewell, the daughter encounters people who are also on their bicycles. The fact that the protagonists, who meet each other, are of different ages supports the illustration of the respective life phase, the daughter is in (see Fig. 3).

At the beginning of the movie the director already establishes another *Topic Line*, which is repeated four times during the movie: The protagonists ride their bicycles uphill or upwind. We see the strain the protagonists are experiencing and we feel part of it ourselves. The scenes illustrate the enormous effort it takes the daughter to deal with the loss of her father [6, 21].

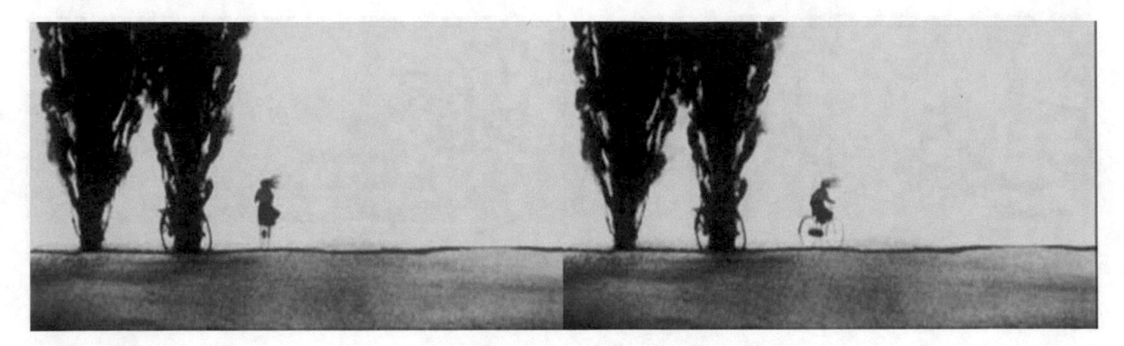

Fig. 2 Example for one of 12 events signifying longing (*one the left side of the figure*) and an example for one of seven events signifying futility (*one the right side*)

Fig. 3 Two examples of nine events showing encounters

In *Father and Daughter* each change-over from one segment to another is reinforced by remarkable changes of the music. These turning-points provide a clear segmentation of the movie and thus help the audience to structure the plot [6, 21].

Humor in *Father and Daughter* allows for distance. The term *humor* stands for funny or spontaneous events, which are characterized by overstatement. Two events in the movie are characterized by humorous elements: In the first scene, the daughter on her bicycle is blown forward by the wind and passes an old woman. In the second scene, a young girl meets the daughter, who is an old woman now. After the encounter, the young girl hops with her bicycle over a stone and the daughter, who is now an old woman, rings the bell of her bicycle [6, 21].

The next event type is called *hole*. Such holes provide the possibility to reflect, because the pictures on screen contain only minor information. For instance, pictures of landscape just indicate that time passes by ("epic distance") [24, 25]. Another form of distance is possible by using symbols. In *Father and Daughter* different symbols interrupt the flow of narration and point to the movie as an esthetic construction. Such pictures are often enigmas, which have to be decoded ("lyric distance") [24, 25]. Finally, distance may be induced by the music. The music swings up and then tilts into a more bright or joyous direction in contrast to a somehow painful event on screen. Or reversely, the music is sad in contrast to the cheerful action displayed, for instance, when children are playing on the shore. The antagonism between picture and music, so-called *counterpoints*, allows for distance [6, 21, 26].

4 Operationalization of Reception Processes

In order to operationalize the reception processes, we refer to data of 30 participants gathered in a previous study [6]. In this study, the movie was preceded by a cinema-like program with commercials and movie trailers in order to allow the participants to adjust to the unfamiliar situation. At the outset participants answered a short questionnaire to obtain information about movie experience and socio-demographical data. During the movie reception heart rate (HR) and skin conductance (SC) were measured. For those participants, who consented to be videotaped, facial behavior was recorded via a camera. Each participant's liking of the movie and self-reported emotional experience was assessed during a personal interview, which was conducted right after the measurement. In sum and in terms of the conceptualization of emotion processes by Scherer [12, 13], the physiological, the motor, and the subjective components of emotions were measured.

The movie was presented on a 19-in. diagonal computer monitor. The camera was positioned behind the monitor. Physiological data were collected with a transportable measurement system. For measurement of HR, three electrodes were attached to the upper part of the body. For the measurement of SC, two electrodes were attached to the palm of the nondominant hand. All dates were arranged for the evening between 5:30 p.m. and 8:30 p.m. in order to control for possible differences caused by the day time. Data recording was conducted individually for each participant. The conditions of the situation were standardized for all participants. Each session was supervised by the same (female) experimenter. During the session the experimenter sat 2 m (about 6 ft) away from the participant. She logged the observed body movements of the participants into the system using a keyboard that is part of the measurement device. With the help of these information artifacts caused by movements could be paralleled exactly with the physiological reactions and be removed for the analyses. One session lasted approximately 45 min (for a more detailed description see [6]). For our analyses we used the data of the 16 participants, ten women and six men, who consented to be videotaped. Table 2 provides an overview of their physiological reactions and body movements.

Table 2
Number of significant physiological reactions and of body movements

	Minimum	Maximum	Mean
Number of significant SCRs	0	84	39.25
Number of significant HR accelerations	3	25	11.81
Number of significant HR decelerations	1	18	10.94
Number of movements	0	8	2.13

The recordings of the facial expressions of the participants during reception were analyzed using the *Facial Action Coding System* (FACS), which was developed by the psychologists Ekman and Friesen [7]. FACS is a reliable method of describing the contraction of each facial muscle. Certified FACS-coders deconstruct an observed facial expression into so-called *Action Units* (AU). Specific combinations of Action Units stand for basic emotions. The coding was done by one coder in a blind-trial, without her knowing, which scene of the movie caused the reactions. The most frequent facial expressions were AU 45 (the eye blink; approximately 60 times for all participants), AU 14 (the lip corner depressor or dimpler; approximately 50 times), and AU 14 in combination with other AUs (approx. 30 times).

5 T-Pattern-Analysis

First of all, we are interested in patterns of events within the film *Father and Daughter*. Such patterns allow for insights into the narrative complexity of movies [27, 28]. Secondly, we want to analyze the dynamic relations between aspects of the movie and the reactions of the spectators. The analysis of T-patterns perfectly meets the requirements of an analysis of the manifold and complex dynamic relations between events in the movie and reactions of the viewers: Using THEME enables us to analyze both, immediate and delayed reactions. Even an anticipatory reaction preceding the event is possible. Moreover, the fact that recipients remember past events with a longer time delay can be taken into account. In Fig. 4 the data input prepared for an analysis with THEME is visualized.

The behavior records of all subjects were joined in a single data set. On the timeline, 16 series of events (one for each of the participants) appear one after another. The event types include the coding of cinematic events (event types 1–22; for each event begin and end were coded), the FACS-coding (23–119), the coding of movements (122) and the coding of physiological data (heart rate deceleration 120, heart rate acceleration 121, and skin conductance response 123).

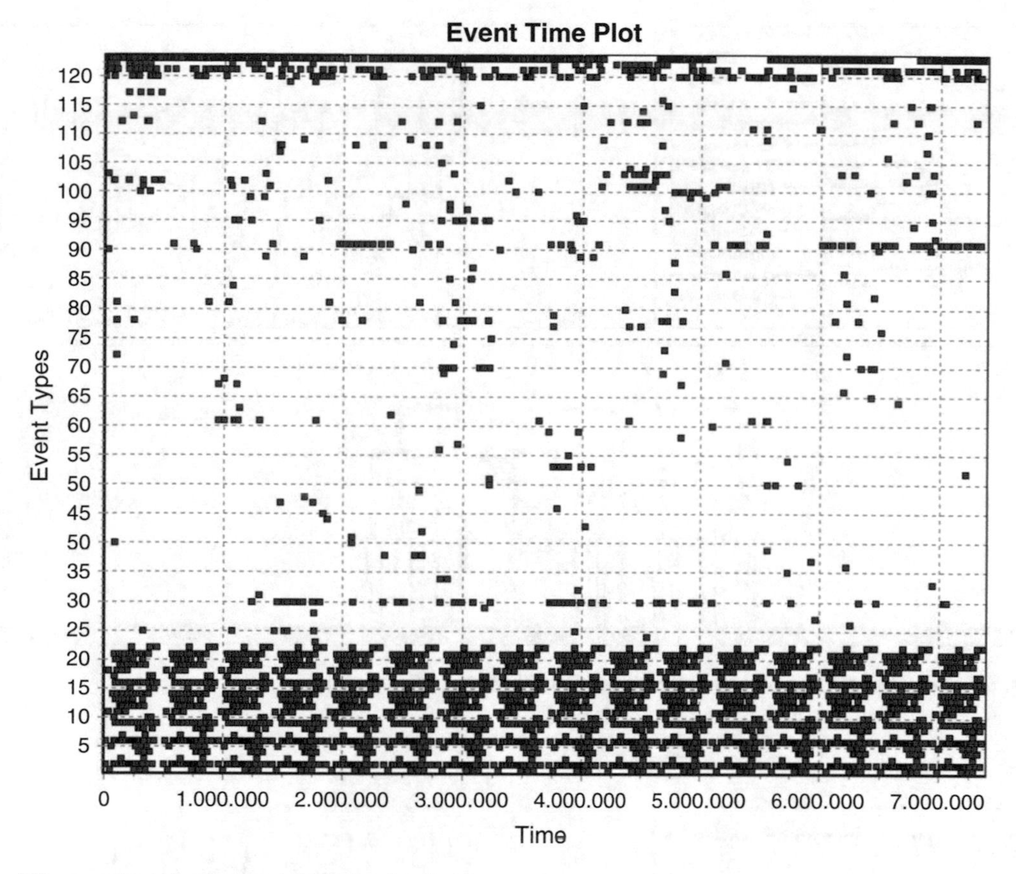

Fig. 4 Time series behavior record of all subjects

5.1 T-Patterns of Events in the Movie

The 11 event types include 83 events overall. We looked only at highly significant patterns ($p < 0.005$), which appear at least three times during the movie. Figure 5 shows the longest pattern we found for the events defined in the movie *Father and Daughter*.

This pattern can be "read" as follows: After showing the futility of the daughter's attempts to deal with the loss of her father the music changes. Then, there is a hole, i.e. pictures with minimal information are shown on screen. Then, the daughter encounters someone. Then she looks out at the place on the shore—symbolizing her longing for her father. Again, the music changes. Again, there is a hole. And again, she encounters someone. This pattern occurs three times in the middle of the movie. A look at the most frequent pattern in the movie (see Fig. 6) reveals that a part of this longest pattern is consistent during eight repetitions: After a scene showing the daughter's longing for her father, the music changes and then she encounters someone (see also [29]).

This most frequent pattern impressively points to the repetitive narrative structure of the short film, and at the same time demonstrates the importance of Topic Lines (longing, futility, encounter). Topic Lines generate sense by the inner-textual variation of important

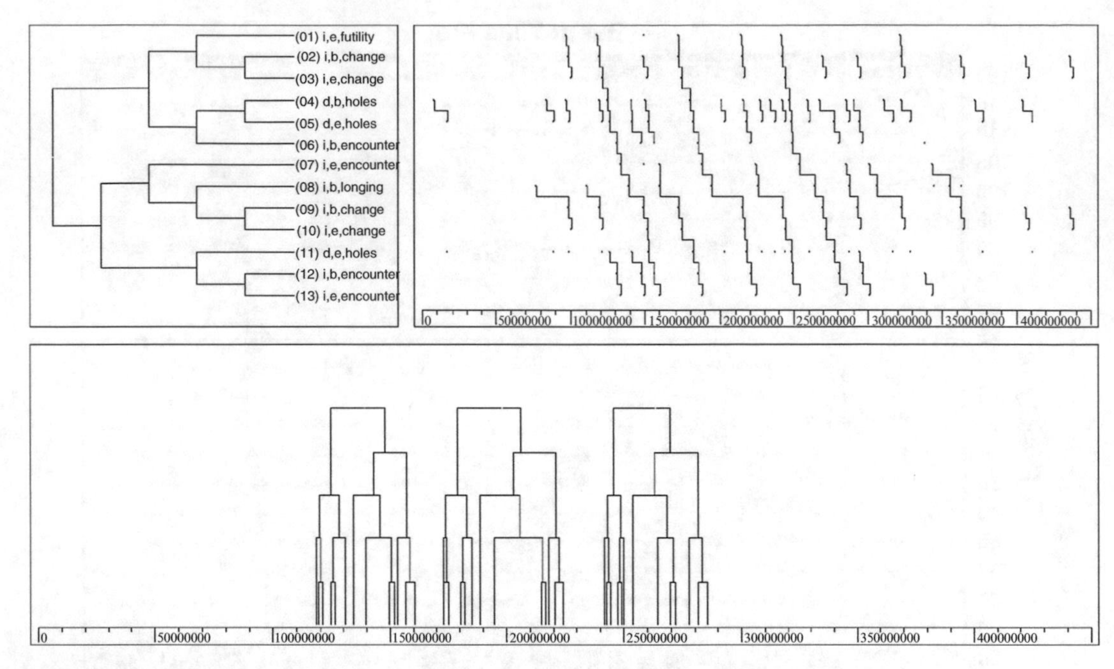

Fig. 5 The longest pattern of events (*i* stands for involvement, *d* for distance)

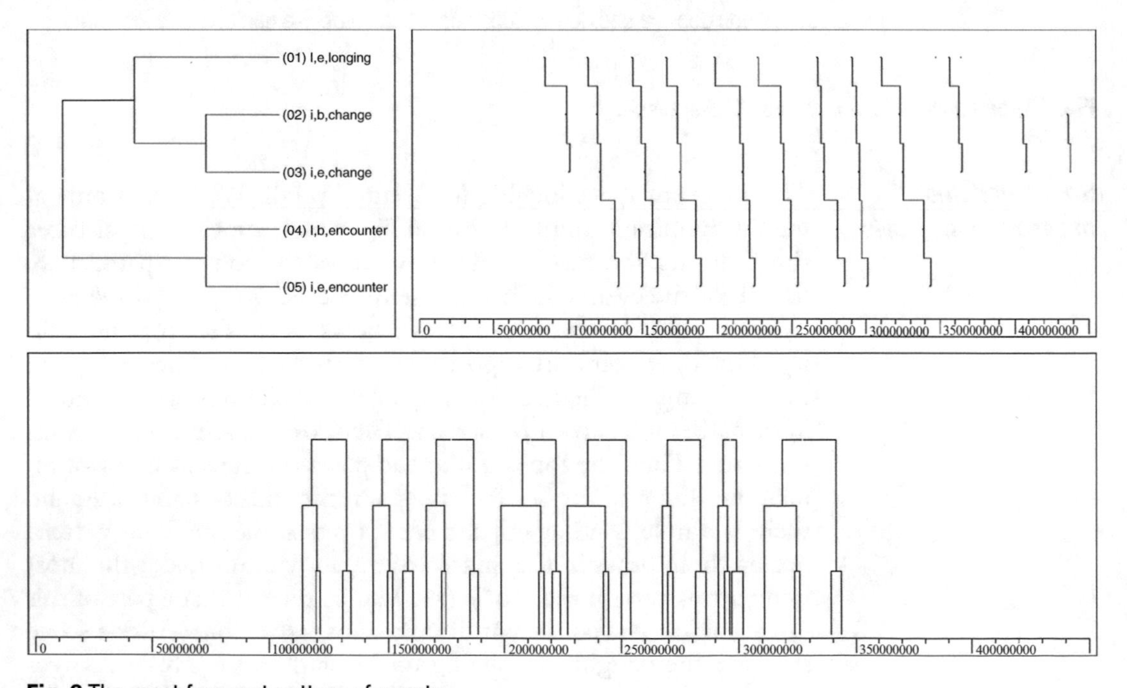

Fig. 6 The most frequent pattern of events

themes or motives. They differ in range, dispersion, and complexity. The term *Topic Line* summarizes quite different circumstances: Often, behavioral manners that characterize the protagonists, are merged into different contexts ([22], p. 68–81). Topic Lines are more inconspicuous and relatively unconsciously perceived cinematic structures. Nevertheless, the repetition can evoke latent expectations in the viewer's mind. The evidence of Topic Lines is low at the beginning of a movie and increases with the frequency of repetition throughout a movie ([22], p. 29, 70; [6], p. 42f.). Combinations of different Topic Lines evoke measurable effects on the spectators [6]. THEME can help to describe and to visualize structural relations in movies, which cannot easily be recognized neither by watching the movie again and again, nor through a detailed dramaturgical analysis, but nevertheless are of relevance for reception processes (for analyses of other movies see [27, 28]).

5.2 T-Patterns of Topic Lines, Physiological Reactions, and Facial Expressions

To further elaborate on the notion of the relevance of Topic Lines for reception processes we conducted analyses concentrating on patterns of Topic Lines, physiological reactions, and facial expressions for the 16 participants, who consented to be videotaped. The main Topic Lines in the short film *Father and Daughter* intensify the inner conflict of the daughter by showing her frequent returns to the place, where her father left, signifying her longing and the futility of her attempts to deal with the loss of her father (see 3). We find peculiar T-patterns that enable us to draw conclusions about the impact of these narrative structures on the recipients: Seven (of 16) participants smile before a scene ends, in which the daughters' longing for her father is shown. Altogether, 14 times a combination of AU 6 (the cheek raiser) and AU 12 (the lip corner puller) was observed (see Fig. 7).

The combination of AU 6 and AU 12 indicates that these scenes in the movie are appraised as intrinsically pleasant and/or as conducive to reach goals [18]. The participants, who smile, may expect that the father will come back now. However, we also find significant patterns that indicate a defensive reaction immediately after the movie scene that shows the daughter's longing for her father: We observe eight significant heart rate accelerations immediately after the longing-scenes (for seven participants). And even more obvious, we observe significant skin conductance reactions for nearly all participants for most of the longing-scenes throughout the movie. HR acceleration and skin conductance reactions indicate a negatively valenced reaction. Looking at patterns of the futility-scenes with facial expressions reveals a more detailed interpretation: At least four participants move their body (after longing- and) during futility-scenes and in their faces appears a combination of AU 10 (the upper lip raiser) and AU 14 (lip corner depressor or dimpler). The combination of AU 10 and AU 14 indicates, that the participants of our study appraise the scene as

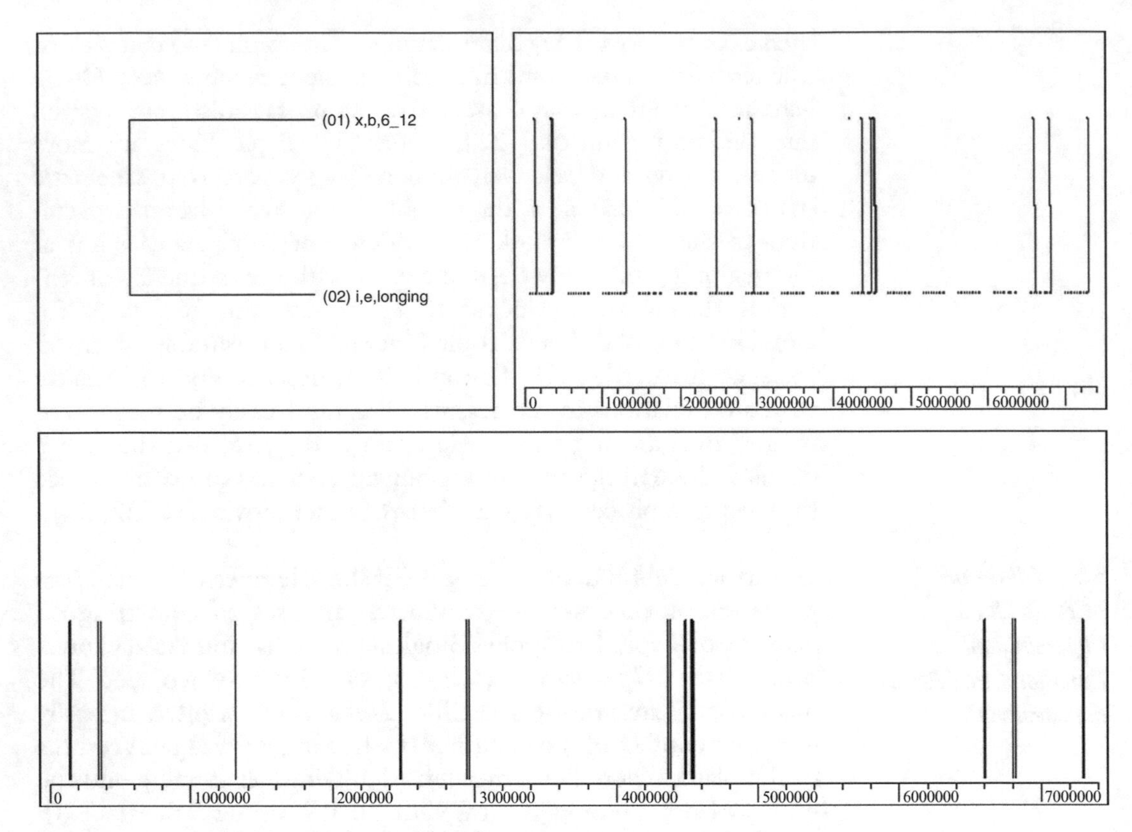

Fig. 7 T-pattern combining longing and a smile

violating external and internal standards [18], when the futility of the daughter's attempts to deal with the loss of her father is shown in the movie. The participants, who show these facial expressions, may think about the inappropriate behavior or the unfairness of a father leaving his daughter alone. The T-patterns illustrate the possibility of stepwise appraisals postulated by Scherer and colleagues [14, 15, 18, 19] and their transferability to movie reception.

AU 14 is one of the most frequent facial expressions observed in the study reported in this contribution as well as in a number of other studies with audiovisual media stimuli [30, 31]. AU 14 thus indicates a very typical emotional appraisal when watching TV or movies (in the laboratory). If this can be interpreted as a kind of "Leit-Affekt" linked to moral evaluations is currently under dicussion [32, 33].

5.3 T-Patterns of Acoustic Events and Facial Expressions

As searching for T-patterns can help to find out more about reactions the participants are not aware of, we decided to also have a look at patterns consisting of the acoustic events in *Father and Daughter* and the facial expressions. The music supports the clear-cut segmentation of the short film. We find four occurrences of a T-pattern combining changes of the music with a combination of AU 1 (the inner brow raiser), AU 2 (the outer brow raiser), and

AU 5 (the upper lid raiser). This combination of AUs signalizes that the situation is appraised as new and kind of sudden [18]. The combination of AUs also indicates the basic emotion *surprise* [7]. Moreover, we found patterns of changes of the music and AU 45 (eye blinks; four occurrences), which can also be interpreted as concomitants of newness and surprise. Even more interesting are other patterns, in which a combination of AU 4 (brow lowerer) and AU 7 (lid tightener) occurs (seven times) shortly *before* the music changes. This combination of AUs is commonly interpreted as an appraisal of an unexpected event [18, 34]. The participants seem to anticipate the change within the plot. The fact that T-patterns help to operationalize expectations makes this method especially interesting for movie impact research.

The results of the T-pattern analysis of the counterpoints (antagonisms between picture and music) are also interesting. Counterpoints allow for distance [26]. In fact, we find patterns in which AU 14 (the dimpler) accompanies scenes during which the music contradicts the content at least for three persons. Basically, AU 14 signalizes that an event is appraised as "something is wrong", i.e. as a kind of irritation. Possibly, this was the exact intention of the movie's music composer. Very little is known about the effects of such acoustic features. The results call for further empirical research at any rate, because we observe that the acoustic events are very often accompanied by body movements. As body movements are not coded in detail in this study we do not report the results here (see also [35]).

5.4 T-Patterns Associated with Moments of Impact

A regression analysis of the physiological data of the 30 participants, conducted by Suckfüll [6], revealed a very strong effect for the humorous scene, in which the daughter is blown forward by the wind and a strong effect of the conflict situation, when the daughter seems to commit suicide (see Fig. 8).

The strong effects of the two scenes can be ascribed to their dramaturgical embedding: Both scenes are prepared by the use of Topic Lines. The motive of effort prepares the recipient by eliciting motor mimicry for the relief effect of the scene in which the daughter is blown forward by the wind. And the longing for her father and the futility of her attempts, repeated eight times during the short film, prepares the recipient for strong emotions, when the final resignation of the daughter is shown.

But, how exactly do the facial reactions of the recipients look like? The humorous scene clearly elicits positive emotions. A number of participants smile, which was indicated by a combination of AU 6 and AU 12 in reaction to the scene, in which the daughter is blown forward by the wind. The pattern occurs seven times in the behavior record of all subjects. The smile indicates a pleasantness-appraisal [18] and the basic emotion *happiness* [7]. Grodal speculates that in movies a sensation of relief may be accompanied by eased laughter ([36]: Chap. 8). Systematic research is

Fig. 8 Scenes causing strong physiological reactions

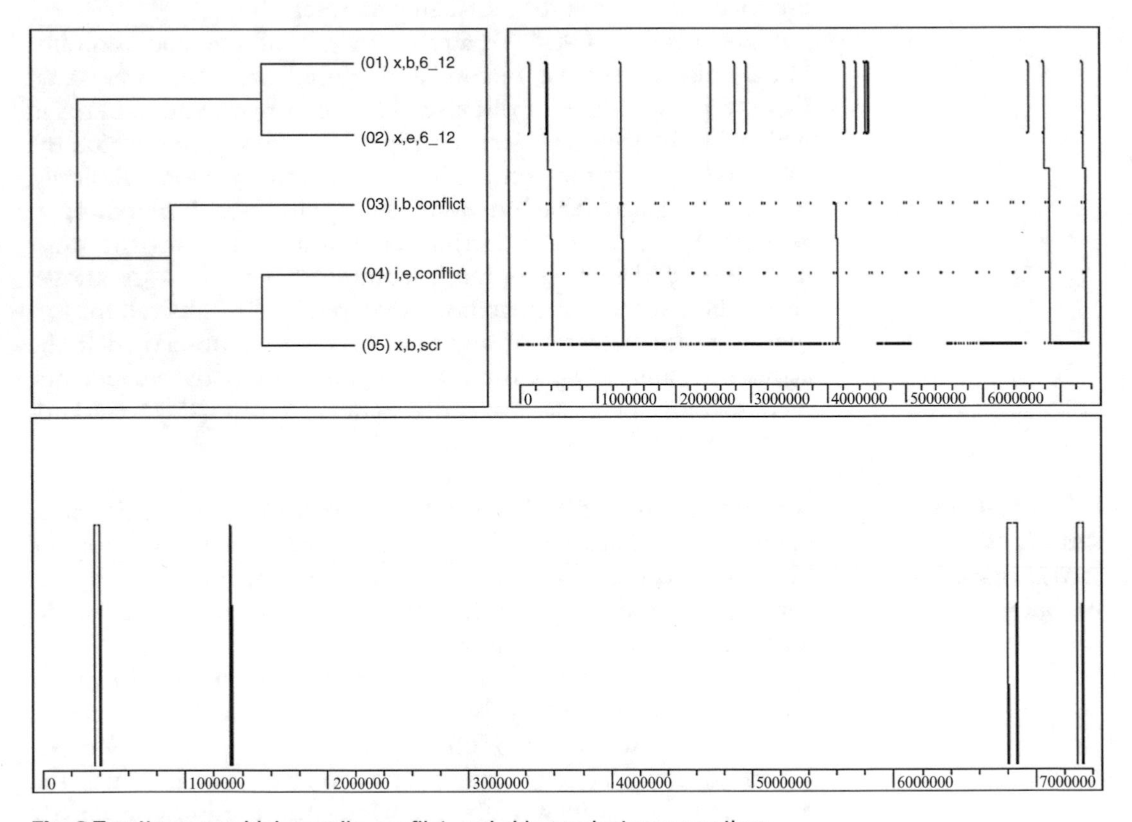

Fig. 9 T-patterns combining smile, conflict, and skin conductance reactions

required, in which other possible indicators of relief (e.g., body movements) are observed, coded, and analyzed.

The patterns associated with the conflict-scene, which causes strong physiological reactions (see Fig. 8, right screenshot), also involve smiles. Yet, the smiles (combinations of AU 6 and AU 12) occur *before* the daughter seems to commit suicide. All participants, who smile *before* the daughter seems to commit suicide, have a significant skin conductance reaction afterwards (four participants, see Fig. 9).

Another pattern, which occurs 14 times (eight participants) illustrates that the conflict-scenes are often followed by a heart rate deceleration. A decrease in HR can be seen as an indicator of the initiation of an attentional state [6]. It is interesting that the first conflict-scene (the father leaves the daughter) is not part of this pattern. In the beginning of the movie, the participants obviously did not yet realize the central conflict the movie is based on.

6 Implications for Future Research

It is impossible to discuss all T-patterns that we found in this paper due to limited space. In summary, the analyses enlighten the dynamic nature of movie reception: Important scenes of a movie are prepared with virtuosity, creating "lines" or repetitions of motives combined with each other. The film maker plays with the expectations and emotions of the viewers—expectations are raised and then disappointed. THEME is perfectly suitable for analyses of iterative and constantly changing processes. One main advantage of adapting THEME for movie reception processes is the possibility to systematically analyze reactions that occur in advance of an event in the movie.

We have already mentioned some suggestions for future research throughout the description of the patterns, but we want to add one more idea for future research: It is especially interesting to look for patterns associated with scenes in a movie, in which *nothing* happens. Thus, future analyses should focus on the so-called *holes*, the scenes in a movie that contain only minor information. It is conceivable that the viewers show intense reactions exactly during these scenes as the film maker offers a break or a pause for reflection about prior scenes or for imagination in the sense of integrating the cinematic events into the viewer's own life. At the same time such scenes may have the function of raising suspense.

The possible emotion processes elicited by such stimuli cannot be analyzed without appropriate data collection methods and tools for analysis. To fully exploit the potential of THEME for an understanding of film art, it is essential to systematically prove hypotheses on the basis of a bigger sample or with manipulated movie sequences in an experiment. The results reported in this contribution support the formulation of concrete hypotheses. The study was already replicated under optimal conditions in a laboratory designed for reception studies with 150 participants watching the short film *Father and Daughter*. The analyses are not yet finalized.

Acknowledgements

Monika Suckfüll wants to thank her research assistants Diana Mirza, who did the data collection, and Flavia Bleuel, who coded the facial expressions.

References

1. Oatley K (1999) Why fiction may be twice as true as fact: fiction as cognitive and emotional simulation. Rev Gen Psychol 3:101–117
2. Tan ES (1996) Emotion and the structure of narrative film: film as an emotion machine. Erlbaum, Mahwah
3. Konjin EA, ten Holt JM (2010) From noise to nucleus: emotion as key construct in processing media messages. In: Döveling K, von Scheve C, Konjin EA (eds) Routledge handbook of emotions and mass media. Routledge, London, pp 37–59
4. Schwab F, Schwender C (2010) The descent of emotions in media: Darwinian perspectives. In: Döveling K, von Scheve C, Konijn EA (eds) The Routledge handbook of emotions and mass media. Routledge, New York, pp 15–36
5. Jennings C, Thijssen W (Producers), Dudok de Wit M (Director) (2001) Father and daughter [Motion Picture]. Netherlands, Belgium, United Kingdom: CinéTé Filmproductie BV, Cloudrunner Ltd.
6. Suckfüll M (2010) Films that move us. Moments of narrative impact in an animated short film. Projections 4(2):41–63
7. Ekman P, Friesen W (1978) Facial action coding system: a technique for the measurement of facial movement. Consulting Psychologists Press, Palo Alto
8. Cosmides L, Tooby J (2000) Evolutionary psychology and the emotions. In: Lewis M, Haviland-Jones JM (eds) Handbook of emotions. Guilford, New York, pp 91–115
9. Scherer KR (1982) Emotion as a process: function, origin, and regulation. Soc Sci Inf 21:555–570
10. Frijda N (1986) The emotions. Cambridge University Press, New York
11. Lazarus RS (1991) Emotion and adaptation. Oxford University Press, New York
12. Scherer KR (2001) Appraisal considered as a process of multi-level sequential checking. In: Scherer KR, Schorr A, Johnson T (eds) Appraisal processes in emotion: theory, methods, research. Oxford University Press, New York, pp 92–120
13. Scherer KR (2005) What are emotions? And how can they be measured? Soc Sci Inf 44(4):695–729
14. Kaiser S, Wehrle T (2000) Ausdruckspsychologische Methoden. In: Otto JH, Euler HA, Mandl H (eds) Emotionspsychologie: Ein Handbuch. Weinheim, Beltz, pp 419–428
15. Kaiser S, Wehrle T (2001) Facial expressions as indicators of appraisal processes. In: Scherer KR, Schorr A, Johnstone T (eds) Appraisal theories of emotion: theories, methods, research. Oxford University Press, New York, pp 285–300
16. Scherer KR (1992) What does facial expression express? In: Strongman K (ed) International review of studies on emotion, vol 2. Wiley, Chichester, pp 139–165
17. Scherer KR (1995) Toward a concept of modal emotions. In: Ekman P, Davidson RJ (eds) The nature of emotion. Fundamental questions. Oxford University Press, New York, pp 25–31
18. Scherer KR, Ellgring H (2007) Are facial expressions of emotion produced by categorical affect programs or dynamically driven by appraisal? Emotion 7:113–130
19. Wehrle T, Kaiser S, Schmidt S, Scherer KR (2000) Studying the dynamics of emotional expression using synthesized facial muscle movements. J Pers Soc Psychol 78(1):105–119
20. Scherer KR (1998) Emotionsprozesse im Medienkontext: Forschungsillustrationen und Zukunftsperspektiven. Medienpsychologie 10:276–293
21. Suckfüll M (2012) Emotion regulation by switching between modes of reception. In: Shimamura AP (ed) Psychocinematics: exploring cognition at the movies. Oxford University Press, New York, pp 314–336
22. Wuss P (2009) Cinematic narration and its psychological impact: functions of cognition, emotion and play. Cambridge Scholars, Newcastle
23. Suckfüll M (2004) Rezeptionsmodalitäten. Ein integratives Konstrukt für die Medienwirkungsforschung. Fischer, München

24. Rabenalt P (1999) Filmdramaturgie. Vistas, Berlin
25. Stutterheim K, Kaiser S (2009) Handbuch der Filmdramaturgie: Das Bauchgefühl und seine Ursachen. Lang, Frankfurt/Main
26. Adorno TW, Eisler H (2006) Komposition für den Film. Suhrkamp, Frankfurt/Main
27. Suckfüll M (2007) Rhythmus im Film. Narrative Zeitmuster, Komplexität und ästhetischer Eindruck. In: Früh W, Wünsch C, Gehrau V (eds) Integrative Modelle in der Rezeptions- und Wirkungsforschung: dynamische und transaktionale Perspektiven. Fischer, München, pp 241–260
28. Suckfüll M (2008) Rhythm in films. Narrative time patterns, complexity and aesthetic experience. In: Bordens K (ed) Proceedings of the twentieth biennial congress of the International Association of Empirical Aesthetics, Chicago
29. Suckfüll M (2010) Time patterns in films. A case for distance editing. Paper presented at 8th international conference of the Society for Cognitive Studies on the Moving Image. Virginia Tech, Roanoke
30. Unz D (2010) Effects of presentation and editing on emotional responses of viewers: the example of TV news. In: Döveling K, von Scheve C, Konjin EA (eds) Routledge Handbook of emotions and mass media. Routledge, London, pp 294–309
31. Unz D, Schwab F, Winterhoff-Spurk P (2008) TV news - the daily horror? Emotional effects of violent TV news. J Media Psychol 20(4):141–156
32. Rozin P, Lowery L, Imada S, Haidt J (1999) The CAD triad hypothesis: a mapping between three moral emotions (contempt, anger, digust) and three moral codes (community, autonomy, divinity). J Pers Soc Psychol 76(4):574–586
33. Suckfüll M (2015) Moral emotions. Paper presented at 13h international conference of the Society for Cognitive Studies on the Moving Image. University of London, London, UK
34. Unz D, Schwab F (2005) Viewers viewed: facial expression patterns while watching TV news. In: Anolli L, Duncan S, Magnusson M, Riva G (eds) The hidden structure of social interaction. From genomics to cultural patterns. Ios Press, Amsterdam, pp 253–264
35. Schütte E, Mirza D, Suckfüll M (2013) Taking the body into account. Developing a coding system for body movement and posture during film reception. In: Freitas-Magalhães A (ed) Emotional expression: the brain and the face, vol 4. Edições Universidade Fernando Pessoa, Porto
36. Grodal T (2002) Moving pictures. A new theory of film genres, feeling, and cognition. Clarendon, Oxford

Chapter 10

Immersive Dynamics: Presence Experiences and Patterns of Attention

Michael Brill, Gudberg K. Jonsson, Magnus S. Magnusson, and Frank Schwab

Abstract

The present chapter addresses the application of T-Pattern Detection to video games research. By their nature as interactive media, games offer great degrees of freedom for their users which can result in very different gaming experiences; strictly speaking, no two sessions are the same. If researchers do not plan to solely rely on summative post-session measures, but also intend to investigate media-use processes in their temporal course, these challenges call for an analysis method which proves to be robust towards such inter-individual differences in the data. We discuss the use of T-Pattern Detection as a method which meets these challenges. We present an example study which merges summative questionnaire data and structural process analysis data in order to investigate the reception processes that underlie the reception phenomenon of presence.

Key words T-pattern, Structure, Game studies, Games, Gaming, Presence, Attention

1 Introduction

Media research in general addresses a wide range of media, especially in a modern information society. With constantly emerging new media capabilities, research has been facing new questions and new challenges. One of these novelties was the feature of interactivity which is one of the key-features of video games. Since pattern detection has already been used successfully for analyses of human interaction and of humans' reactions towards media, we argue that this method begs to be used for the analysis of humans' interaction with media, in this case video games. We will present a study in which a special kind of behavior of gamers, blinking, is analyzed and will demonstrate how pattern detection offers new possibilities for analysis and interpretation of data.

Magnus S. Magnusson et al. (eds.), *Discovering Hidden Temporal Patterns in Behavior and Interaction: T-Pattern Detection and Analysis with THEME™*, Neuromethods, vol. 111, DOI 10.1007/978-1-4939-3249-8_10,

1.1 Games as Interactive Media

Why do a myriad of people go to the movies and watch million-dollar blockbusters such as *Titanic* or *Lord of the Rings* over and over again? Why do users of an online game like *World of Warcraft* or racing games like *Need for Speed* spend countless hours visiting virtual worlds? What happens to our minds when we use media or play games on our computers, or are absorbed by an exciting book or movie [1]? Nowadays, in our so-called information age, the media have become a central part of our daily life. Current research shows that individuals spend an average of 4 h a day watching TV, 3 h a day listening to the radio, and almost 1.5 h a day browsing the Internet [2]. In face of this almost omnipresent media use, media psychology aims at describing and explaining behavior and experiences concerning the usage of mass and individual media [3]. Mass media include the press, radio, TV, and cinema, whereas individual media encompass landline telephones, mobile devices, and social network services and video games. In addition to learning and knowledge acquisition, the one focus of media psychology is to find an answer to the question why humans are willing to invest such a great amount of time and money into (media) entertainment and playful computer usage. One special aspect within this field is the phenomena which relate to the user's experience of the mediated environment. While there are several adjacent concepts dealing with this aspect using different stimuli media, we focus on the so-called presence experiences in the domain of video games.

Especially for the video games domain, interactivity is one key aspect of interest: Do video games compared to other media offer more immersion and presence, due to their interactive challenges? Do they make their users more aggressive because they let their users be active? So in fact, the key aspect of games—and also the one that ultimately led to the application of the T-Pattern Detection method in this example—is interactivity. In most studies the dependent variables are summative outcome measures like post-session questionnaires for presence or aggressiveness. Consequently, those outcome measures are used to assess effects of gaming, but don't focus on the process of gaming itself. A more fine-grained observation is delivered by studies incorporating objective measures for e.g. body posture [4, 5] or facial expressions [6]. In these studies, process measures are collected during a period of media use, are then summed up and compared across different experimental conditions. But strictly speaking, in these cases there is still an observation of behavior in its sum, not of the temporal process itself and its structural characteristics. At this point we suggest that one may gain valuable insights about the emergence of effects when considering the structural aspects of the user-media-interaction, as well. Patterns of user-media-interactions that unfold over the course of time could be used rather than cumulated parameters of isolated user behavior. As described by Magnusson [7], this approach follows the idea of identifying event structures in data rather than pure event frequencies or durations.

However, analysis gets more complicated when the subject's individual behavior in the course of time is of interest. Due to the high degrees of freedom and interactivity in modern video games, data may become more complex and may differ considerably among subjects. One tool to meet these analytic challenges is the method of T-Pattern Detection [8, 9].

1.2 T-Pattern Detection

The method of T-Pattern Detection is the connecting link of the researchers inside the MASI network. The algorithm has frequently been deployed for analysis of social interaction among animals and humans and there are numerous studies about this most natural form of interaction, be it for example in the context of psychotherapy settings [10] or human interaction with dogs and robots [11]. But the method's versatility allows it to be also used—besides many more applications—in similar contexts like reactions of humans towards classical media, for example which facial expressions occur during viewing of TV news [12].

Aside from these situations' common element—behavior occurs which can be analyzed with T-Pattern Detection—there are obviously different levels of possible interaction. In a social face-to-face situation, interaction is only limited by the social rules which determine what is appropriate to say, do and display. If instead of face-to-face communication a medium is used as a channel of communication, well-known additional constraints apply according to the medium's unique capabilities. As mentioned above, when watching TV or movies almost none of the recipient's actions alter the mediated content. Speaking of video games and their key-feature interactivity, they could be seen situated in between the noninteractive media and the perfectly interactive personal dyads. They allow a certain degree of interaction, however, not in such a broad spectrum and in such a natural way as human face-to-face interaction. In general, for single and multiplayer settings the interaction done via the game's controls is limited to certain actions and channels which have been incorporated by game designers.

So due to these media limitations, strictly speaking it would be inaccurate to describe for example the last study mentioned above, the observation of TV viewers' facial expressions, as an interactive situation. But this last example also shows that some behavior occurs despite the lack of a proper backchannel and thus without apparent (media-related) use: while in a dyad facial expressions can serve relational regulation or for illustration of spoken content [13], there is no obvious use for displaying joy, anger, or contempt while watching TV since it does not influence the TV program. So these recipient actions don't serve a purpose in the communication process, but when they can be seen as indicators for internal processes or as remnants of social interaction then they can be a valuable source of information for researchers.

2 Example Study

2.1 Background

In the presented study we used a racing video game to investigate an internal process during media use which is called the experience of presence. Presence is often referred to as the sense of "being there" or a "perceptual illusion of nonmediation" [14], i.e. during reception of media content the user becomes unaware of the fact that the content is brought to him/her by a medium. While there are several concepts of presence we focus on spatial presence and use the conceptualization by Wirth et al. [15]. They propose a two-stage model for presence experiences: On the first level, users need to focus their attention on the medium and create a mental model of the depicted situation, the spatial situation model. This leaves users with two rivaling mental models: one for the real environment and one for the mediated environment. Following the model, presence should occur on a second level if users choose the model from the mediated environment as their so-called primary ego-reference-frame. According to the model, there are two facets of spatial presence experiences: the users' feeling that their self is located in the mediated environment, and the users' feeling that their possible actions are determined by the mediated environment.

In our case, subjects were sitting in a darkened booth and were asked to play a modern racing game. The booth was equipped with a steering wheel for controls, surround sound, and a screen capable of stereoscopic 3D. According to the theory, the participants needed to build a spatial situation model of the simulated setting "car on the race track"; presence should have been experienced when they neglected the fact that they were sitting in a lab.

When it comes to measurement of the subjective presence experiences, the most common approach is to let users report their experiences in questionnaires. Depending on the conceptualization at hand, there is available a number of questionnaires. For our study we used the MEC-SPQ [16] by the same workgroup who proposed the presence model at hand. Because questionnaires as subjective measures are by design prone to subjective distortions such as memory effects, there have also been approaches to use objective indicators for presence experiences. They would have the advantage of being less prone to distortions and—in the best case—would be able to be recorded continuously without interrupting the media reception. In contrast, questionnaires can only sum up certain intervals of experiences after they took place. Approaches to objective measurement of presence include behavioral aspects [4, 5], brain-imaging methods [17], and measures for attention allocation [17].

In the latter, there is already the rationale of determining the degree to which subjects focus their attention on the stimulus by using eye movement. We would like to also follow this rationale by using eye blinks as indicators.

For our study we considered the so-called spontaneous eye blinks. While reactive eye blinks serve as protection of the eye, e.g. against close or approaching objects, the function of spontaneous eye blinks is seen in the moistening of the cornea [18]. One might think that this process occurs at random or is just determined by the eyes' level of humidity, but research findings show that there is more behind the ubiquitous closing of our eyes.

Nakano et al. [19] investigated the spontaneous eye blink rate during TV viewing. They first state that each blink means an inevitable loss of visual information. They thus hypothesized that blinking does not occur in a random manner but is somehow controlled to minimize loss of information. In their study, the researchers found that viewers of a comedy show tended to blink during certain breaks in the story telling, i.e. viewers inhibited their blinking when relevant story information was presented. This behavior did not occur during viewing of a non-story telling show or an audio story. More recent work by Nakano et al. [20] investigates possible functions of eye blinks in attention regulation.

A very similar direction is followed by the proposal of a two-component model for spontaneous eye blinks by Galley [18], in which blinking is seen as determined by two antagonist processes: on the one hand, excitatory processes that depend on arousal and increase spontaneous eye blink rate, and on the other hand inhibitory processes related to attention which can suppress blinking to a certain degree until it can be done safely. The proposal also claims that the inhibition could serve as a kind of optimizer of visual intake and avoid the loss of relevant visual information.

In an exploratory part and further analysis of the study we aimed at finding clues towards eye blinks as a continuous online-measure for presence experiences: Would players with different levels of presence apply their attention to the game in different ways? We used spontaneous eye blinks as an indicator for attentional processes and related them to post-session questionnaire measures of presence. The first step focused on intra-individual differences in blinking behavior: Since the racetrack consists of straight sections and curved sections, there is varying difficulty and thus varying demand for attention along the racetrack. We first derived the assumption from the described theoretical background that the rate of spontaneous eye blinks should be lower for the demanding curved sections when compared to the easier straight sections. In further analysis of the data we then considered the research question of whether participants reporting more or stronger presence experiences would fit their blinking behavior more distinctly to the racetrack's demands. This may serve as an indicator for being focused on the medium rather than on the real surroundings, a necessary condition for presence.

3 Methods

The whole study used two experimental factors: in the first factor we varied the mode of presentation, i.e. whether subjects played in regular 2D or with 3D glasses in stereoscopic 3D. For the second factor we varied the measurement of eye blinks via derivation of an electrooculogram by applying or not applying electrodes. We regarded this as necessary to control for the possibility that our method of measurement, electrodes, interfered with the measured construct of presence. Because both factors were fully crossed, four experimental conditions resulted. Participants were randomly assigned to these conditions. For analysis we could rely on the data of 48 subjects aged 19–45 ($M=24.2$, $SD=4.3$) with normal or corrected to normal vision.

The experimental sessions included six laps of training, followed by six laps of racing during which data was collected, all done on the same race track (see Fig. 1). After the gaming period we collected the post-session data including the presence questionnaire. As output we had several data sets for each participant, some of which are relevant for our present research question: the presence questionnaire ratings as a summarizing post-session measure, and the game's racing data and collected eye blinks as event time series data. Because racing and blinking data was recorded in sync, it was possible to determine when and where on the racetrack the subjects had blinked.

3.1 Summative Analysis: Frequencies and Questionnaire Data

To first create one artificial lap for each subject, the blinks from all six laps of each subject were condensed into one lap. These artificial laps from each subject were in turn condensed into one artificial lap including all blinks of all subjects. This way, the pooled distribution of blinks along the racetrack becomes visible (see Fig. 2); χ^2-tests were conducted to statistically test the distribution's equality. The first binning distinguished between straight sections versus curved sections, i.e. race track sections between corner entrance and corner exit. This first approach did not yield significant results ($\chi^2(1)=1.806$, $p=0.179$). For a second analysis

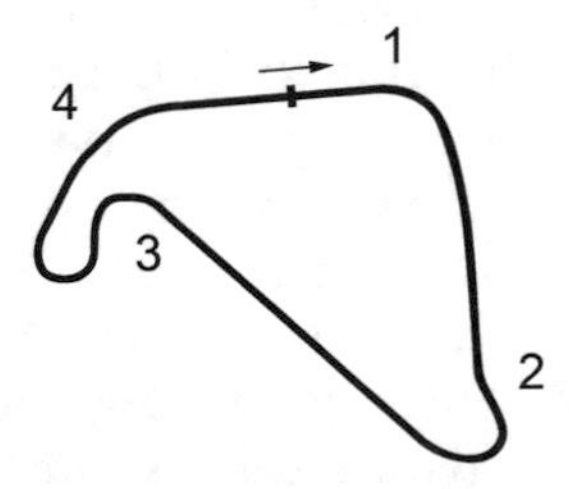

Fig. 1 Race track used in the study. *Numbers* indicate corner entries: 1 (at 6.5 % of race track), 2 (30.5 %), 3 (63.5 %), 4 (86 %)

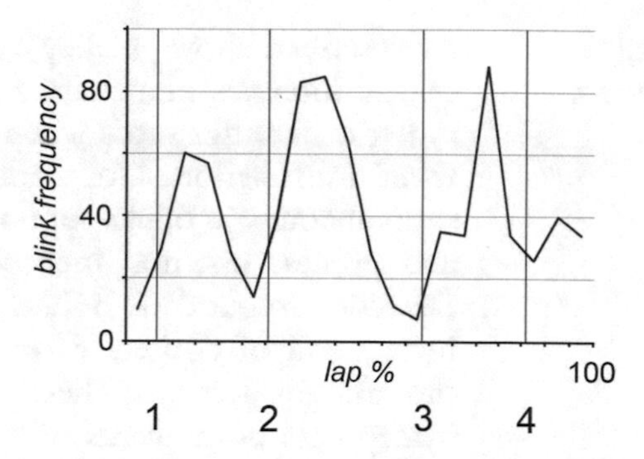

Fig. 2 Frequencies of all subjects' blinks for each section of race track. *Numbers* below indicate corner entries: 1 (at 6.5 % of race track), 2 (30.5 %), 3 (63.5 %), 4 (86 %)

we shifted the limits for curved sections forward to account for the fact that the entry area of a corner actually is more challenging than the corner exit. This layout led to a χ^2-test that showed a significant result ($\chi^2(1)=27.616$, $p<0.0001$). This confirms that the blinks are not distributed equally along the racetrack. Apparently, the spontaneous eye blink rate drops considerably in intervals around the curve entries. This may be explained by the fact that these intervals are the critical phases in which the users need to adjust speed and steering to properly pass the corner. The distribution suggests that, according to the theory, subjects try to avoid loss of visual information in these critical intervals: blinking is inhibited right before corners and is rather performed after the corners' apex.

For the assessment of inter-individual differences we first correlated individual blinking rates with presence experiences: if a reduced eye blink rate due to more focused attention leads to more presence experiences then there should be a significant negative correlation. Correlational analyses showed no significant correlation of eye blink rate to presence questionnaire data (scale "spatial presence—self-location": $r=0.284$, $p=0.190$; $N=23$; scale "spatial presence—possible actions": $r=.082$, $p=0.710$; $N=23$). In additional analyses of inter-individual differences we use the method of T-Pattern Detection to gain insights into the temporal structures of the interactive gaming process. We try to assess possible relations between blinking on the one hand and questionnaire based presence scores on the other hand. Additionally, the question is investigated if there is a relation between blinking and different performance levels of players measured by total racing time; e.g. low performing players may already be challenged by the controls, whereas high performing or experienced players can better focus on racing.

3.2 Structural Analysis: T-Pattern Detection

As described above, Galley's proposal [18] suggests that for such critical intervals blinks are more likely to be inhibited and may rather occur afterwards when the situation allows a certain loss of visual information. Our data of intra-individual distributions of spontaneous eye blinks supported this notion. As a more sensitive and process oriented method of analysis we applied T-Pattern Detection to our data: Because blinks can be done more hazard-free in the end of corners when the participants accelerate towards the straight sections, there should be identifiable patterns that incorporate both events of "passing the corner exit" and "blinking." In preparation of this approach data was recoded and two extreme outliers in blinking frequency were excluded from further analyses. Data were then imported into the THEME T-pattern Detection software.

Subjects were categorized into two groups along the median of their scores on the scale "spatial presence—self-location." Thus event patterns of subjects who reported lower presence experiences could be compared to subjects who reported higher presence experiences. A χ^2-test showed a significant difference among the two very heterogeneous groups with regard to blink numbers ($\chi^2(1) = 9.03$, $p < 0.005$), so in our limited sample participants who reported more presence experiences had also been blinking more during the race. The search parameters were set for detection of T-Patterns in the individual subject files and included THEME's "burst detection" to detect rapidly timed events. Only two types of events were specified for the search: "passing the curve exit" and "blinking", so an example for a possible pattern would be that a blink follows within a critical time interval after the player has left the corner. A level of significance of $p = 0.005$ was chosen for both general patterns and burst detection; furthermore patterns were only considered when they occurred at least three times in a subject's data set. For the low presence group a total of 62 patterns, for the high presence group a total of 139 patterns were detected. Another χ^2-test was conducted to test the difference for its statistical significance. Under the assumption that a larger number of blinks also allows more patterns, the expected values for the χ^2-test were estimated from the previously found distribution of eye blinks between the two groups. Under these premises the test yielded a significant result ($\chi^2(1) = 12.67$, $p < 0.001$).

The two groups were split again along the median of their total racing time, dividing the participants into four groups. In this 2×2 split it appears that the number of blinks is distributed rather equally among slower and faster drivers and that there appears to be some difference between low- and high-presence players (see Table 1). These numbers are again used to estimate the numbers of expected T-Patterns for each cell. In the numbers of patterns detected by the THEME software there are apparently slight differences for slow players (low-presence: 52 pattern occurrences,

Table 1
Number of eye blinks in median-split groups

	Low presence	High presence
Fast racing time	113	154
Slow racing time	116	144

Table 2
Number of pattern occurrences in median-split groups

	Low presence	High presence
Fast racing time	10	91
Slow racing time	52	48

high presence: 48 pattern occurrences), but apparently remarkable differences for the faster players: In the group of faster players, in sum 10 pattern occurrences had been identified for the low-presence group, whereas a sum of 91 pattern occurrences had been identified for high-presence players (see Table 2). A χ^2-test was conducted to compare the observed number of pattern occurrences with the blink-number-based expected numbers and showed a significant result ($\chi^2(3) = 45.03$, $p < 0.001$).

With due care because of limited sample size, large inter-personal differences in eye blink frequency and our explorative approach, some findings can be derived for informing hypotheses in future studies. First, the finding of more pattern occurrences in eye blinks of more present players is in line with the assumption that players who feel more present may orient their blinking behavior more towards the game's demands. Second, THEME extends an invitation for further research into the differences in blinking behavior between the faster and the slower race drivers which may contribute to research about the influence of expertise; this will have to be addressed again with appropriate sample sizes.

4 Conclusion

We tried to use spontaneous eye blinking behavior as an observable indicator of unobservable cognitive processes. The rationale was to determine the degree to which subjects focus their attention on the virtual world by using blinking behavior and relate this to established questionnaire measures of presence. Regarding frequency and distribution of blinks, the participants showed differential blink distributions along the race track; since this distribution falls in line with theoretical assumptions on attention allocation, this may be a result of the players' attentional processes that take place when using the virtual race track. Beyond that, we used the T-Pattern analysis of THEME to investigate structural aspects in the event-time-series of blinking and racing events. These patterns were in turn used for further analyses and uncovered previously unnoticeable relations in the data. Presence literature supposes attention to be a necessary, but not sufficient prerequisite for presence experiences [15], so this analysis alone does not lead to ultimate insights on the formation of presence. Nevertheless, the example shows that T-Pattern Detection offers the chance to get a deeper insight into the mental processing of interactive media. The phenomena within the research fields of immersion, presence, or transportation are not yet fully understood and are mostly addressed with questionnaires on a summative, attributional level of description. T-Pattern Detection enhances the researcher's possibilities and offers a chance to depict the dynamics of the events that let us get immersed in mediated environments.

References

1. Schwab F, Schwender C (2010) The descent of emotions in media: Darwinian perspectives. In: Konijn E, Dövelin K, Scheve CV (eds) The Routledge handbook of emotions and the mass media. Routledge Chapman & Hall, London
2. ARD-ZDF-Onlinestudie (online study conducted by German public television) (2012). www.ard-zdf-onlinestudie.de
3. Winterhoff-Spurk P (2004) Medienpsychologie. Eine Einführung, 2nd edn. Kohlhammer, Stuttgart
4. Freeman J (1999) Subjective and objective approaches to the assessment of presence. Dissertation, University of Essex
5. Freeman J, Avons SE, Meddis R et al (2000) Using behavioural realism to estimate presence: a study of the utility of postural responses to motion-stimuli. Presence 9:149–164
6. Kaiser S et al (1998) Emotional episodes, facial expression, and reported feelings in human-computer interactions. In: Fischer AH (ed) Proceedings of the Xth Conference of the International Society for Research on Emotions. ISRE Publications, Würzburg, pp 82–86
7. Magnusson MS (2005) Understanding social interaction: discovering hidden structure with model and algorithms. In: Anolli L, Duncan S Jr, Magnusson MS, Riva G (eds) The hidden structure of interaction: from neurons to culture patterns. Emerging communication: studies in new technologies and practices in communication, vol 7. Ios Press, Amsterdam
8. Magnusson MS (1996) Hidden real-time patterns in intra- and inter-individual behavior: description and detection. Eur J Psychol Assess 12(2):112–123
9. Magnusson MS (2000) Discovering hidden time patterns in behavior: T-patterns and their detection. Behav Res Methods Instrum Comput 32(1):93–110
10. Merten J, Schwab F (2005) Facial expression patterns in common and psychotherapeutic situations. In: Anolli L, Duncan S, Magnusson MS, Riva G (eds) The hidden structure of interaction. From neurons to culture patterns. Ios Press, Amsterdam
11. Kerepesi A, Jonsson GK, Kubinyi E, Miklósi A (2007) Can robots replace dogs? Comparison of temporal patterns in dog-human and robot-human interactions. In: Lazinica A (ed) Human-robot interaction. I-Tech Education and Publishing, Vienna, Austria
12. Unz D, Schwab F (2005) Viewers viewed: facial expression patterns while watching TV news. In: Anolli L, Duncan S, Magnusson MS, Riva G (eds) The hidden structure of social interaction. From neurons to culture patterns. Ios Press, Amsterdam

13. Merten J (1997) Facial-affective behavior, mutual gaze and emotional experience in dyadic interactions. J Nonverbal Behav 21(3):179–201
14. Lombard M, Ditton T (1997) At the heart of it all: the concept of presence. J Comput Mediat Commun 3(2) http://jcmc.indiana.edu/vol3/issue2/lombard.html
15. Wirth W, Hartmann T, Böcking S et al (2007) A process model of the formation of spatial presence experiences. Media Psychol 9(3):493–525
16. Wirth W, Schramm H, Böcking S et al (2008) Entwicklung und Validierung eines Fragebogens zur Entstehung von räumlichem Präsenzerleben. In: Matthes J, Wirth W, Fahr A, Daschmann G (eds) Die Brücke zwischen Theorie und Empirie: Operationalisierung, Messung und Validierung in der Kommunikationswissenschaft. Halem, Köln
17. Böcking S, Wirth W, Hartmann T et al (2008) Zur Messung von räumlichem Präsenzerleben: Ein Vergleich von vier alternativen Messmethoden. In: Matthes J, Wirth W, Fahr A, Daschmann G (eds) Die Brücke zwischen Theorie und Empirie: Operationalisierung, Messung und Validierung in der Kommunikationswissenschaft. Halem, Köln
18. Meinold PE (2005) Psychologie des Lidschlags: Eine literatur- und methodenkritische Studie. Dissertation, University of Cologne
19. Nakano T, Yamamoto Y, Kitajo K et al (2009) Synchronization of spontaneous eyeblinks while viewing video stories. Proc R Soc B 276:3635–3644
20. Nakano T, Kato M, Morito Y et al (2013) Blink-related momentary activation of the default mode network while viewing videos. Proc Natl Acad Sci U S A 110:702–706

Chapter 11

Accessing Individual Style through Proposed Use of *THEME* Associates

Liesbet Quaeghebeur and David McNeill

Abstract

The phenomenological concept of personal "style," being the individual's idiosyncratic mode of being, is registered in speech and gesture during narrative discourse. We propose THEME is able to empirically pinpoint instances of this largely intuitive concept. Gestural catchments are stylistic elements of language use which THEME could recover by identifying them as associates.

Key words Phenomenology, Gesture, Style, Theme, T-pattern, Associates

1 Introduction

"Individual style" is a typical example of a concept that is transparent to us in everyday life, yet which turns opaque when we try to capture it in scientific terms. You and I need not explicitly agree that a third person's style, say Mary's, is "earthy, on the verge of blunt." Yet, discussing her actions, a shared and quite specific know-how of Mary's manner of being enables us to make sense of why Mary did this or that or to predict how she will respond to certain situations in the future. Hampson [1] (quoted in Brunas-Wagstaff [2]) sees in these everyday practices an "amateur approach to personality psychology" or a "Lay Perspective" on (cognitive) style: it is based on experience[1] and is related to predictability. In other words, there is something in the way that individuals behave, which creates, in others, the expectation of just this kind of behavior.

[1] We use the term experience where Hampson [1] uses the term observation. In everyday social reality, people most of the time do not observe each other's behavior, but rather acquire know-how of another person's manner of being through interactively engaging with them. We will use experience to denote a more general form of "information pick-up," subsuming also those moments in which we solemnly retreat to observe and deliberate over other people's behavior.

Magnus S. Magnusson et al. (eds.), *Discovering Hidden Temporal Patterns in Behavior and Interaction: T-Pattern Detection and Analysis with THEME™*, Neuromethods, vol. 111, DOI 10.1007/978-1-4939-3249-8_11,

We will suggest that one way of making the implicit everyday reality of personal style scientifically visible is to look at individual differences in so-called gestural "catchments." To this end, we will link up our interpretation of the work on style by the French phenomenological philosopher Merleau-Ponty [3–5] to the gestural theory of the Growth Point as elaborated by McNeill [6, 7]. Using these theoretical concepts, we can turn to the possibility of THEME extraction of patterns, and in particular the recovery of "associates" of the patterns, to greatly expand their empirical reach.

2 The Concept of "Cognitive Style"

The topic "cognitive style" was introduced into personality psychology by Gordon Allport [8] and can be defined loosely as "a person's typical or habitual mode of problem-solving, thinking, perceiving, and remembering" (Riding and Cheema [9]: 194). In the 1950s and 1960s theoretical research on cognitive style took off, with applied research and significant shifts in theoretical interest following from the 1970s onwards. We will follow Kozhevnikov's [10] literature review to survey the historical evolutions in the field of cognitive style research.

Kozhevnikov abstracts the following definition of the field's object of investigation to withstand its different interpretations over the years:

> *Cognitive style* historically has referred to a psychological dimension representing consistencies in an individual's manner of cognitive functioning, particularly with respect to acquiring and processing information. (p. 464).

Klein [11] proposed the "sharpeners–levelers" distinction to relate individual differences in perception and personality. Witkin et al. [12] identified *field dependent* persons (those whose judgments were highly influenced by their surrounding field) and *field independent* (those whose judgments were not). These individual differences in perception were explained as the outcome of different modes of adjustment to the world. In the 1950s and 1960s Klein and Witkin's ideas about the measurement of cognitive styles in terms of bipolar dimensions attracted many followers, resulting in a proliferation of proposed bipolar cognitive style dimensions.

While the early theoretical research related cognitive style to basic perception tasks, the research in applied fields focused on more complex tasks, such as problem-solving, decision-making, learning, and the individual's causal explanations of life events. Proposed dimensions in this research were very much related to practical use—such as Kirton's [13] introduction of the *adaptors–innovators* dimension as preferred modes of tackling problems;

research in psychotherapy resulted in "personal styles"; and the largest number of applied studies were carried out for the advancement of education, proposing "learning styles." Gregorc [14], for example, localized subjects on two dimensions, perception and ordering, with each of these dimensions consisting of two poles: perception is concrete or abstract, ordering is sequential or random. Learning styles such as abstract-random were then concluded to relate to the whole system of thought. Kolb [15] proposed four learning styles, *diverging*, *assimilating*, *converging*, and *accommodating*, based on the way experience is grasped (concrete or abstract) and the way it is transformed (reflective or active).

More recent trends in theoretical cognitive style research have attempted to come up with complex models to accommodate some apparent contradictions present in earlier research. Allinson and Hayes [16], for example, proposed a cognitive style measure called the Cognitive Style Index which was based on the unified structure *analytical–holistic*. Riding and Cheema [9] proposed two major style families, based on their review of different cognitive styles: *analytic–holistic* and *verbalizer–imager*. Sternberg [17], on the other hand, proposed a whole new theory of thinking styles based on the metaphor of government: individual differences in the regulation of intellectual activity, the legislative, executive, and judicial. The third recent trend in theoretical cognitive style research, which Kozhevnikov connects to, is the integration of a hierarchical organization of styles into information processing theory. Nosal [18] proposes a theory that connects different styles to different levels of information processing: *field dependence–field independence* become styles at the perceptual level, while the *rigidity–flexibility* dimension operates at the level of metacognition.

2.1 Some Problematic Aspects of Cognitive Style Theory

2.1.1 Pigeonholing

Immediately striking about Kozhevnikov's [10] literature review is the unidirectionality of cognitive style research over the years: the universally assumed intended result of an investigation of cognitive style seems to be a model, consisting of poles, which can be used to divide individuals up into categories. After her review of the initial stage of theoretical cognitive style research, Kozhevnikov remarks that "[t]his [kind of research] led to a situation in which as many different cognitive styles were described as there were researchers who could design different tasks" [10: 465]. Yet, frustration surrounding the infinite proliferation of bipolar cognitive style dimensions did not lead to a reassessment of the concept's theoretical foundations or the field's purpose: "new" strategies of approaching the phenomenon attempted to unify dimensions proposed earlier, by subsuming them in a "metastyle," a unifying theory, or a hierarchical structure. We will call this the "pigeonholing" tendency of personality psychology regarding cognitive style (Fig. 1).

Fig. 1 Pigeon holes

We are not necessarily suggesting here that pigeonholing regarding cognitive style is an automatically sinful activity. Pigeonholing is useful for practically oriented social psychology research yet it does not shed light on the universality of the phenomenon that is being pigeonholed, in this case, individual style.

2.1.2 Theoretical Foundations of Cognitive Style Research, and the Lack Thereof

Also for a second problem related to cognitive style research, we may initially refer to Kozhevnikov's discussion of the literature:

> A (…) problem, as noted in many reviews (…) was the lack of a theoretical foundation for identifying cognitive style dimensions. Most studies of cognitive styles were descriptive, did not attempt to elucidate the underlying nature of the construct or relate styles to information processing theories, and were designed according to the assumption that styles are limited to only very basic information processing operations. [10: 469]

The concept of "cognitive style" as it is used in the pigeonholers' investigations is far removed from the general concept of style as we know it from everyday experience. The type of research carried out on cognitive style pre-defines its object as a mental function: it is related to such "basic" and "complex" activities (to use Kozhevnikov's terms) as perception, concept formation, sorting, categorization, problem-solving, decision-making, learning, and causal explanation. Two assumptions are at work here: (1) stylistically marked activities such as perception or decision-making are mental, cognitive functions, and (2) a meaningful reduction is possible from an individual's general behavior to her cognitive functioning, i.e. cognitive style is a meaningful entity that can be studied independently from other entities. It will be of no use here to extract from these assumptions the theory or theories of mind that underlie the existing cognitive style research. What does need to be noted is that explicit references to such theories of mind and arguments in favor of the possibility of the reduction are lacking.[2] It is in everyday experience that we (as living, perceiving, and (inter)acting human beings) acquire the basic, implicit starting knowledge of the phenomena that our alter-egos (the scientific investigators of human behavior) decide to study. From everyday reality we know individual style as a phenomenon that overarches all of an individual's actions, from her doodling to making life choices. It seems that this basic phenomenon should be elucidated *before* an argumentation for an investigative focus on cognitive style can be set up.

Furthermore, due to the pigeonholing tendency and the lack of theoretical foundation, it seems that cognitive style researchers miss two major points embedded in the original definition of their object:

> (...) a psychological dimension representing consistencies in an individual's manner of cognitive functioning (...) [10]

While the point of cognitive style research is to capture an *individual's* manner of functioning, the pigeonholing tendency aims at categorizing individuals into larger entities, i.e., cognitive types, thus removing and losing, in the end, the individual's particularity which was the investigation's point of interest in the first place. The lack of theoretical foundation, on the other hand, makes it unclear how the consistency of cognitive style relates to the results of the investigation. Does "consistency" imply "predictability?" If so, how far does the predictability derived from one task stretch out? How is consistency "stored?"

2.2 Merleau-Ponty on Style

The human subject of the philosophy of mind has been called many names: a "cognitive system," an "intentional system," a "symbolic system," an "intentional agent," an "organism," a "cognizer," a "perceiver," etcetera. In an investigation of perception or cognition, it seems only natural to define your subject in terms of

---

[2] Sternberg and Grigorenko [19: 700], for example, argue that thinking styles, i.e. the preferred ways of using certain abilities that will be the object of their paper, are elements of cognitive style. Cognitive styles, in their turn, being characteristic modes of processing information, are elements of style, being a characteristic manner of acting. The meaningfulness of the reduction from style to cognitive style, or from cognitive style to thinking style is not argued for. Can thinking styles or cognitive styles be studied in separate, while the data we have to go on does not consist of "the use of abilities" or "the processing of information," but human actions?

your views on how her perceptual or cognitive functions work. However, this "outward" focus on what happens between perceiver and perceived leads the theorist's attention away from her original starting point: the person who does the real, nontheoretical kind of perceiving and cognizing, i.e. a human individual. Jacob and Jeannerod [20: 22], for example, characterize two people in social interaction as "(…) conspecifics, who, unlike inanimate targets of action, can act back". When reading something like this, the layman will wonder why we talk about people as if they're close to the most basic organism you can imagine, defined solely by being alive and by being able to move its limbs. And the layman has a point. Of course, when doing philosophy of mind, it is impossible in words to avoid reducing your subject to her cognitive functioning—this is what we do too. The question is rather whether this reduction is made explicit and whether its consequences have been weighed.

Merleau-Ponty's [3–5] œuvre can be interpreted as an ongoing attempt to thoroughly account for the facticity of human being (its concrete situatedness) in the dialogue between phenomenology and cognitive science. He walks a middle ground by arguing that the biological body (which we expect to provide us with universal features of perception) is fundamentally malleable, so that the distinction between body and mind, nature and nurture, essence and contingency disappears. The software, as it were, gives shape to the hardware and vice versa. The "internal," bodily aspect, or the way the subject has been formed by her life history, is increasingly being accounted for in socalled "embodied" cognitive science. Merleau-Ponty's own analysis of the body's "formedness" undergoes a transformation from his early to his later works, from body schema to style. We will argue that although philosophers of mind and embodied cognitive scientists have picked up on the body schema (or a variation on the theme inspired by a different philosophical background), the added value which (our interpretation of) style has to offer over the body schema has not been accounted for. While the body schema is a general form of situatedness, style adds an aspect of individual situatedness to the human body. We perceive and behave very much like other members of our community, yet we do it in a consistently idiosyncratic way. We argue that this aspect of idiosyncrasy is nontrivial and that it has consequences for some traditional problems in cognitive science.[3]

Merleau-Ponty mainly opposes the rationalist account of human behavior, in which the different parts of the body are considered in objective space, as elements of anatomy and as existing

[3] Merleau-Pontian phenomenology is not to be confused with phenomenological methods applied in some approaches to style, i.e. self-image or analysis and cataloguing of overt behavior in style types (see [10]: 9).

independently and next to each other (as "partes extra partes"). In such an account, the body's harmonious functioning has to be organized by an external factor, a "command central" which overlooks and synchronizes action. The action is orchestrated visàvis the objective features (a certain shape, size, position, etcetera) of the object of my intention. This command central quite literally navigates the body through space, by planning, recording, and controlling the course of her "ship": climbing the stairs, for example, is a complex task involving the maintenance of body equilibrium, the estimation of distances and proportions, the organization of a smoothly alternating amount of pressure applied by left and right upper legs, calves, and feet, a constant monitoring of the effect of the whole and a repeated recalculation of the optimal strategy in light of new information. The navigator does have a memory and can get used to climbing a specific flight of stairs, yet, although perhaps less calculations need to be made or she could even execute the routine blindfolded, each and every command still needs to be given to the different body parts, thus guiding the process. In either way, in the rationalist framework, the organization of even the simplest action requires an enormous (but unconscious) mental effort.

In an embodied framework, action is not guided by the objective features of the perceived object, but by a "motor intentionality." This means that the object (e.g. a glass of water I reach for) does not passively relate to me as a thing waiting to be approached, but instead it elicits from my body a certain action because of its pragmatic value: for me, this glass opens up the possibility that I might drink from it. Gibson [21], giving the concept a "biological" twist, calls these values "affordances": they are properties of objects that are related to the organism's interests.

In spite of their internal differences,[4] philosophers of embodied cognitive science agree that something like the body schema, i.e. a system of acquired interactive worldly knowhow, is at the core of everyday perception and action. It is a direct correlation of our being embodied and, therefore, of our situatedness. Importantly, the schema is literally a property of the body: worldly knowledge lies entrenched in our muscles and these muscles act appropriately to situations without being supervised by a navigator.

When Merleau-Ponty talks about style, he by no means intends to give a qualitative evaluation of a certain form of expression. Style is not something that must be acquired in order to possess it (versus, for example, Robinson's [22] interpretation). Merleau-Ponty's

[4] Gallagher [24: 24] defines the body schema as "(...) a system of sensory-motor capacities [Thompson [25: 411] calls these "dynamic sensorimotor principles"] that function without awareness or the necessity of perceptual monitoring."

concept of style can be described as the subject's mode of bodily being, or, as Singer [23] puts it particularly well: "Style is the (...) modal consequence of being an embodied point of view." For Merleau-Ponty, we do not just coincide with our bodies, but each and every one of us coincides with a particular body, a unique mode of bodily being. Due to our bodies, we are situated in the world in a general and in an individual manner. I behave like other members of the communities I make part of now and in the past (cultural, linguistic, socioeconomic class, education, family, partner, friends...), yet at the same time I am a unique individual, a unique constellation of such memberships. Style is reflected in everything we do: the way we walk, cook, drive, reason, joke, etcetera. We deal with the world in our idiosyncratic, yet largely shared way. This largely shared but particular way of being is important for what we perceive, what we do and how we do what we do.

The reader might wonder: how does style then differ from the body schema?

Strictly speaking, body schema and style are the same thing. This allows Gallagher [26: 549] to say that "the body schema (...) is precisely the style that organizes the body as it functions in communion with its environment." However, while the former description does allow for an interpretation which accounts for the particularity of the molded body (we all behave in a slightly different manner), the latter description stresses this aspect of unique individuality. Body schema refers more to form, while style refers to content.

The visualization in Fig. 2 is far from exhaustive, yet it will help understand the position of style in cognitive science. The vertical axis represents the communities we make and have made part of,

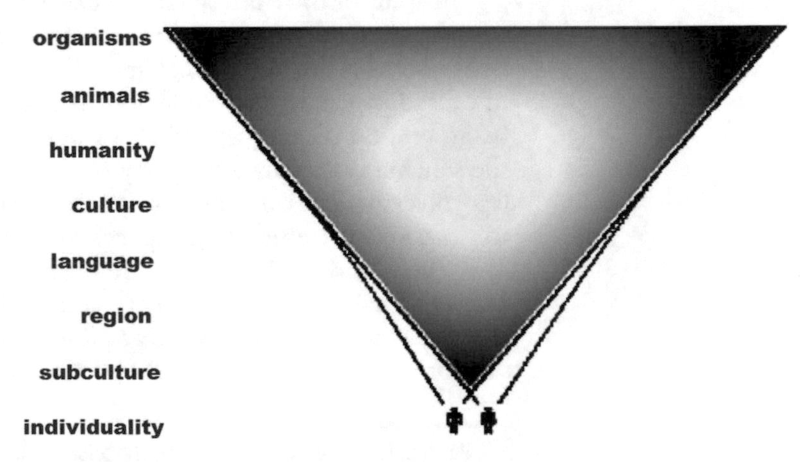

Fig. 2 "Style Funnel 1" showing total situatedness at any time and how two individuals find separate styles (differing slightly in language, more in region, still more in subculture, and most of all, not surprisingly, in individuality.) From Quaeghebeur [27]

with at the lower extreme the individual, a "community" of one (of which we picture two specimens). The horizontal axis represents the set of bodily knowhows we have acquired because we are or have been a member of this or that community. The gray area represents the bodily knowhow we share. At the level of the organism, we share, for example, the fact that we are autonomous agents, as Thompson put it, or the fact that each and every one of us (organisms) has a spatiotemporal location. In a very basic sense, I differ from other people due to the fact that mine and their bodies can never take up the same space at the same time (or actually, this feature belongs to worldly objects, and therefore also to organisms). The difference between an organism and an animal, i.e. the question what all animals share, is a complex one. With a view to perception and action, however, we can state that, at the level of the animal, what seems relevant is that we are capable of movement (although there may be a few exceptions) and we are capable of sexual reproduction (although some animals are capable of asexual reproduction). The question of what it is that differentiates humans from animals may be an even more difficult one. We share a bodily makeup, that's for sure. We all seem to have language and a set of existential beliefs, and contrary to animals, we are capable of blushing. From the level of humanity downwards is where people start differing from each other. We differ biologically with respect to age, gender, skin color, and sexuality. We are part of a culture, i.e. a very large group of people who share certain beliefs and practices having to do with the family, society, food, economy, male–female relations, religion, etcetera. We speak a certain language and belong to a social class. We have attended certain schools and have developed certain interests that we share with friends. We come from a family and have shared a certain upbringing with our siblings. We share a relationship with our partner, and finally we ourselves are the unique conglomerate of all these memberships.

In this paper we discuss personal style in the form in which it is scientifically most easy to access: as multimodal expression (i.e. language use as embodied action, including not only speech, but also, among other modalities, the use of gesture). Stylistically speaking, we express our intentions like we carry out all our activities: much like those close to us, but with a personal twist. The language (including gesture), dialect and genre I use, or am capable of using, is dictated by what I learned during childhood, at school, at work, etc., in other words, by my memberships in different linguistic groups. In this sense, I am like others. But I differ from them in the way in which I mix all of these influences together in a specific phrasing of a specific intention at a specific point in time: though mostly consistent, I have my very own linguistic style.

3 Style and Gesture

Given the history of cognitive style research sketched earlier, and crucially in the light of Merleau-Ponty's concept of style as an all-encompassing mode of being in a given situation, we adopt a two-pronged empirical approach based on new concepts of thought and action, and how they intertwine. This approach we then suggest motivates an extension by use of one feature of *THEME*, the associate.

3.1 Multimodal Expression: Thought in Action

Gestures are a unique source of information for the empirical investigation of individual styles. They are closely linked to spoken language but largely outside conscious regulation and linguistic convention. Thus the speaker can display aspects of individually marked behavior without the filter of knowing adjustment. The kinds of gestures we mean are not the familiar "emblems" such as the "OK" sign, but the unconscious hand, arm, and head movements seen with speech. Such gestures open a window onto thought beyond that provided by speech itself. Gestures have both a content and a mode of semiosis that, with training and experience, can be "read." The gesture, though closely linked to speech, differs from speech and this is the basis of many new insights. The most crucial of these is the concept of the *growth point* or GP. GPs are considered to be the smallest dynamic units of language and thought. In a GP gesture synchronizes with speech to form an unbreakable unit, a unit inherently multimodal—at the same time, on its verbal side, semiotically segmented, analytic and combinatoric and, on the gestural, semiotically global, synthetic, and non-combinatoric. Having the same meaning simultaneously in opposite semiotic modes is unstable, and fuels verbal-conceptual change on a microgenetic time-scale. Seeking closure the GP is "unpacked" into a static, i.e., a stable, grammatical form (a grammatical construction is the dialectic's "stop-order"). The GP and its unpacking effect is an output of the "style funnel" resolved down to a fraction of a second. Figure 3 is a depiction. More broadly, a GP starts with an individual's differentiation of a significant opposition in a context or field of oppositions, and ends with a socially constituted "unpacking." Still more broadly, it starts with creativity and ends with conformity, and perhaps in this is a microcosm of the individual's position as an agent in a social framework. The dual semiotic can also be likened to Mead's "I"/"me"—the self at the verge of creativity and conformity, poles alternating endlessly (see Mead [28]).

3.2 Mimicry as Method

To access styles, we use *mimicry as a method of observation*. By replicating the speech and gesture of another you can recreate the person's GP. The method rests on the insight that gestures are

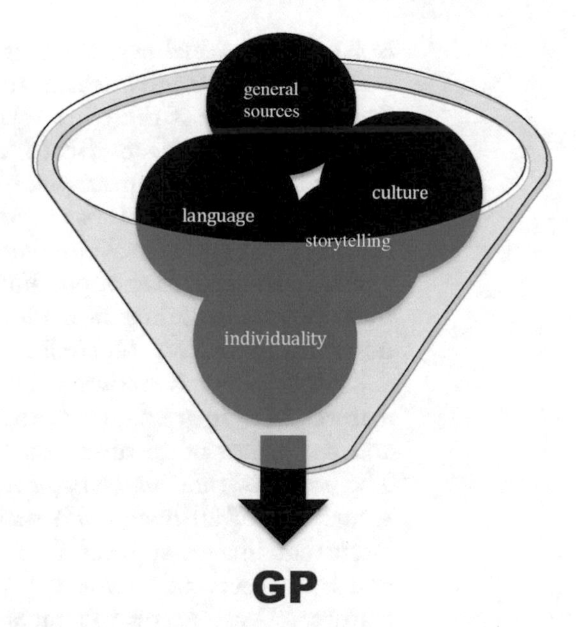

Fig. 3 "Style Funnel 2" showing situatedness as a momentary state of cognitive being in speaking. Also shows the motivation of THEME "associates." The growth point idea unit is the temporal focus point of individual style. The speaker differentiates a field of oppositions that she has in part constructed to make this differentiation meaningful. The idea unit consolidates everything flowing down the funnel—culture, language, storytelling, individuality, all else that comprises the speaker's individual being—and focuses it "all at once" (see [27])

actual components of speech, not accompaniments or "add-ons" (Adam Kendon's [29] term), but integral parts of it. Mimicry fosters a kind of mind-merging whereby one mind (ours) melds with another. If you further ask in what context this gesture and speech combination could comprise a newsworthy point of differentiation, you also get at the context. Mimicry is a kind of *borrowed embodiment*. Gesture coders spontaneously mimic to get at meaning. Here, we mimic entire narrative passages, speech, and gesture, keeping as close to the model as possible. In the exercise we experience forms of organizing meanings not our own, a kind of validation (we are not projecting our own experience into the video).

3.3 The Catchment

In addition to mimicry, which taps our own experience of the speaker's field of oppositions, we can use the speaker's own gestures to identify context via the *catchment*. The catchment metaphor is the land area draining into (being differentiated by) a body of water (the idea unit). Visualize someone speaking during some more or less extended stretch of discourse. The first thing to notice is that gestures occur more or less continuously. Gestures are not isolated, occasional events—they take place roughly one per second over long periods. It thus makes sense to examine bundles of gestures and see if there are relations among them. This kind of examination has given rise to the catchment concept. A given catchment could, for example, be defined by the recurrent use of the same trajectory and space with variations of hand shapes within the trajectory and space. This would suggest a discourse unit within which meanings are grouped but with contrasts among them. All these points, including how idea units and catchments relate, are described in detail in McNeill [7].

A catchment is recognized when two or more of such gesture features recur in at least two (not necessarily consecutive) gestures, and the group of gestures relates to a cohesive discourse theme. The logic is that *imagery generates the gesture features.* Thus, recurrence of an image will generate recurrent gesture features. Recurrent images suggest a common discourse theme. In other words, a discourse theme will produce gestures with recurring features. These gesture features can be detected. Then working backwards, the catchment of recurring features offers clues to the cohesive linkages in the text with which it co-occurs. A catchment is, in other words, a kind of thread of visuospatial imagery that runs through a discourse to reveal the separate parts that cohere into larger discourse units. By discovering the catchments created by a given speaker, we can see what this speaker is combining into larger discourse units—what meanings are being seen as similar or related and grouped together, and what meanings are being put into other catchments or are being isolated, and thus are seen by the speaker as having distinct or less related meanings. Individuals differ in how they link up the world into related and unrelated components. Catchments give us a way of detecting these individual characteristics, or individual styles.

4 Experiments on Style

Using a quasi-experimental approach of animated cartoon narrations, we observe catchments and styles at work at the level of individual utterances. In theory, these are specific moments in a speaker's cognitive being. Our method is to show Canary Row, a 1950 Warner Brothers Tweety and Sylvester classic, and have the participant immediately recount the story from memory to a second participant who has not seen the cartoon. The participants were told the experiment was a study of storytelling. Emphasis was on accuracy and completeness (the narrator was told her listener would retell the story based on what she recounted). There was no mention of gesture (cf. McNeill [6] and references therein for details of coding). Our narrators, identified as Speaker 1 and Speaker 2, were describing (independently) an event from the cartoon in which one character (Sylvester, a cat) attempted to reach another character (Tweety, a canary) by climbing the inside of a drainpipe that conveniently topped off next to the window ledge where Tweety had perched. Tweety however sees Sylvester coming, rushes off screen and reappears with an improbably large bowling ball that only in the cartoon world he could support, which he drops into the pipe. Halfway down bowling ball and Sylvester meet explosively and Sylvester is next seen shooting out the bottom of the pipe, the bowling either inside him or under him. Speaker 1 and Speaker 2 have different catchment organizations of this episode. For Speaker 1 the main catchment is interrupted to make an ironic comment. For Speaker 2 the whole episode is a metaphor of morally opposed forces. Each adds something extra, but this extra differs between then and is an aspect of each speaker's individual style.

4.1 "Ironic" Style

The example picks up where Speaker 1 describes the bowling ball and its fall (Fig. 4), the last of what had been until then a straight narrative description and right before the ironic aside. She interrupts herself to make the aside—"you can't tell if the bowling ball is under Sylvester or inside of him"—which occupies the rest of the figure until the end, when she returns to straight narrative. The aside has its own catchment. The first panel of Fig. 4, the narrative account before irony began, occupies the central pole of the gesture space. The ironic aside is literally "to the side"—the gestures are all in the lower left corner of the gesture space; this is its new catchment. The final panel, the resumption of the narrative line, is also back to the central pole, and the narrative catchment returns.

Using mimicry, we evoke a vivid experience of the ironic style. It is ironic in the sense that the speaker, in the midst of describing Sylvester's catastrophic encounter with the bowling ball, pauses to identify what she considers to be an epistemological problem—whether, after the bowling ball and Sylvester collide, it is inside or under him—a matter she felt the cartoon makers had left unclear. Other descriptions with this kind of ironic aside occur elsewhere in Speaker 1's narrative. The style seems characteristic. We analyze this interlude to show how speech and gesture actualized irony.

The hesitations in (2) (silent "/" and filled "ah") are key elements—they mark the moment Speaker 1 shifts from straight storytelling to ironic detachment. As she poses the ambiguity her left hand rises slightly and her gaze tilts up into the visual space shared with her listener, two features that suggest a new discursive unit.

[/ and it **goe**s <u>dOWn</u>] Narrative catchment. Two hands, cupped in shape of either the bowling ball or the pipe (or both), move down – the panel shows the beginning of the gesture. Gaze directed into lower central space, for the speaker her "thought" space.	
and it* [/ <u><ah></u> The irony catchment. One hand (rather than two) in lower left corner (circled). Gaze shifts out of "thought" space, marking start of ironic aside. This possibly is more an "interaction" space, as the speaker shifts to the aside, a kind of bookend of the ironic style.	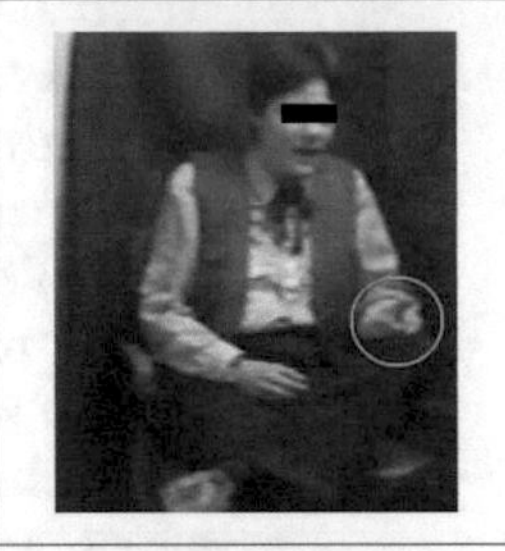
<u>you ca</u>] [[**an't tell**] [if the **bOW**<u>ling bAll /</u>]] [[is **UNd**<u>er Sylvester</u>] Left hand in "irony" apace; right hand goes to rest. Gaze shifts back to lower center "thought" space. The bowling ball in its <u>under</u> Sylvester location within the "irony" aside.	
[or **ins**<u>Ide</u> of him]] The bowling ball in its '<u>inside</u> location. This is anti-iconic, in that the hand moves outward, but is metaphorically appropriate as imagery of contrast – "inside" versus "under" – that is the ambiguity she proclaims. Gaze shifts back to interlocutor space: marks end of ironic aside, the second bookend.	
but [$_8$[it **rOll**] $_9$[s **him OUt***]] Returns to narrative line. Back to central catchment, and gaze back into "thought" space.	

Her gaze (3 "you can't tell") then returns to the lower-center narrative space of (1) where it remains through (4 "if the bowling ball," 5 "is un..." and 6 "is under Sylvester"), whereupon it shifts briefly again to the shared space (7 "or inside of him"), and then back to lower-center for the end of our example, (8) "but it rolls" and (9) "him out." Thus, two spatially defined catchments, one of which is an ironic aside. The details of space, gaze, and phonology are items we propose that *THEME* could detect in narrative gesture and speech transcripts.

4.2 "Moral" Style

Again using the mimicry method, Speaker 2, in her narrative approach, emerges as unswervingly sequential. Still, in her speech and gesture, we detect a kind metaphoric thinking in which the episode is cast in moral terms, "The Good" (here the bowling ball as Tweety's surrogate) versus "The Bad" (Sylvester). The clue to this perspective is how she synchronizes a specific gesture with two segments of speech. The gesture depicts Tweety launching the bowling ball into the pipe. At the same time she is saying, "Tweety Bird runs and gets a bowling ba[ll and drop<u>s</u> **it do<u>wn</u>** the drainpipe]," with the gesture phrase and stroke as indicated—the stroke timing exactly with "it down" and skipping over its nearby lexical affiliate, "drops," which identifies this very action. This verb was uttered during the continuing preparation for the gesture stroke and may even have been touched with a brief prestroke hold, as if the stroke, now cocked, was waiting until "drops" had completed and "it down" could begin. This overshoot, if that is what it is, is our clue. Suppose the "it" and "down" were the core meaning of the idea unit. Then the stroke was not an overshoot but a precise synchronization. What meaning could make this idea unit, the downward thrusting gesture and "it" and "down" the core? Speaker 2's catchments suggest an answer: the bowling ball was not just an object but was a force against Sylvester in what for

Fig. 4 Ironic gesture style. Gesture and speech notation in Susan Duncan's standard transcription: Square brackets indicate the beginning and end of the gesture phrase (Kendon [30]), the interval where the hands are actively engaged in performing the gesture. '/' indicates fraction-of-second silence, "#" indicates an audible breath pause. Font size indicates prosodic emphasis. Boldface indicates the gesture stroke, the meaningful phase of the gesture, here the hands moving down (the interval from the '[' and the start of boldface is the preparation phase of the gesture. Underlining is a gesture hold, a temporary cessation of movement, here a poststroke hold where the hands freeze in midair holding the stroke shape and final position (not present in this example, but an interval from the end of boldface or underlining and the ']' is the retraction phase (in this case, the hands went directly to rest)

Table 1
Speaker 2's catchment themes

Catchments	Utterances
C1 ("Sylvester solo") One-handed gestures tie together references to Sylvester as a solo force.	(1) he tries going up the inside of the drainpipe and (6) [and he comes out the bottom of the drainpipe
C2 ("moral") Two-handed symmetrical gestures group descriptions where the bowling ball is the antagonist, the dominant force. The 2-handed form echoes the shape of the bowling ball.	(2) Tweety Bird runs and gets a bowling ball and drops it down the drainpipe (7) and he's got this big bowling ball inside him (8) and he rolls on down into a bowling alley (9) and you hear a strike
C3 ("inside the pipe") Two-handed asymmetrical gestures group items in which the bowling ball (LH) and Sylvester (RH) are equals differing only in their placement and direction of motion.	(3) and as he's coming up (4) and the bowling ball's coming down (5) he swallows it

Speaker 2 was a moral battle of The Good versus The Bad. She had several gestures with this theme based on hand use. The moral catchment we term "two similar hands," in which the hands outline the shape of the bowling ball. The first of these two-similar hands gestures accompanied "it down."[5]

Table 1 lists the catchments of this episode description, the "moral" catchment (**C2**, "two-handed symmetrical") and two others (**C1**, "one-handed" and **C3**, "two-handed asymmetrical"). The "moral" catchment conveyed instances where the bowling ball was acting as an antagonistic force in Tweety's interests, and occurred with each of the following:

(2) ba[ll and drops **it do**wn the drainpipe]

(7) […and he's **got thi**s big bowling ball inside h]im

(8) [and h**e rolls on down**]

(8 continued) [into **a bow**ling all]

(9) [ey and then **you hear a** sstri]ke

Thus Speaker 2 was not bothered by Speaker 1's epistemological puzzle, just as Speaker 1 did not regard the episode as a moral drama.

When the experiment was over and the tape recorder about to be shut off, Speaker 2 called out to the experimenter, "Do you want analysis?", an instance of her own irony but an irony not

[5] For ease of exposition we describe this example in an order other than Speaker 2's actual order. The spoken order can be recovered from Table 1. See McNeill [7], Fig. 2.8, for the entire narration.

where Speaker 1 had produced it, in the midst of an episode. The fact that Speaker 2 put irony at the end while Speaker 1 dropped it into the middle of a scene, indicates different styles of discourse with possible attendant "associated" features in the *THEME* sense. For Speaker 1 irony was allowed to intrude; for Speaker 2 sequence was sacrosanct and irony was allowed only after the story was done. Thus one negative associate of irony for Speaker 2 is temporal sequence. This in turn affects what can be the target of the irony, the epistemology of a narrative event or "analysis" as a kind of practiced universal metapragmatics (cf. Silverstein [31]).

5 The Upshot

We have seen two selective effects of style, influencing both what is included and what is not—Speaker 2's Good vs. Evil, which had no place in Speaker 1's sense of irony and, vice versa, Speaker 1's irony which would have neutralized, were it included, Speaker 2's moral drama (not to say that Speaker 2 did not also add ironic touches, but they were not present in this scene).

5.1 THEME "Associates"

We are not equipped to carry out a *THEME* analysis but we can note how it could empirically pinpoint instances of personal style in this kind of investigation. It can show in detail what a given style selects and what it excludes. A wider array of features than the unaided eye can see can be accessed. We imagine inputs of good speech plus prosody and gesture transcripts of the kind demonstrated, grammatical analyses at the clause level, finer descriptives for functions of a speech-act kind, and glosses like "irony," "moral drama," "metaphor," "narrative sequence" and others. We ask *THEME* what positive and negative associates does a given a narrative sequence carry? With the kinds of examples we have shown, we could direct *THEME* to look for associates of patterns such as narrative interlarded with irony versus narrative followed by it, etc. *THEME* finds both positive and negative associates—those that occur with a catchment, and those that tend to avoid it. Both seem close to the idea of personal style, what the person, over time, tends to gravitate toward or tends to avoid.

Figure 3 captures our sense that the moment of expression of a personal style is the funneling of a host of determinants. We spot some of these through normal transcription, but we suspect that many are hidden from unaided view and that *THEME* will particularize more over a much wider range of behavioral particulars..

Magnus Magnusson on the THEME "associate" (notes from University of Chicago lecture in 2010): A positive or negative (+/−) associate of a T-pattern is: some behavior that is not a part of that pattern, but occurs within or around significantly more of its occurrences (positive) or less (negative) than expected by chance. These we propose are among the style "balls" of Fig. 4 funnel.

Using this idea, some likely associates with the ironic and moral style speakers, based on observation:

Ironic narrator: detachment and amused irony permeate every moment. For her, irony is a + associate of the narrative scheme.

Moral narrator: ironic comment is separated from narrative events. As noted, this form of irony is a – associate to the narrative scheme.

The two styles in turn open the door to different ***kinds*** of irony—individualized events, by necessity with Speaker 1; dissociated from any single event with Speaker 2. Each style is typified by a + associate that is the – associate of the other. So, overall, there is "irony" for both but its form differs.

With an actual *THEME* analysis of narrative and other forms of discourse, we expect to see many more such plus and minus features, and from them gain new insights into the embodiment of personal style that avoids pigeonholing, rests on uniform theoretical-empirical foundations, and builds on Merleau-Ponty's insights into the moment-by-moment situatedness of being.

References

1. Hampson SE (1988) The construction of personality: an introduction. Routledge, London
2. Brunas-Wagstaff J (1998) Personality: a cognitive approach. Routledge, New York
3. Merleau-Ponty M (1962) Phenomenology of perception. Colin Smith (trans.), Routledge, London
4. Merleau-Ponty M (1964) Signs, Richard C. McCleary (trans. and introduction). Northwestern University Press, Evanston
5. Merleau-Ponty M (2006) The structure of behavior. Alden L. Fisher (trans.), John Wild (forward.), Duquesne Press, Pittsburgh
6. McNeill D (1992) Hand and mind: what gestures reveal about thought. University of Chicago Press, Chicago
7. McNeill D (2012) How language began: gesture and speech in human evolution. Cambridge University Press, Cambridge
8. Allport GW (1937) Personality: a psychological interpretation. Holt, New York
9. Riding R, Cheema I (1991) Cognitive style—an overview and integration. Educ Psychol 11:193–215
10. Kozhevnikov M (2007) Cognitive styles in the context of modern psychology: toward an integrated framework of cognitive style. Psychol Bull 133:464–481
11. Klein GS (1951) A personal world through perception. In: Blake RR, Ramsey GV (eds) Perception, an approach to personality. Ronald Press Company, New York
12. Witkin HA, Lewis HB, Hertzman M, Machover K, Bretnall PM, Wapner S (1954) Personality through perception, an experimental and clinical study. New York, Harper & Brothers
13. Kirton MJ (1976) Adaptors and innovators, a description and measure. J Appl Psychol 61:622–629
14. Gregorc AF (1979) Learning/teaching styles, potent forces behind them. Educ Leadersh 36:234–236
15. Kolb DA (1976) Learning style inventory: technical manual. Prentice Hall, Englewood Cliffs
16. Allinson J, Hayes C (1996) The cognitive style index: a measure of intuition-analysis for organizational research. J Manag Stud 33:119–135
17. Sternberg RJ (1988) Mental self-government: a theory of intellectual styles and their development. Hum Dev 31:197–224
18. Nosal CS (1990) Psychologiczne modele umyslu [Psychological models of mind]. PWN, Warsaw
19. Sternberg RJ, Grigorenko EL (1997) Are cognitive styles still in style? Am Psychol 52:700–712

20. Jacob P, Jeannerod M (2005) The motor theory of social cognition: a ritique. Trends Cogn Sci 9:21–25
21. Gibson JJ (1979) The ecological approach to visual perception. Houghton Mifflin, Boston
22. Robinson J (1985) Style and personality in the literary work. Philos Rev 94:227–247
23. Singer L (1981) Merleau-Ponty on the concept of style. Man World 14:153–163
24. Gallagher S (2005) How the body shapes the mind. Clarendon, Oxford
25. Thompson E (2005) Sensorimotor subjectivity and the enactive approach to experience. Phenomenol Cogn Sci 4:407–427
26. Gallagher S (1986) Body image and body schema: a conceptual clarification. J Mind Behav 7:541–554
27. Quaeghebeur L (2010) A philosophy of everyday, face-to-face conversation. University of Antwerp, Antwerp
28. Mead GH (1974) Mind, self, and society from the standpoint of a social behaviorist, C. W. Morris (ed. and introduction). University of Chicago Press, Chicago
29. Kendon A (2008) 'Some reflections on the relationship between 'gesture' and 'sign.''. Gesture 8:348–366
30. Kendon A (1980) Gesticulation and speech: two aspects of the process of utterance. In: Key MR (ed) The relationship of verbal and nonverbal communication. Mouton and Co., The Hague, pp 207–227
31. Silverstein M (2003) Indexical order and the dialectics of sociolinguistic life. Lang Commun 23:193–229

Part II

Animal and Neuronal Behavior (Non-human Behavior)

Chapter 12

Application of T-Pattern Analysis in the Study of Rodent Behavior: Methodological and Experimental Highlights

Maurizio Casarrubea, Magnus S. Magnusson, Giuseppe Di Giovanni, Vincent Roy, Arnaud Arabo, Andrea Santangelo, and Giuseppe Crescimanno

Abstract

In our laboratories we use T-pattern analysis to study rat behavior in different and well-known experimental assays widely employed as rodent models of anxiety: the open field, the hole board and the elevated plus maze. By using Theme software and T-pattern analysis, we have observed that numerous events, characterizing rodent behavior in each experimental model, occurred sequentially and with significant constraints on the interval lengths separating them. In this chapter, for each test, we highlight some key aspects of our behavioral analyses, with a twofold attempt: first to provide the researcher with useful information concerning the application of T-pattern analysis in the study of rodent behavior and, second, to present and discuss various results of our studies.

Key words Multivariate analysis, T-pattern analysis, Anxiety, Rat, Open field, Hole board, Elevated plus maze

1 Introduction

The first step in the experimental study of human or animal behavior is commonly represented by the construction of a formal list containing descriptions of individual behavioral components. Such a formal list, namely an ethogram, dismounts the observed behavior

This chapter is dedicated to the memory of my father Prof. Giuseppe Casarrubea (March 4th 1946—June 7th 2015), historian and writer. His meticulous and patient work has clarified numerous and obscure aspects in the modern history of Italy.

As historian, he was fascinated by concepts and theories concerning T-patterns and the related possibility to study repetition of events from a scientific point of view.

As father, he would have been happy to read this book and proud to see his son Coeditor with such prestigious Colleagues and Authors.

Maurizio Casarrubea
Palermo, June 15th 2015

Magnus S. Magnusson et al. (eds.), *Discovering Hidden Temporal Patterns in Behavior and Interaction: T-Pattern Detection and Analysis with THEME™*, Neuromethods, vol. 111, DOI 10.1007/978-1-4939-3249-8_12,

into discrete components that, in turn, can be characterized by means of latencies, durations, per cent distributions, and so on. In other terms, by using a quantitative approach, each component of a given behavioral repertoire can be quantified and described with numbers. In the analysis of behavior, quantitative evaluations are useful because changes of specific parameters do often provide valuable information; for instance, it could be useful to appreciate the frequency of a specific behavioral element and its modifications following the administration of a drug. On the other hand it goes without saying that the behavior of a living being is much more than simple frequencies, latencies, or durations of individual elements, disjointed from the comprehensive behavioral structure: the meaning of the behavior lies in the relationships among its constitutive components [1, 2]. As a consequence, a thoughtful approach to the study of behavior should take into consideration suitable analytical tools able to assess these relationships. Such a crucial aspect calls for different means of detection, data handling, and analysis.

The terms "multivariate analyses" are used to indicate a set of techniques aimed at the assessment of data sets with more than one variable. These methods were greatly developed only along the last three decades because they often require the computational support of modern computers and specific software. The great advantage of a multivariate approach is the possibility to assess the behavior in terms of underlying interrelationships among the behavioral elements. In addition, all multivariate techniques share the possibility to describe behavioral dynamics otherwise undetectable by means of quantitative assessments. On the basis of these features, multivariate analyses have been considered essential tools to study the structure of animal behavior in several experimental assays such as the hot plate [3–6], the open field [7], the hole board [8–10], the elevated plus maze test [11, 12], or the forced swimming test [13].

Different multivariate approaches are available. For instance, cluster analysis, stochastic analysis, or adjusted residuals analysis are multivariate techniques based on the elaboration of transition matrices. A characteristic of these methods is that they explore the comprehensive observational time window providing little information on the temporal structure of the behavior. Actually, the lack of information concerning the temporal characteristics is a common aspect of various multivariate analyses applied to the experimental study of behavior. To fill this gap the T-pattern analysis can be used. Such a multivariate technique has been developed to determine whether two or more behavioral events occur sequentially and with statistically significant time intervals [14, 15]. T-pattern analysis has been successfully used to study behavioral modifications in neuro-psychiatric diseases [16], route-tracing stereotypy in mice [17], interaction between human subjects and animals or artificial agents [18], hormonal-behavioral interactions [19], feeding behavior in broilers [20], patterns of behavior associated with emesis [21] and, in our laboratories, to investigate exploration and anxiety-related behaviors in rodents [2, 22–24]. The present article will discuss the application of T-pattern analysis in the study of rodent's anxiety-related behaviors in three different behavioral assays. Various aspects of our behavioral analyses will be highlighted with a twofold aim: first to provide the reader with methodological information concerning the application of T-pattern analysis in the study of rodent behavior and, second, to discuss various results of our studies.

2 Methods

2.1 Apparatus

In the present chapter the T-pattern analysis has been carried out to assess rodent's behavior in three different experimental assays: the open field, the hole board, and the elevated plus maze. Common characteristics of these tests are the relatively low cost and simple testing procedures.

The *Open Field* (OF) needs little introduction being the most used experimental assay in laboratories of behavioral sciences and animal psychology. This experimental apparatus generally consists of an enclosed circular, square, or rectangular perimeter where freely moving rodents are observed for a limited period of time. The OF is commonly employed to study exploration [25] and anxiety-related behaviors [26, 27]. The rationale supporting the utilization of OF in the study of anxiety lies in the natural rodents' aversion for novel environments and unprotected areas. Indeed, once placed in the OF, rats spontaneously prefer the periphery, remaining close to the surrounding walls (a phenomenon known as "thigmotaxis"). Increase of time spent in the central zone as well as the increase of the ratio central/total locomotion or the decrease of the latency to enter the central zone represent widely accepted indexes of anxiolysis [26, 27]. The OF used in the present study consisted of a square arena 50×50 cm made of white opaque Plexiglas floor surrounded by three white opaque walls and a front transparent one.

The *Hole Board* (HB), similarly to the OF, is an exploration-based assay commonly used to examine various features of anxiety-related behaviors in rodents [8, 28–34]. This experimental apparatus consists of a square or rectangular arena with a variable number of holes in the ground [28, 35, 36] where a rat (or a mouse) can insert its head. Excluding modified HBs [37], the presence of the holes represents the essential difference between an OF and a HB. Changes of head-dipping behavior (frequency, latency, duration) reflect the anxiogenic and/or anxiolytic state of animals: anxiogenic drugs decrease both the number and duration of head-dips [38], on the contrary, if anxiolytic drugs are administered, increases in the number and duration of head-dips are observed [38]. The HB used in the present study consisted of a square 50×50 cm arena made of white opaque Plexiglas with a raised floor positioned 5 cm above a white opaque Plexiglas subfloor and containing four equidistant holes, 4 cm in diameter. Each hole center was 10 cm from the two nearest walls so that holes were equidistant from adjacent corners. The arena was surrounded by three white opaque Plexiglas walls and a front transparent one.

The *Elevated Plus Maze* (EPM), introduced by Handley and Mithani [39], is a widely used model to assess anxiety-related behaviors in rodents. Basically, the apparatus consists of an elevated plus-shaped platform characterized by the presence of two open and two enclosed arms. EPM usefulness has spread towards the understanding of the biological basis of emotionality related to learning and memory, hormones, addiction, and withdrawal [40]. The rationale underlying the utilization of EPM in the study of rodents' anxiety-related behaviors is based on the assumption that rodents exposed to the apparatus will respond to a conflict elicited by the presence of safe parts of the maze that are closed and protected, and aversive parts of the maze that are open, unprotected, and more brightly lit [40]. The apparatus we used was elevated at a height of 50 cm above the floor [24, 41]. The closed arms were surrounded by a 50 cm wall while open arms presented 0.5 cm edges in order to maximize open-arm entries [42]. The floor of the maze was covered with grey plastic.

2.2 Subjects

Observations have been carried out on 30 specific pathogen-free male Wistar rats divided in three groups. Each group, encompassing 10 animals, was utilized for the observations in one experimental apparatus. All subjects were housed in a thermoregulated room, maintained at constant temperature. In their home cages all animals had free access to food (standard laboratory pellets) and water.

2.3 Experimental Procedures

To minimize transfer effects and avoid possible visual or olfactory influences, rats were transferred from housing room to testing room inside their own home cages and allowed to acclimate for 30 min far from the experimental apparatus. Environmental temperature in testing room was maintained equal to the temperature measured in the housing room. Concerning the OF and the HB, each rat was placed in the center of the arena and allowed to freely explore for 5 and 10 min respectively. Concerning the EPM, each rat was placed in the central platform facing an open arm and allowed to freely explore for 5 min.

All rats, experimentally naïve, were observed only once. After an observation, the apparatus was cleaned with ethylic alcohol to remove possible scent cues left by the animal. Rodents' behavior was recorded through a digital camera, and video files were stored in a personal computer for following analyses. Concerning HB and

OF the camera was placed in front of the apparatus. As to EPM the camera was placed above the apparatus.

2.4 Ethical Statement

All efforts have been carried out to minimize the number of animals used. All the experiments here described have been conducted in accordance with the European Communities Council Directive 86/609/EEC concerning the care and use of animals for scientific purposes.

2.5 Ethogram and Coding

In the present research behavioral observations have been carried out on the basis of the ethograms we have employed in our recent studies [2, 22–24]. From a methodological point of view it is important to underline that establishing an ethogram is always a critical moment because an error (e.g. a behavioral element not described or, worst, misinterpreted) is potentially able to negatively influence the comprehensive analysis. Such a statement might appear exaggerated until one does not consider that the "raison d'être" of a multivariate behavioral analysis lies in its ability to describe interrelationships among individual components. Notably, this is even more important if the multivariate approach used is the T-pattern analysis: actually, an event can be uncommon (e.g. occurring only few times for each subject and/or not in all subjects) nonetheless the temporal relationships it establishes can be extremely important for the behavioral architecture. However, once video files have been collected and the ethogram is ready/available, the following step is normally represented by the coding process, i.e., the utilization of specific software that allows the researcher to record the occurrences of all the behavioral elements performed by the actor. The result of the coding process is an event log file that is, in its simplest form, a text file containing a sequence of behavioral events occurring at specific time points (milliseconds, seconds or, even, video frames). In the present study all video files have been coded using The Observer (Noldus Information Technology BV, The Netherlands).

2.6 Search Procedure

T-pattern analysis can be carried out by means of Theme™ software (PatternVision Ltd, Iceland; Noldus Information Technology BV, The Netherlands). This software, by means of a sophisticated detection algorithm, processes event log files evaluating possible significant relationships among the events in the course of time [14]. The search advances following a bottom-up process. In brief, being A, B, C three hypothetical events occurring in a given event log file, the algorithm compares the distributions of each pair of the behavioral elements A and B searching for a time window after A such that more occurrences of A contain B than expected by chance. In this case A and B are indicated as (A B) and form a T-pattern. After that, such first level T-patterns are marked and considered as potential A or B terms in higher patterns, for example, ((A B) C). Thus, more complex patterns may be created, step by step, following this bottom-up detection process. The search is completed when no more patterns are found. More details concerning theories and concepts behind T-pattern analysis can be found in various chapters of this book and/or in our previous articles [14, 15, 22–24].

2.7 Search Parameters

To perform a search for T-patterns, Theme™ requires specific parameters. Crucial is the "significance level" (i.e. the maximum accepted probability of any critical interval relationship to occur by chance). Extremely small values of this parameter (e.g., $p = 0.0000001$) are often useless because will lead to the detection of very few and short patterns (that is, for instance, T-patterns encompassing only two events) or, more probably, no patterns at all. On the contrary, higher values (e.g. 0.05) may produce many more and longer patterns. Thus the selected significance level strictly depends on the available data that need to be analyzed. In the analysis of rodent behavior we've found that values of 0.0001 and 0.005 work very well. Additional and important parameters with a substantial impact in the detection of T-patterns are the "minimum occurrences" (i.e. the minimum number of times a T-pattern must occur to be detected), the "lumping factor" (i.e. forward and backward transition probability above which A and B of a T-pattern (A B) are lumped, that is, A and B are not considered separately but only as the (A B) pattern) and the "minimum samples" (i.e. the minimum percent of subjects in which a pattern must occur to be detected). It is important to remember that the "minimum samples" parameter has a particular relevance when samples have been concatenated. Indeed, Theme™ is able to concatenate all event log files into a single file. Such a joining procedure is very useful because it makes possible the detection of patterns that may occur only once in each event log-file and/or, possibly, not in all samples. After the concatenation of individual log files, by setting the appropriate value in "minimum samples", uncommon but possibly interesting patterns (non detectable by analyzing each individual log file) may be detected. On this subject, since behavioral observations with rodents are normally carried out with a reasonable number of subjects (e.g. 10 or 15 rats per group) it is clear that the coding process will produce a certain number of event log files that can be concatenated to search for uncommon patterns. In the present research, coherently with our previous studies [2, 22–24], the following search parameters have been employed:

- "Significance level" = 0.0001 (0.005 in OF observations)
- "Lumping factor" = 0.90
- "Minimum samples" = 50 % (100 % in EPM observations)

2.8 Statistics

Albeit each critical interval implies the existence of a statistical significance, the enormous number of possibilities of such relationships in data with several occurrences of behavioral events might raise the question whether the detected T-patterns are there only by chance. Theme™ deals with this important issue by randomizing and analyzing the original data: using the same search parameters as with the real data, the average number of patterns detected in the randomized data is then compared with that obtained from the original data. Such a randomization process is essential because if during the assessment of one or more event log files the detected T-patterns in the real data are not significantly different from the number of patterns detected in the randomized data, then it is likely that too permissive search parameters have been used and, in brief, the detected T-patterns are not at all representative of specific behavioral dynamics but, simply, have been detected only by chance.

3 Results

3.1 Quantitative Analyses

Tables 1, 2, and 3 present the ethograms used in OF, HB, and EPM respectively. Per cent distribution of behavioral elements are presented in Fig. 1. In OF (Fig. 1a) the behavioral elements more represented are immobile sniffing, walking, immobility, climbing, and front paw licking, together reaching 81.93 % of the behavior; in HB (Fig. 1b) hole exploratory activities (HD and ES) do encompass a noticeable slice of behavior and, together with walking, immobile-sniffing, and climbing, represent the 80.79 %; finally, concerning the EPM (Fig. 1c), due to the more complex ethogram used, several behavioral elements range from 1 to 10 %. However, sniffing (-Sn), walking (-Wa, -Ent), and vertical exploration (-Re, -HDip), taken together, represent more than 90 % of the total number of behavioral events.

3.2 T-Pattern Analysis

Results from T-pattern analysis demonstrated that numerous events, characterizing rodent's behavior in each experimental model, occurred sequentially and with significant constraints on the interval lengths separating them. Figure 2 presents T-patterns length distribution in open field (Fig. 2a), hole board (Fig. 2b) and elevated plus maze (Fig. 2c).

Concerning OF, 28T-patterns of different composition have been detected (Fig. 2a); as to HB, 22 different T-patterns have been detected (Fig. 2b); finally, concerning the observations in the EPM, 197T-patterns of different composition have been detected (Fig. 2c).

Figure 2 shows also the average number of patterns detected in the randomized data + 1 standard deviation (for 5 random runs).

Table 1
Ethogram of rat's behavior in the open field

Behavioral element	Description
Walking (Wa)	The rat walks around sniffing the environment
Climbing (Cl)	The rat maintains an erect posture leaning against the Plexiglas wall, usually associated with sniffing
Rearing (Re)	The rat maintains an erect posture without leaning against the wall, usually associated with sniffing
Immobile Sniffing (IS)	The rat sniffs the environment, firmly standing on the ground
Front Paw Licking (FPL)	The rat licks or grooms its forepaws
Hind Paw Licking (HPL)	The rat licks or grooms its hind paws
Face Grooming (FG)	The rat rubs its face with the forepaws
Body Grooming (BG)	The rat rubs the body combing the fur by fast movement of the incisors
Immobility (Im)	The rat maintains a fixed posture
Chewing (Ch)	The rat produces rapid jaw movements

Table 2
Ethogram of rat's behavior in the hole board

Behavioral element	Description
Walking (Wa)	The rat walks around sniffing the environment
Climbing (Cl)	The rat maintains an erect posture leaning against the Plexiglas wall, usually associated with sniffing
Rearing (Re)	The rat maintains an erect posture without leaning against the wall, usually associated with sniffing
Immobile Sniffing (IS)	The rat sniffs the environment, firmly standing on the ground
Edge Sniffing (ES)	The rat sniffs the border of the hole without inserting the head inside
Head Dip (HD)	The rat puts its head into one of the four holes
Front Paw Licking (FPL)	The rat licks or grooms its forepaws
Hind Paw Licking (HPL)	The rat licks or grooms its hind paws
Face Grooming (FG)	The rat rubs its face with the forepaws
Body Grooming (BG)	The rat rubs the body combing the fur by fast movement of the incisors
Immobility (Im)	The rat maintains a fixed posture

Table 3
Ethogram of rat's behavior in the elevated plus maze

Behavioral element	Description
Closed Arm Entry (CA-Ent)	The rat moves from the central platform to a closed arm
Open Arm Entry (OA-Ent)	The rat moves from the central platform to an open arm
Closed Arm Return (CA-Ret)	The rat from a closed arm puts its head and forepaws in the central platform then rapidly re-enters in the closed arm
Closed Arm Walk (CA-Wa)	The rat walks in a closed arm
Open Arm Walk (OA-Wa)	The rat walks in an open arm
Central Platform Entry (CP-Ent)	The rat moves from an open or a closed arm to the central platform of the maze
Immobile Sniffing (p/u-ISn)[a]	The rat sniffs the environment standing on the ground
Corner Sniffing (p/u-CSn)[a]	The rat sniffs the entrance border of a closed arm
Stretched Attend Posture (p/u-SAP)[a]	The rat stretches its head and shoulders forward
Head Dip (p/u-HDip)[b]	Scanning movements over the sides of the maze in the direction of the floor
Rearing (p/u-Re)[a]	The rat maintains an erect posture
Defecation (p/u-Def)[a]	Excrements are produced
Grooming (p/u-Gr)[a]	The rat licks/rubs its face and/or body
Paw Licking (p/u-PL)[a]	The rat licks its paws
Immobility (p/u-Imm)[a]	An immobile posture is maintained

[a]The behavioral element is considered protected (p-) if occurring in the central platform or in a closed arm, unprotected (u-) if occurring in an open arm
[b]The head dip can be protected (p-) only in the central platform, or unprotected (u-) in an open arm

In EPM, among the 24 elements of the ethogram (Table 3) only 11 elements are encompassed in the structure of detected patterns: six protected elements (CA-Ent, CA-Wa, CP-Ent, p-Csn, p-ISn, p-Re) and 5 unprotected ones (OA-Ent, OA-Wa, u-Csn, u-HDip, u-ISn). In addition all the 197 patterns can be divided in three different groups on the basis of their composition: T-patterns occurring in central platform—open arms, in central platform—closed arms and in all the three zones of the EPM.

Figure 3 illustrates the tree structure and the connection diagram of three different T-patterns occurring 11, 41, and 22 times in OF, in HB and in EPM respectively. Their terminal strings are:

- ((Wa IS)(Cl Im)) (Fig. 3a)
- (ES (HD IS)) (Fig. 3b)

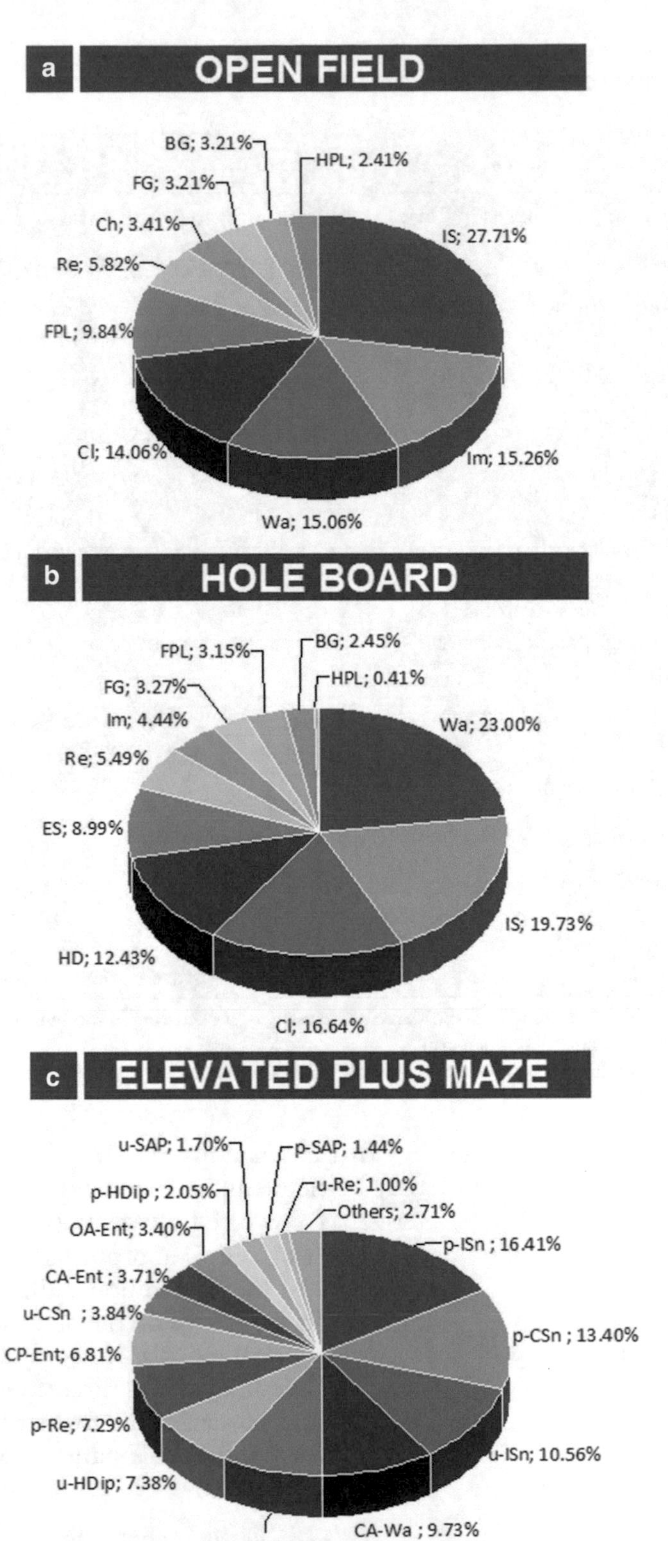

Fig. 1 Per cent distribution of behavioral elements in open field (**a**), hole board (**b**), and elevated plus maze (**c**). For abbreviations see Tables 1, 2, and 3

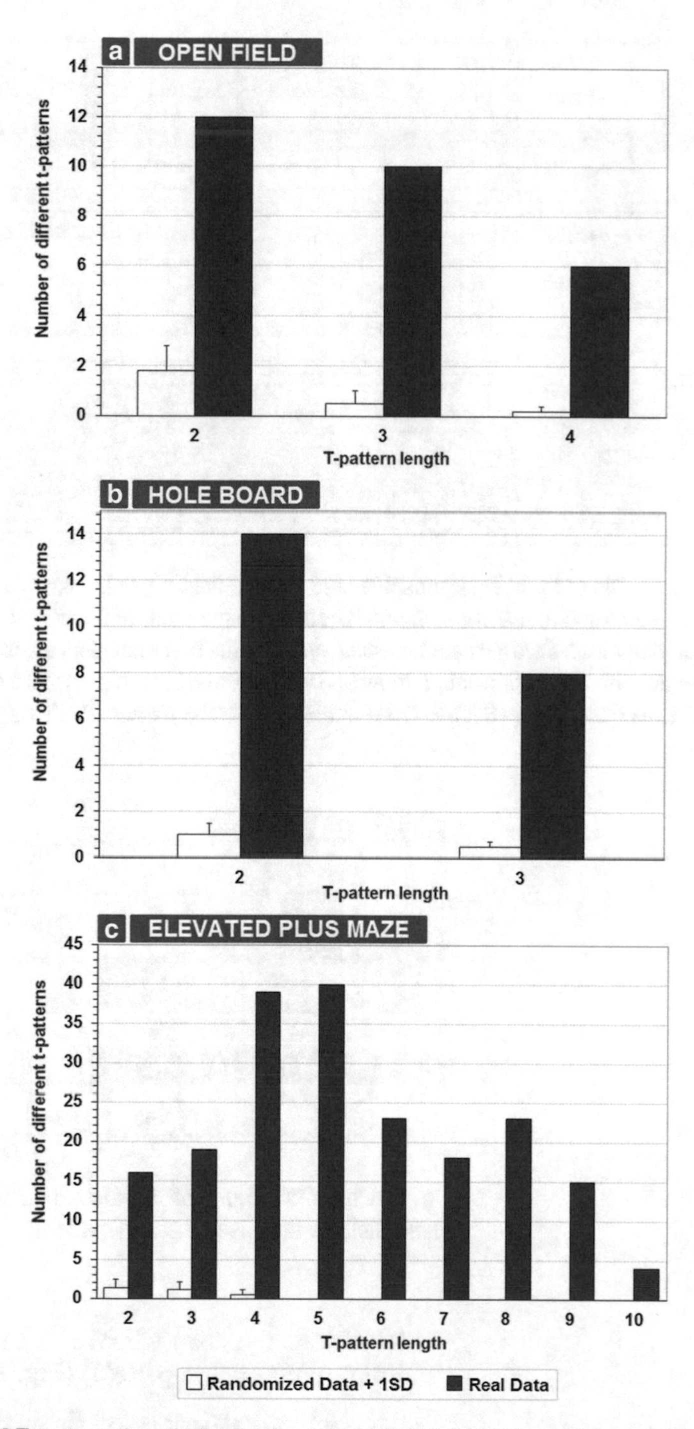

Fig. 2 T-patterns length distribution in open field (**a**), hole board (**b**), and elevated plus maze (**c**). *X*-axis = number of events encompassed in the structure of the T-pattern; *Y*-axis = number of T-patterns of different composition. *Dark columns* = real data; *White columns* = randomized data +1 SD

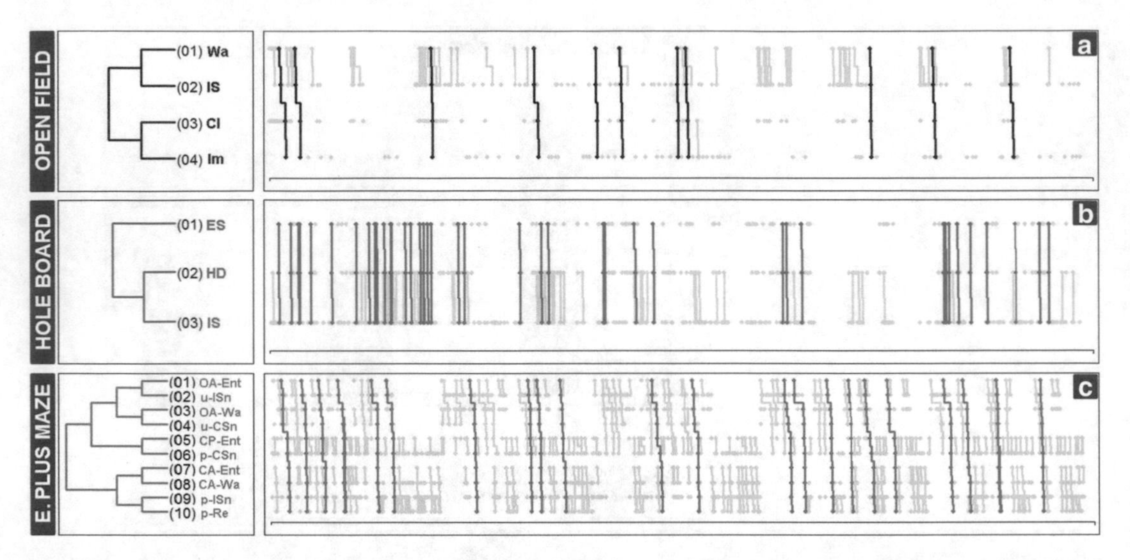

Fig. 3 Example of three T-patterns detected in open field (**a**), hole board (**b**), and elevated plus maze (**c**). *Left boxes*: tree structures. *Number in brackets* indicate the order of appearance of each event. *Right boxes*: connection diagrams. *Dots* indicate the occurrences of the corresponding events indicated in the *left boxes*. Lines connecting the *dots* represent patterns and subpatterns. Search procedure carried out on concatenated event log files, as described in Sect. 2.7. See Tables 1, 2, and 3 for abbreviations

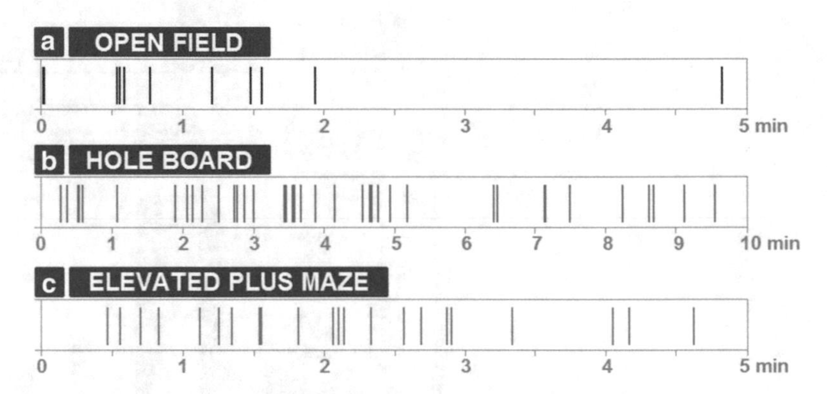

Fig. 4 Behavioral stripes of the three T-patterns illustrated in Fig. 3. *Vertical marks* indicate the onset of each T-pattern

– ((((OA-Ent u-ISn)(OA-Wa u-CSn))(CP-Ent p-CSn))((CA-Ent CA-Wa)(p-ISn p-Re))) (Fig. 3c)

The behavioral stripes of these three patterns are shown in Fig. 4.

Finally, Table 4 and Fig. 5 present the terminal strings and the behavioral stripes of 30 different patterns detected in the elevated plus maze.

Table 4
Terminal strings of 30 different T-patterns taken from the total amount of 197 detected in EPM

#	Terminal strings	A (%)	B (%)
1	(oa-wa ((u-hdip u-csn)(cp-ent oa-ent)))	100.00	0.00
2	(u-hdip (u-isn ((oa-wa u-csn) oa-ent)))	92.00	8.00
3	((u-isn ((oa-wa u-csn)(cp-ent oa-ent)))u-hdip)	90.63	9.38
4	(oa-ent (u-hdip (oa-wa u-csn)))	90.00	10.00
5	((u-hdip u-csn)(cp-ent oa-ent))	89.66	10.34
6	((u-isn ((oa-wa u-csn) oa-ent)) u-hdip)	89.66	10.34
7	((u-csn (cp-ent oa-ent))(oa-wa u-hdip))	89.47	10.53
8	((u-csn (cp-ent oa-ent)) u-hdip)	88.89	11.11
9	((oa-wa u-csn)((cp-ent oa-ent) u-hdip))	88.00	12.00
10	(u-hdip ((oa-wa u-csn)(cp-ent oa-ent)))	87.50	12.50
11	(u-hdip u-csn)	86.96	13.04
12	(((u-isn u-hdip)(oa-wa u-csn))(cp-ent oa-ent))	85.71	14.29
13	((u-isn u-hdip)((oa-wa u-csn)(cp-ent oa-ent)))	85.19	14.81
14	(((cp-ent p-csn) oa-ent)(oa-wa u-hdip))	85.19	14.81
15	(oa-ent ((u-isn u-hdip)(oa-wa u-csn)))	85.00	15.00
16	(cp-ent ca-ent)	38.27	61.73
17	((cp-ent p-csn)((ca-ent p-isn)(ca-wa p-re)))	38.10	61.90
13	(cp-ent ca-ent) p-re)	38.03	61.97
19	((cp-ent ca-ent)(p-isn p-re))	37.88	62.12
20	(p-csn ca-ent)	37.50	62.50
21	((ca-ent p-isn) p-re)	37.14	62.86
22	(ca-ent (ca-wa p-re))	36.99	63.01
23	(ca-ent ca-wa)	36.90	63.10
24	((ca-ent ca-wa)(p-isn p-re)	36.51	63.49
25	(ca-ent p-re)	36.49	63.51
26	(ca-ent (p-isn p-re))	36.23	63.77
27	((ca-ent p-isn)(ca-wa p-re))	36.17	63.83
28	(ca-ent p-isn)	35.71	64.29
29	(ca-wa p-re)	32.73	67.27
30	(p-isn p-re)	32.50	67.50

#1–#15 = T-patterns occurring in the central platform and open arms; #16–#30 = T-patterns occurring in the central platform and closed arms. %A and %B per cent distribution in the first and in the second part of the observation. For abbreviations see Table 3

Fig. 5 Behavioral stripes of 30 different T-patterns taken from the total amount of 197 detected in EPM. Numbers on the *left* indicate the corresponding string presented in Table 4. #1—#15 = T-patterns occurring in the central—platform and open arms; #16–#30 = T-patterns occurring in the central—platform and closed arms. *Vertical marks* indicate the onset of each T-pattern

4 Discussion

Present results demonstrate that rat's behavior in OF, HB, and EPM is organized on the basis of behavioral events which occur sequentially and with significant constraints on the interval lengths separating them.

4.1 Quantitative Analyses

In the experimental study of behavior a possible synergy between quantitative analyses and multivariate approaches should always be taken into consideration. In addition it is important to consider that results from a multivariate approach might be quite difficult to interpret, even for an experienced researcher, without a preliminary outline of the studied behavior. Such a preliminary outline can be provided by means of "conventional" evaluations such as the assessment of latencies, durations, frequencies, per cent distributions etc. In this chapter we have presented, for illustrative purposes, per cent distributions of the behavioral elements. Various simple information can be appreciated: for instance, sniffing and walking activities, both in OF, HB and EPM, do encompass the largest extent of the behavioral repertoire (Fig. 1). This is not a surprising result since these activities are essential for the environmental exploration and it is well known that rodents have the innate tendency to explore novel environments/objects remaining, at the same time, as protected as possible [43]. Of course, sniffing and walking must be analyzed and interpreted taking into consideration the experimental protocol. It is also interesting to notice the peculiar distribution of grooming- and immobility-related activities, more infrequently observed in EPM than in OF or HB (Fig. 1): since these behavioral elements do require, to be performed, a fixed position, it is possible to suggest that naïve rats in the EPM have a behavioral repertoire more heavily oriented toward locomotion and exploration if a comparison with results from the open field and the hole board is carried out. Hence, on the basis of relatively simple observations of quantitative results (per cent values in this case), it is possible to provide a general outline of what a following multivariate approach may be able to better illustrate in terms of underlying behavioral dynamics.

4.2 T-Pattern Analyses

Results from Theme™ provide various interesting topics of discussion. First of all a comparison of the distribution of T-patterns on the basis of their length (Fig. 2) shows that 28 and 22 different T-patterns have been detected in OF and HB respectively (Fig. 2a, b). Concerning the EPM, Fig. 2c illustrates that 197 different patterns are present and that more complex patterns are also more numerous. Such results gain even more emphasis taking into consideration the more rigid values used for the search parameters in EPM (namely, a search run carried out using a minimum samples of 100 %). These data demonstrate a higher complexity of the temporal structure of rodent's behavior in the EPM if compared with OF or HB. The reason could be the different impact of the EPM, in comparison with OF and HB, in terms of risk assessment and approach-avoidance conflict. Actually, the presence, in the EPM, of different zones (i.e. open arms, closed arms, and central platform), characterized by different levels of aversion [40], makes this apparatus quite different from other assays utilized to study anxiety-related behavior. For instance, during an open field or a hole board test, the rodent explores first the perimeter and only in a second moment the central zones of the arena [44, 45]. Therefore, early during the exploration of an open field, the rat collects adequate information concerning, at least, the boundaries of the novel environment. In EPM, due to its structural features, all visual and somatosensory cues, originating from the different zones of the apparatus, cannot be readily accessible. It is clear that the interaction of the animal is possible only with the surrounding environment while, at the same time, other parts/zones of the apparatus remain unapproachable and will be explored when physically reachable. On the basis of these considerations, it is possible to hypothesize that the higher structural complexity of the EPM, necessarily limiting the rodent in specific zones, elicits more complex and structured behavioral patterns arising from the interaction of the subject with each zone. Coherently, all the 197 patterns detected in EPM are organized in three different groups on the basis of their composition: T-patterns occurring in central platform—open arms, in central platform—closed arms and in all the three zones of the EPM. The exemplificative T-pattern illustrated in Fig. 3c highlights such a third circumstance. In addition even at a first glance it is clear that this pattern encompasses three different subpatterns, each occurring in one of the three zones of the maze: ((OA-Ent u-Isn)(OA-Wa u-CSn)), (CP-Ent p-CSn), and ((CA-Ent CA-Wa)(p-ISn p-Re)).

4.3 T-Patterns' Stripes

The classical tree representations and the connection diagrams (Fig. 3) have the great advantage to show the structure of the patterns detected and their distribution along the observational window; moreover, these illustrative approaches are very intuitive. The drawback is the huge amount of space required. For instance, concerning present results (see Sect. 3.2), the representation of all the different patterns detected in OF, in HB and in EPM by means of tree structures and connection diagrams would be very difficult. Actually, the detection of large amount of different T-patterns each occurring even hundreds of times is not uncommon [22–24]. Last but not least, if a concatenation procedure has been carried out (see Sect. 2.7), the resulting connection diagram concerns all the concatenated log files. For these reasons we have developed the representation of T-patterns by means of behavioral stripes, that is, the illustration of the onset of each T-pattern, along the *x*-axis timeline, by means of vertical marks [22–24]. An example of this representation is illustrated in Fig. 4.

To avoid misunderstandings, it is important to underline that each mark is not an individual behavioral element but the first event of a given T-pattern. So, taking into consideration Fig. 4, each vertical mark indicates the onset of the patterns ((Wa IS) (Cl Im)), (ES (HD IS)), and ((((OA-Ent u-ISn)(OA-Wa u-CSn)) (CP-Ent p-CSn))((CA-Ent CA-Wa)(p-ISn p-Re))) illustrated in Fig. 3. For clarity and completeness, the stripes should be partnered with information concerning the structure of each occurring pattern. We suggest the utilization of a separate table containing the corresponding terminal strings. Data concerning the onset of detected patterns and the terminal strings can be obtained by using the appropriate saving/export options available in Theme™. Figure 5 illustrates the onset of 30T-patterns among the 197 detected in EPM. The composition of each pattern is presented in Table 4 by means of corresponding terminal strings.

The strings in Table 4 and the stripes in Fig. 5 show 15T-patterns occurring in the central platform + open arms (from #1 to #15) and 15T-patterns occurring in the central platform + closed arms (from #16 to #30). Notably, T-patterns from #1 to 15 do occur for the largest extent within the first part of the observation; on the other hand, T-patterns from #16 to 30 have a more homogeneous distribution but with a prevalence during the second part of the test. On the basis of these results it is possible to conclude that the structure of rat behavior in the EPM has a complex temporal organization dependent on the zone of the maze explored and, importantly, on the moment of the exploration. Since it is very well known that naïve rodents in novel environments have the strong innate tendency to avoid open and illuminated areas, the presence of numerous T-patterns in central platform—open arms during the first part of the observation could be explained by a fear-related urgency to find an escape route rather than by a simple curiosity-related exploration [41].

5 Conclusion

The behavior is much more than simple latencies, durations, and per cent distributions of behavioral elements disjointed from the comprehensive behavioral structure. A given behavioral repertoire, in its natural completeness, can be literally dismounted into single pieces, namely the behavioral units of a given ethogram. Of course this is an obligatory step if a behavioral analysis must be carried out. On the other hand, if only a quantitative approach is used, the "risk" is to overemphasize each behavioral element in its individuality. It is our contention that the possibility to reduce a behavior into single "pieces", describing each individual element through even thousand of numbers does not imply the possibility to use those numbers to reconstruct the behavior and/or to figure out what the behavior is in its wholeness. If by means of suitable approaches, such as multivariate analyses, all the behavioral elements are studied in terms of their reciprocal relationships, new behavioral phenomena, otherwise undetectable, could emerge. In this chapter, by means of the multivariate T-pattern analysis, we have demonstrated, in three different and well-known experimental assays, the existence of significant patterning among the behavioral elements in the course of time. From a temporal point of view, it has been demonstrated that rodents' behavior has more complex and structured features in the elevated plus maze than in open field and/or in hole board. Such a higher complexity has been suggested to be linked with the different impact of the plus maze in terms of risk assessment and approach-avoidance conflict. In addition several methodological highlights, concerning the application of T-pattern analysis in the study of rodent behavior, have been presented.

References

1. Spruijt BM, Gispen WH (1984) Behavioral sequences as an easily quantifiable parameter in experimental studies. Physiol Behav 32:707–710
2. Casarrubea M, Sorbera F, Crescimanno G (2009) Multivariate data handling in the study of rat behavior: an integrated approach. Behav Res Methods 41:772–781
3. Espejo EF, Mir D (1993) Structure of the rat's behaviour in the hot plate test. Behav Brain Res 56:171–176
4. Espejo EF, Mir D (1994) Differential effects of weekly and daily exposure to the hot plate on the rat's behavior. Physiol Behav 55:1157–1162
5. Casarrubea M, Sorbera F, Crescimanno G (2006) Effects of 7-OH-DPAT and U 99194 on the behavioral response to hot plate test, in rats. Physiol Behav 89:552–562
6. Casarrubea M, Sorbera F, Santangelo A, Crescimanno G (2011) Learning influence on the behavioral structure of rat response to pain in hot-plate. Behav Brain Res 225:177–183
7. Casarrubea M, Sorbera F, Crescimanno G (2008) Multivariate analysis of the modifications induced by an environmental acoustic cue on rat exploratory behavior. Physiol Behav 93:687–696
8. Casarrubea M, Sorbera F, Crescimanno G (2009) Structure of rat behavior in hole-board: I) multivariate analysis of response to anxiety. Physiol Behav 96:174–179
9. Casarrubea M, Sorbera F, Crescimanno G (2009) Structure of rat behavior in hole-board: II) multivariate analysis of modifications induced by diazepam. Physiol Behav 96:683–692
10. Casarrubea M, Sorbera F, Santangelo A, Crescimanno G (2010) Microstructure of rat behavioral response to anxiety in hole-board. Neurosci Lett 481:82–87
11. Cruz AP, Frei F, Graeff FG (1994) Ethopharmacological analysis of rat behavior on the elevated plus-maze. Pharmacol Biochem Behav 49:171–176
12. Espejo EF (1997) Structure of the mouse behaviour on the elevated plus-maze test of anxiety. Behav Brain Res 86:105–112
13. Lino-de-Oliveira C, De Lima TC, de Pádua Carobrez A (2005) Structure of the rat behaviour in the forced swimming test. Behav Brain Res 158:243–250
14. Magnusson MS (2000) Discovering hidden time patterns in behavior: t-patterns and their detection. Behav Res Methods Instrum Comput 32:93–110
15. Casarrubea M, Jonsson GK, Faulisi F, Sorbera F, Di Giovanni G, Benigno A, Crescimanno G, Magnusson MS (2015) T-pattern analysis for the study of temporal structure of animal and human behavior: a comprehensive review. J Neurosci Methods 239:34–46
16. Kemp AS, Fillmore PT, Lenjavi MR, Lyon M, Chicz-Demet A, Touchette PE et al (2008) Temporal patterns of self-injurious behavior correlate with stress hormone levels in the developmentally disabled. Psychiatry Res 157: 181–189
17. Bonasera SJ, Schenk AK, Luxenberg EJ, Tecott LH (2008) A novelmethod for automatic quantification of psychostimulant-evoked route-tracing stereotypy: application to Mus musculus. Psychopharmacology (Berl) 196:591–602
18. Kerepesi A, Kubinyi E, Jonsson GK, Magnusson MS, Miklósi A (2006) Behavioral comparison of human-animal (dog) and human-robot (AIBO) interactions. Behav Process 73:92–99
19. Hirschenhauser K, Frigerio D, Grammer K, Magnusson MS (2002) Monthly patterns of testosterone and behavior in prospective fathers. Horm Behav 42:172–181
20. Hocking PM, Rutherford KMD, Picard M (2007) Comparison of time-based frequencies, fractal analysis and t-patterns for assessing behavioral changes in broiler breeders fed on two diets at two levels of feed restriction: a case study. Appl Animal Behav Sci 104:37–48
21. Horn C, Henry S, Meyers K, Magnusson MS (2011) Behavioral patterns associated with chemotherapy-induced emesis: a potential signature for nausea in musk shrews. Front Neurosci 5:1–17
22. Casarrubea M, Sorbera F, Magnusson M, Crescimanno G (2010) Temporal patterns analysis of rat behavior in hole-board. Behav Brain Res 208:124–131
23. Casarrubea M, Sorbera F, Magnusson MS, Crescimanno G (2011) T-pattern analysis of diazepam-induced modifications on the temporal organization of rat behavioral response to anxiety in hole-board. Psychopharmacology (Berl) 215:177–189

24. Casarrubea M, Roy V, Sorbera F, Magnusson MS, Santangelo A, Arabo A, Crescimanno G (2013) Temporal structure of the rat's behavior in elevated plus maze test. Behav Brain Res 237:290–299
25. Drai D, Kafkafi N, Benjamini Y, Elmer G, Golani I (2001) Rats and mice share common ethologically relevant parameters of exploratory behavior. Behav Brain Res 125:133–140
26. Choleris E, Thomas AW, Kavaliers M, Prato FS (2001) A detailed ethological analysis of the mouse open field test: effects of diazepam, chlordiazepoxide and an extremely low frequency pulsed magnetic field. Neurosci Biobehav Rev 25:235–260
27. Prut L, Belzung C (2003) The open field as a paradigm to measure the effects of drugs on anxiety-like behaviors: a review. Eur J Pharmacol 463:3–33
28. File SE, Wardill AG (1975) The reliability of the hole-board apparatus. Psychopharmacologia 44:47–51
29. Rodriguez Echandia EL, Broitman ST, Foscolo MR (1987) Effect of the chronic ingestion of chlorimipramine and desipramine on the hole board response to acute stresses in male rats. Pharm Biochem Behav 26:207–210
30. Adamec R, Head D, Blundell J, Burton P, Berton O (2006) Lasting anxiogenic effects of feline predator stress in mice: Sex differences in vulnerability to stress and predicting severity of anxiogenic response from the stress experience. Physiol Behav 88:12–29
31. Harada K, Aota M, Inoue T, Matsuda R, Mihara T, Yamaji T, Ishibashi K, Matsuoka N (2006) Anxiolytic activity of a novel potent serotonin 5-HT2C receptor antagonist FR260010: a comparison with diazepam and buspirone. Eur J Pharmacol 553:171 184
32. Saitoh A, Yamada M, Yamada M, Kobayashi S, Hirose N, Honda K, Kamei J (2006) ROCK inhibition produces anxiety-related behaviors in mice. Psychopharmacology 188:1–11
33. Kalueff AV, Wheaton M, Murphy DL (2007) What's wrong with my mouse model? Advances and strategies in animal modeling of anxiety and depression. Behav Brain Res 179:1–18
34. Kamei J, Hirose N, Oka T, Miyata S, Saitoh A, Yamada M (2007) Effects of metylphenidate on the hyperemotional behavior in olfactory bulbectomized mice by using the hole-board test. J Pharmacol Sci 103:175–180
35. File SE, Wardill AG (1975) Validity of head-dipping as a measure of exploration in a modified hole-board. Psychopharmacologia 44:53–59
36. Hughes RN (2007) Neotic preferences in laboratory rodents: Issues, assessment and substrates. Neurosci Biobehav Rev 31:441–464
37. Ohl F, Holsboer F, Landgraf R (2001) The modified hole board as a differential screen for behavior in rodents. Behav Res Methods Instrum Comput 33:392–397
38. Takeda H, Tsuji M, Matsumiya T (1998) Changes in head-dipping behavior in the hole-board test reflect the anxiogenic and/or anxiolytic state in mice. Eur J Pharmacol 350: 21–29
39. Handley SL, Mithani S (1984) Effects of alpha-adrenoreceptor agonists and antagonists in a maze exploration model of "fear"-motivated behavior. Naunyn Schmiedeberg's Arch Pharmacol 327:1–5
40. Carobrez AP, Bertoglio LJ (2005) Ethological and temporal analyses of anxiety-like behavior: The elevated plus-maze model 20 years on. Neurosci Biobehav Rev 29:1193–1205
41. Roy V, Chapillon P, Jeljeli M, Caston J, Belzung C (2009) Free versus forced exposure to an elevated plus-maze: evidence for new behavioral interpretations during test and retest. Psychopharmacology (Berl) 203:131–141
42. Treit D, Menard J, Royan C (1993) Anxiogenic stimuli in the elevated plus-maze. Pharmacol Biochem Behav 44:463–469
43. Augustsson H, Dahlborn K, Meyerson BJ (2005) Exploration and risk assessment in female wild house mice (Mus musculus musculus) and two laboratory strains. Physiol Behav 84:265–277
44. Simon P, Dupuis R, Costentin J (1994) Thigmotaxis as an index of anxiety in mice. Influence of dopaminergic transmissions. Behav Brain Res 61:59–64
45. Metz GA, Kolb B, Whishaw IQ (2005) Neuropsychological tests. In: Whishaw IQ, Kolb B (eds) The behavior of the laboratory rat. Oxford University Press, Oxford

Chapter 13

Using Hidden Behavioral Patterns to Study Nausea in a Preclinical Model

Charles C. Horn and Magnus S. Magnusson

Abstract

Nausea is a common clinical symptom reported by many patients experiencing cytotoxic anticancer therapy, gastrointestinal disease, and postoperative recovery. Although the neurological basis of vomiting is reasonably well established, an understanding of the physiology of nausea is lacking. The primary barrier to mechanistic research on nausea is the lack of appropriate animal models. Indeed investigating the effects of anti-nausea drugs in preclinical models is difficult because the primary readout is suppression of vomiting. However, animals often show a behavioral profile of sickness, associated with reduced feeding and movement, and possibly these general behavioral measures are signs of nausea. Here we applied t-pattern analysis to determine patterns of behavior associated with emesis as a potential measure of nausea. Musk shrews were used for these experiments because they have a vomiting reflex and other laboratory animals (rats and mice) do not. Standard emetic test agents were used in Study 1, the chemotherapy agent cisplatin (intraperitoneal), and Study 2, nicotine (subcutaneous) and copper sulfate (intragastric). Emesis and other behaviors were coded and tracked from video files. T-pattern analysis revealed patterns of behavior associated with emesis. Long-term tests (3 days of continuous video recording) showed that eating and drinking, and other larger body movements, including rearing, grooming, and body rotation, were significantly less common in cisplatin-induced emesis-related behavioral patterns in real versus randomized data. Short-term experiments (30 min) using nicotine and copper sulfate showed no difference between male and female shrews but there were more behavioral patterns associated with nicotine compared to copper sulfate treatment. The current approach to behavioral analysis in a preclinical model for emesis using t-pattern analysis could be used to assess the effects of drugs used to control nausea and its potential correlates, including reduced feeding and changed activity levels.

Key words Emesis, Vomiting, Nausea, Conditioned taste aversion, Pica, Avoidance, Anorexia, Cancer

1 Introduction

Antiemetic drugs control vomiting better than nausea and a primary focus of current research is to understand the biology of nausea [1, 2]. Nausea, as a subjective experience, cannot be directly measured in nonhuman animals. Surrogate markers of nausea in animal studies have included conditioned flavor aversion, pica

Magnus S. Magnusson et al. (eds.), *Discovering Hidden Temporal Patterns in Behavior and Interaction: T-Pattern Detection and Analysis with THEME™*, Neuromethods, vol. 111, DOI 10.1007/978-1-4939-3249-8_13,

(ingestion of clay), and physiological measures [3–8]. These approaches have major limitations, including the need for specific testing protocols, lack of face validity, and the use of invasive procedures. An alternative approach is the assessment of behavioral indices of sickness as a potential measure of nausea. In theory, animals should show behavioral changes that correlate with the level of nausea or malaise [9]. Vomiting is an obvious indication of sickness but there might also be behavioral patterns leading up to and/or following an emetic episode. In this way the occurrence of vomiting could be used as an unequivocal anchor for evaluating the behavioral changes that occur with sickness, malaise, and potentially nausea. Reports suggest the existence of some species-specific emesis-related behavioral responses. For example, lip-licking, backwards walking, and burrowing behaviors have been observed in association with emesis in ferrets [10, 11]. However, there has been little effort focused on a quantitative analysis of the patterns of behavior related to emesis.

In the current studies we tested the hypothesis that emesis is associated with behavioral patterns of reduced feeding and movement, i.e., a profile of sickness or nausea. To test for behavioral patterns we employed temporal pattern (t-pattern) analysis [12] to determine statistically significant relationships between emetic events and other behaviors. T-pattern analysis can assess subtle patterns of behavior that can be difficult to detect by other methods [12–14]. To create the time-stamped event data needed for these analyses, we acquired digital video of behaving animals, manually coded behaviors (e.g., emesis, eating, drinking), and automatically tracked body movement with computer software. In these experiments, we injected musk shrews with standard emetic test stimuli, including the chemotherapy agent cisplatin (ip), nicotine (sc), and copper sulfate (po) [15–17]; all of these agents produce high levels of nausea and vomiting in humans [18–20]. The musk shrew (*Suncus murinus*) is a well-established small animal model (40–80 g) used for emesis research [8, 15, 21–24].

2 Materials and Methods

2.1 Subjects

The subjects were adult musk shrews (>35 days of age; Fig. 1) with body weights of 39–91 g for males and 33–53 g for females. Animals were derived from breeding stock acquired from Professor John Rudd, the Chinese University of Hong Kong (a strain originating from Taiwan), and housed individually in clear plastic cages (28 × 17 × 12 cm), with a filtered air supply, using a 12 h light/12 h dark cycle (0700–1900 h light period); animals had free access to food and water. Food consisted of a mixture of 15 % Purina Cat Chow Complete Formula and 25 % Complete Gro-Fur mink food

Fig. 1 A musk shrew (*Suncus murinus*)

pellets [25]. All experiments were approved by the University of Pittsburgh Institutional Animal Care and Use Committee.

2.2 Data Collection Procedures for Study 1: Cisplatin (72 h Test)

The behavior of six male musk shrews was video-recorded for 96 h (24 h before and 72 h after injection of 30 mg/kg of cisplatin, ip). A camera (Sony, DCR-SR300; internal sensor and IR light for recording dark phase activity) was placed above the cage and attached to a computer via a USB port (USB 2.0 Video Capture Cable; StarTech.com). Only one animal was tested each week (4 days in the observation chamber). The videos were captured using Movie Maker software (Microsoft). At ~945 h animals were transferred to the test cages (Fig. 2), which were kept under an animal transfer hood. One day after the acquisition day (baseline), animals were injected with cisplatin. The cage contained a food cup, and water was provided from a sipper tube attached to a graduated cylinder (Fig. 2). Each morning at ~945 h, the musk shrews were weighed and food and water containers were checked and refilled. Because of the need to make these measurements, on the baseline day and days 1 and 2 post-injection there was an average of 16 min less video time for each 24 h segment. One of the baseline days (the 24 h before cisplatin injection) from one animal was recorded for only 14 h due to a software malfunction (these data are not included in t-pattern analysis).

Videos (MPEG-2) were imported into behavioral coding software (The Observer XT 10.1; Noldus Information Technology, www.noldus.com, The Netherlands). Emetic episodes and other behaviors were recorded manually with keystrokes by a trained observer viewing a computer monitor (Table 1; Study 1). A second trained observer validated the occurrence of all emetic episodes and random sections of data for other behaviors. An emetic episode was recognized as a sequence of contractions of the abdomen and head movements (retching). An emetic episode can occur with or without the expulsion of gastric contents [23]; therefore, episodes without expulsion were also counted. The Observer software allows users to slow down, reverse, and check the coding of behaviors stored in a computer file containing the codes and timestamps.

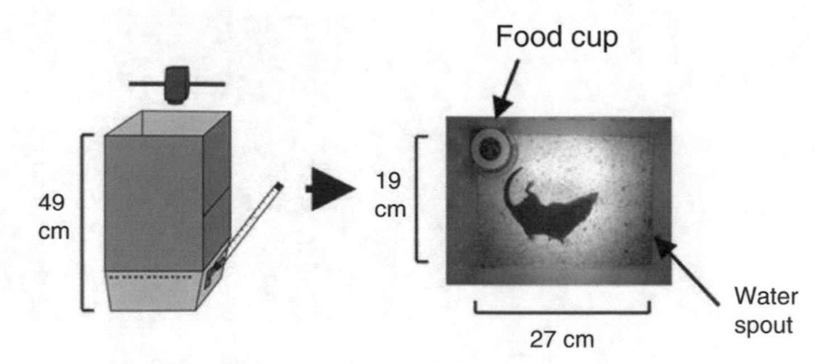

Fig. 2 The behavioral test chamber used in Study 1, cisplatin injection (72 h test). The same chamber was used in Study 2, nicotine and copper sulfate injections (30 min tests), but without food and water access

Videos were imported into animal tracking software (Ethovision 7.1; Noldus). It was not possible to load daily videos (~20 GB each) into the Ethovision software without compression and we determined that conversion to PAL (25 frames/s) using the AVI-MPEG4 format worked well (432 × 720). In the longer term videos, there were significant shadow effects, which affected the tracking of body contour (particularly as the animal approached the sides of the chamber) and therefore only the center body point was tracked for each daily video (15 frames/s using model-based tracking with medium pixel smoothing, 4 pixel erosion, and 2 pixel dilation). In these 24 h videos, we used difference thresholding to track the body contour, i.e., a reference image was automatically updated over the course of the video tracking to correct for changes in illumination and position of bedding material.

2.3 Data Collection Procedures for Study 2: Nicotine and Copper Sulfate (30 min Tests)

The behavior of 28 male and 24 female musk shrews was video-recorded for short-term (30 min) testing. Two tests were conducted with at least 1 week separating each test: (1) subcutaneous nicotine injection (5 mg/kg), and (2) intragastric copper sulfate injection (120 mg/kg). Animals were placed in test chambers (Fig. 1) without food or water access (with a clear plastic bottom without wood chips) for 15 min before injection. Videos were either collected as in Study 1 and recorded to video camera or stored on a PC using Media Recorder software (Noldus). Manual recording of emesis, abdominal contraction, and swaying was conducted at the time of testing by a trained observer sitting outside the clear plastic test chambers. Manual entries were collected on a notebook computer using JWatcher (http://www.jwatcher.ucla.edu/). Tests were conducted between 0830 and 1230 h.

2.4 Data Processing and Analysis

Manually scored events were timestamped and automatically tracked measures were exported from Noldus and JWatcher software. Custom scripts written in Matlab (Version 7.1; Mathworks) were used to process text files for import to T-pattern software.

Table 1
Musk shrew behavioral event types recorded in Studies 1 and 2

Manually coded		Included Study 1	Included Study 2
emesis	A sequence of contractions of the abdomen (retching)	x	x
eat	Putting the head into the food container for ≥2 s and visible movement of head and jaw	x	
drink	Placing the snout on the sipper tube for ≥1 s	x	
sniff	Moving the snout with elongation of the body	x	
rear	Standing on back legs, against the wall, sometimes jumping	x	
abcon	Abdominal contraction; a single retch		x
sway	Swaying of the abdominal region from side to side		x
Automatically tracked			
Locomotion			
dchi dclo	Distance moved, body center, high Distance moved, body center, low	x	x
mc	Movement (velocity, begin > 2 cm/s, end < 1.75 cm/s)	x	x
Turning			
rot rotc	Rotation—clockwise (a turn of 360°) Rotation—counterclockwise	x	x
tanhi tanlo	Turn angle of the nose, high Turn angle of the nose, low		x
tachi taclo	Turn angle of the body center, high Turn angle of the body center, low	x	x
Movement in place			
con norm long	Contracted Normal Elongated body contour		x
immob mob mobhi	Immobile (<1 %) Mobile (>1 %, <8 %) And highly mobile (>8 %; change in body contour)		x
dnhi dnlo	Distance moved, nose, high Distance moved, nose, low		x

Continuous variables that were automatically tracked, e.g., distance moved and velocity, were converted to discrete events using a threshold cutoff of 2 standard deviations (SD) above or below the mean values to generate timestamps (Table 1). In preliminary analyses we determined that relatively low cutoffs (e.g., ±1 SD) were too computationally intensive for the t-pattern analysis software.

T-pattern analysis was conducted using computer software (Theme 6; Noldus Technologies). These algorithms have been tested on numerous data sets [26–29], including several reports using laboratory mice, rats, and musk shrews [13, 14, 16, 30].

We conducted statistical comparisons using analysis of variance (ANOVA) and *t*-tests (Statistica 11.0; StatSoft). When an ANOVA was statistically significant, we used Least Significance Difference test (LSD-test) or Tukey's HSD test for group mean comparisons. For all ANOVA and mean comparisons, $p < 0.05$ was used to detect statistical significance.

3 Results

3.1 Study 1: Cisplatin (72 h Test)

Cisplatin injection produced the predicted outcomes, including emesis, reduced feeding and drinking, and loss of body weight [Fig. 3; $F(3,15) \geq 3.9$, $p \leq 0.03$]. Feces production was not significantly changed (ANOVA). Hence, these animals would be defined as sick on multiple dimensions by day 3 after cisplatin injection. Cisplatin treatment (30 mg/kg) induced acute (<24 h) and delayed (>48 h) phases of emesis (Figs. 3, 4, and 5). One animal showed 3 emetic episodes during the baseline day (prior to cisplatin treatment). On measures of food and water intake and body weight this animal was not significantly different from the other five animals. These few emetic episodes in one animal might be attributable to stress.

Figures 4 and 5show behavioral events per hour for manually and automatically tracked behaviors. Distance moved and turn angle are represented as continuous data prior to detection of discrete events using ± 2 standard deviations. There were small but statistically significant changes in behavior over the 4 days [$Fs(69,345) \geq 1.4$, $ps \leq 0.05$, ANOVA, day by hour interaction effect for measures of eat, drink, groom, sniff, rear, mc, rot, rotc, distance moved, and velocity; $F(3,15) = 5.1$, $p < 0.05$, ANOVA, main effect of day for emesis]. There were no statistically significant effects for turn angle.

There was a nearly perfect correlation between distance moved and velocity ($r = 0.99$, Pearson); therefore we did not use velocity as a metric in the t-pattern analysis. Furthermore, unlike the short-term experiment, there were no low distance moved events (dclo; 2 standard deviations below the mean). T-pattern analysis included 12 event types: emesis, groom, rear, sniff, eat, drink, mc, dchi, rotc, rot, tachi, and taclo (see Table 1).

The t-pattern analysis was focused on Days 1 and 3 post-injection with cisplatin since these represent acute and delayed phases of emesis. Even with only 12 event types it became clear that t-pattern analysis software was not capable of evaluating large daily data sets using all variables simultaneously, which produced malfunctioning of the program. To solve this issue we divided the analysis into sets of event types and also ran analyses on data from 1 h before and 1 h

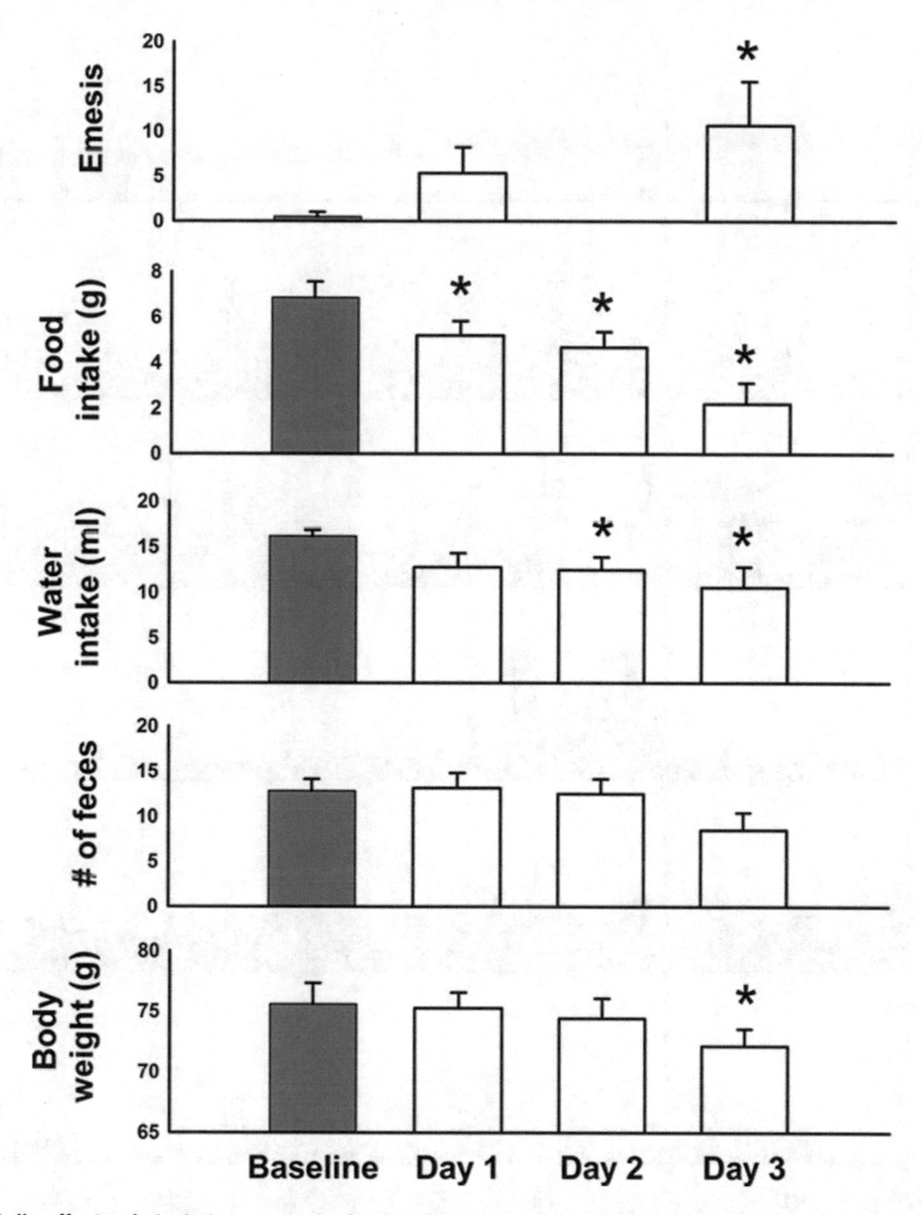

Fig. 3 Daily effects of cisplatin on emesis, food and water intake, number of feces, and body weight in Study 1 (72 h after injection with cisplatin, 30 mg/kg, ip, $n=6$). Values were recorded for one day before (baseline control) and 3 days after injection of cisplatin (30 mg/kg; i.p). Results represent the mean ± SEM. * $P<0.05$ versus baseline, LSD-test [see reference 16]

after each cluster of emetic episodes. Emetic episodes that occurred with intervals greater than 1 h were considered as new clusters. There were six independent analyses (always including emesis as one event type): (a) 24 h data, including groom, sniff, eat, and drink events; (b) 24 h data, with groom, rear, eat, and drink events; (c) 24 h data, using dchi events; (d) 2 h data, including groom, rear, sniff, eat, drink, and mc events; (e) 2 h data, using groom, rear, sniff, eat, drink, and dchi events; and (f) 2 h data, with groom, rear, sniff, eat, drink, rot, rotc, tachi, and taclo events (see Table 1).

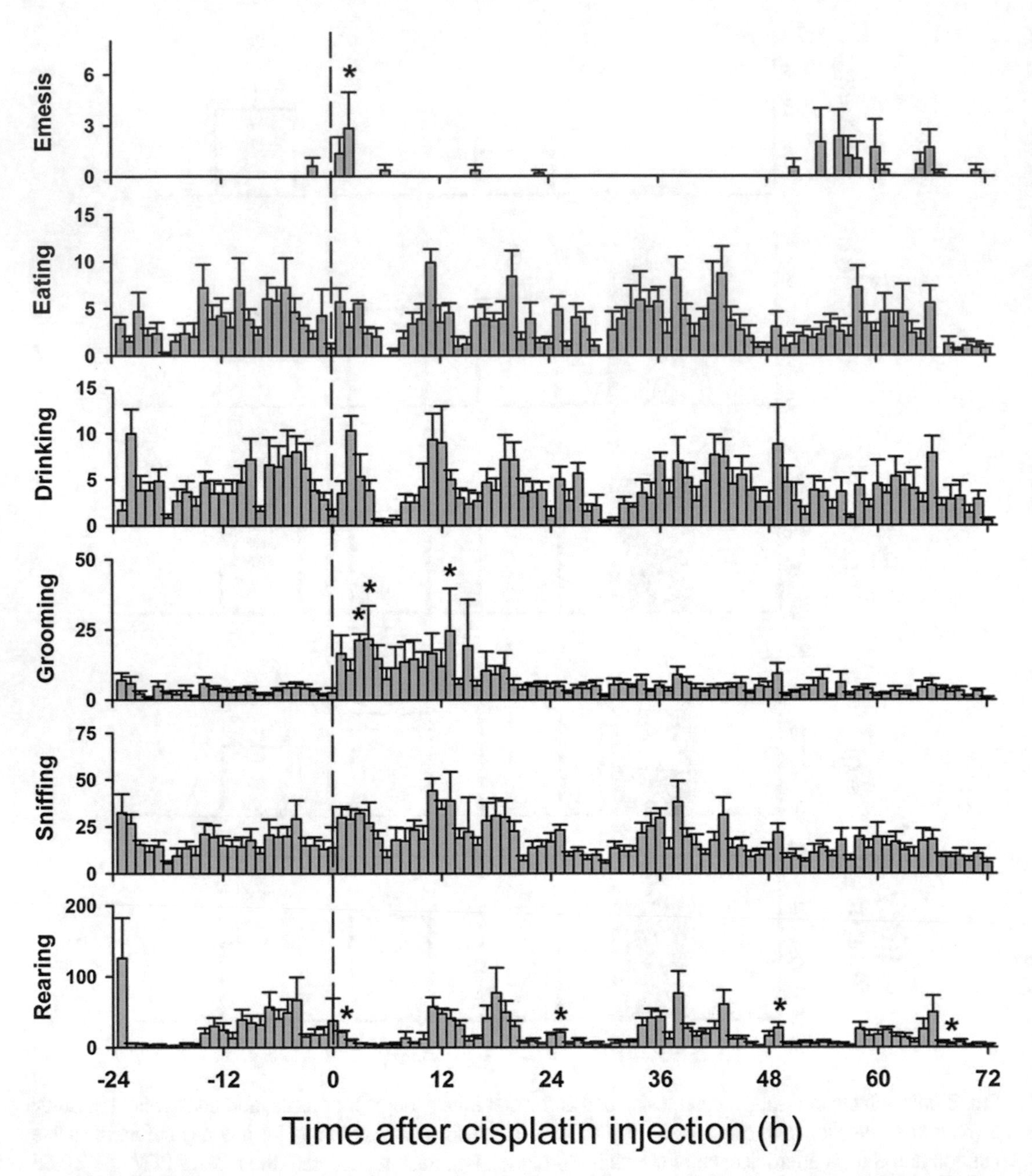

Fig. 4 Hourly effects of cisplatin on emesis, eating, drinking, sniffing, grooming, and rearing/jumping in Study 1 (72 h after injection with cisplatin, 30 mg/kg, ip, $n=6$). Behaviors were recorded for 1 day before (baseline control) and 3 days after the injection of cisplatin (30 mg/kg; i.p). Results represent the means ± SEM ($n=6$). * $P<0.05$ versus corresponding hour from baseline before cisplatin injection, Tukey's HSD test [see reference 16]

Our analyses revealed a large number of statistically significant behavioral patterns with a subset containing emesis: (a) total detected patterns in Day 1 = 6507 ± 1837 (mean ± SEM) and Day 3 = 3335 ± 975, with 46 ± 14 and 227 ± 110 emesis-related patterns, respectively; (b) total Day 1 = 4838 ± 1160 and Day 3 = 2665 ± 737,

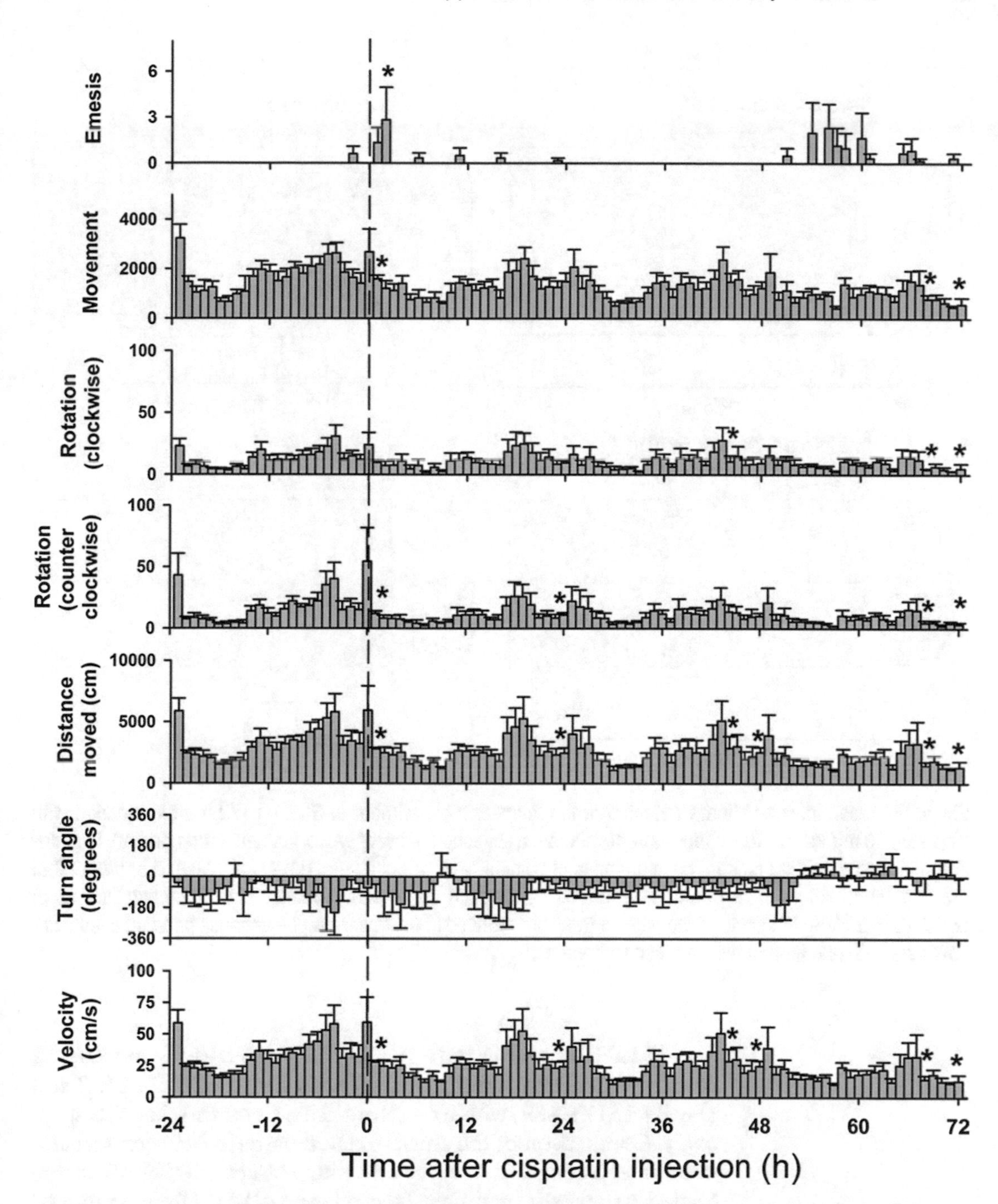

Fig. 5 Hourly effects of cisplatin on emesis and multiple tracked behaviors in Study 1 (72 h after injection with cisplatin, 30 mg/kg, ip, $n = 6$). Behaviors were recorded for 1 day before (control) and 3 days after the injection of cisplatin (30 mg/kg; ip). Results represent the means ± SEM ($n = 6$). *$P < 0.05$ versus corresponding hour from baseline before cisplatin injection, Tukey's HSD test [see reference 16]

with 41 ± 16 and 80 ± 42 emesis-related, respectively; (c) total Day 1 = 22 ± 5 and Day 3 = 29 ± 6, with 9 ± 4 and 19 ± 6 emesis-related, respectively; (d) total Day 1 = 5684 ± 4330 and Day 3 = 2040 ± 885, with 77 ± 46 and 40 ± 17 emesis-related, respectively; (e) total Day

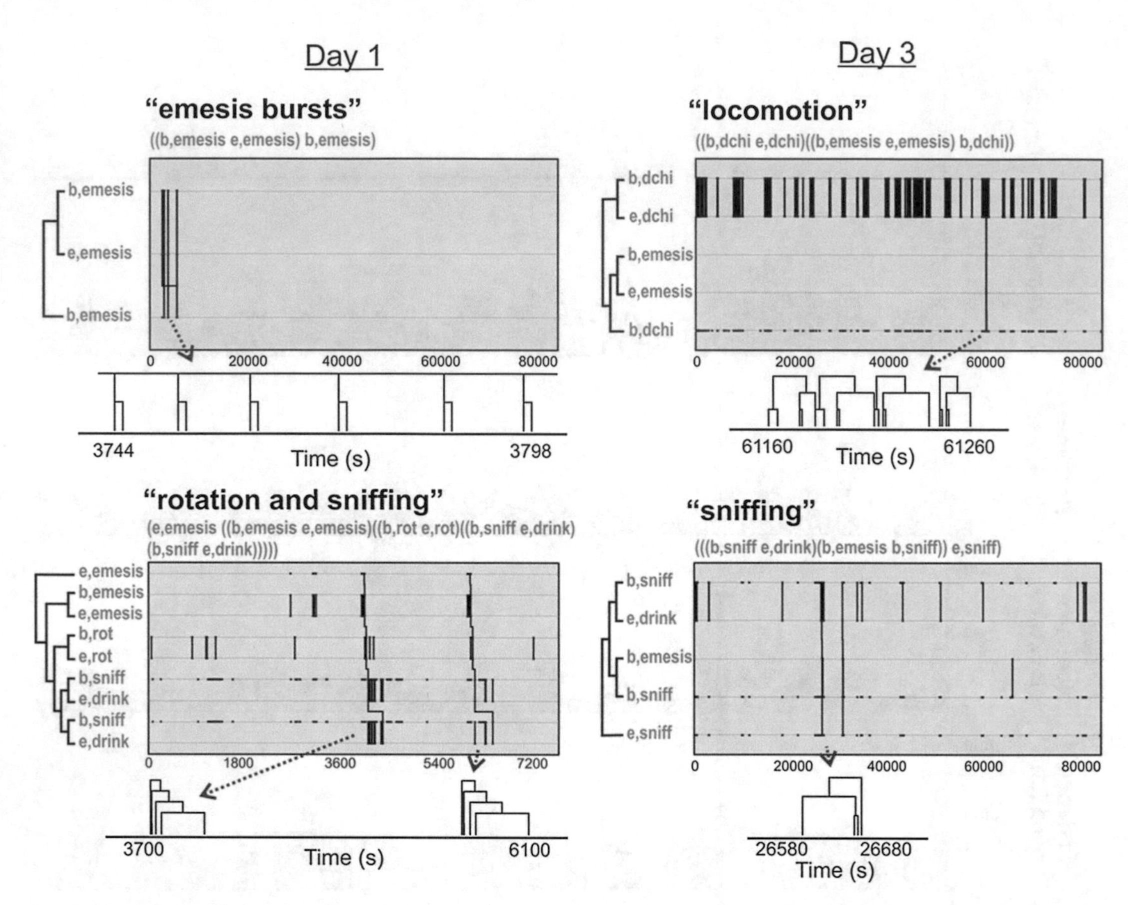

Fig. 6 Representative t-patterns detected in the long-term experiment in Study 1 (72 h after injection with cisplatin, 30 mg/kg, ip). These four patterns contain a collection of event types, including rotation (rot), locomotion (dchi), and sniffing (sniff). The sequences of "b,dchi" and "e,dchi", etc. represent patterns of bursts that are contained within larger patterns. Diagrams below each figure show smaller time scale subsections of pattern trees. Note that some of the pattern trees are reduced in the lower graphs because bursting events are collapsed into one limb of the tree [see reference 16]

1 = 4252 ± 3141 and Day 3 = 1957 ± 752, with 84 ± 53 and 40 ± 12 emesis-related, respectively; and (f) total Day 1 = 3667 ± 2653 and Day 3 = 1317 ± 439, with 67 ± 39 and 13 ± 4 emesis-related, respectively. Comparison of the emesis-related patterns between real and randomized control data revealed a much larger number of emesis-related patterns in real data [see reference 16]. Distance moved (dchi), movement (mc), sniff, drink, groom, rear, and rot occurred in these patterns more often in Day 3 compared to Day 1 [see reference 16, Table 3]. Representative samples of patterns from the six analyses are shown in Fig. 5. None of these analyses showed statistically significant negative associates of emesis-related patterns but positive associates were present for several of the event types that had also occurred as part of detected patterns (Fig. 6).

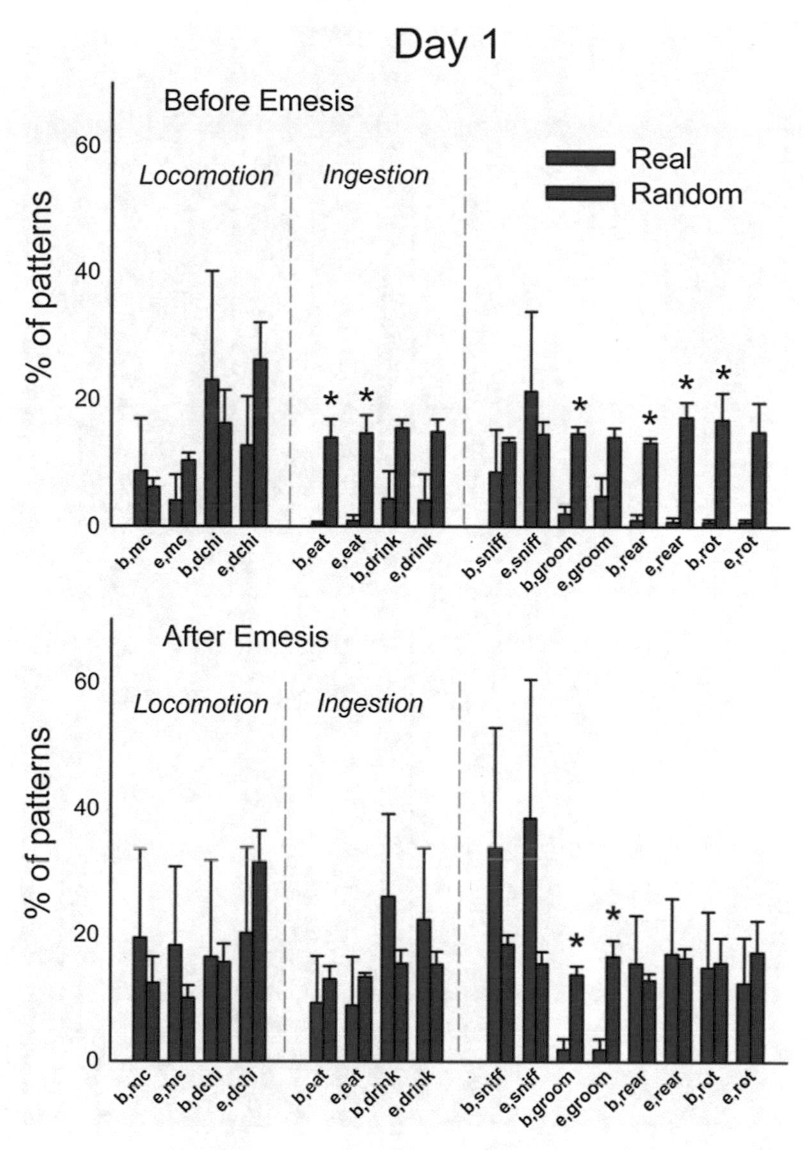

Fig. 7 Comparison of event types in emesis-related t-patterns detected in Study 1, Day 1 (24 h after injection with cisplatin, 30 mg/kg, ip, $n=6$). Results represent the mean ± SEM. $* = p < 0.05$, two-tailed t-test, real vs. random data [see reference 16]

Figures 7 (Day 1) and 8 (Day 3) show the occurrence of event types before or after emesis comparing real and randomized data. Locomotion (mc) occurred significantly more often before emesis on Day 3 than on Day 1 in real versus random emesis-related t-patterns. Ingestive behaviors before and after emesis were less common in real compared to random patterns. Furthermore, animals showed less complex movements, i.e., grooming, rearing, and rotation, in emesis-related patterns from real versus randomized data.

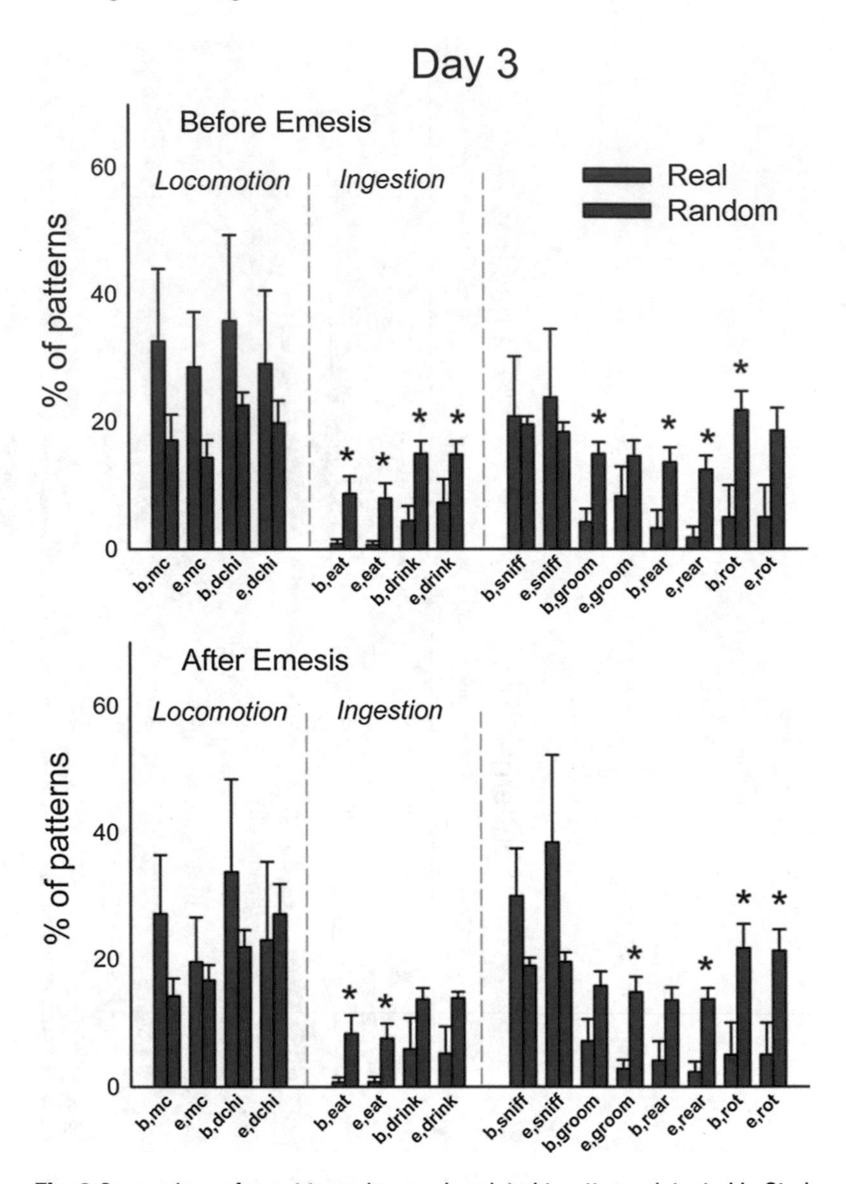

Fig. 8 Comparison of event types in emesis-related t-patterns detected in Study 1, Day 3 (24 h after injection with cisplatin, 30 mg/kg, ip, $n=6$). Results represent the mean ± SEM. * = $p<0.05$, two-tailed t-test, real vs. random data [see reference 16]

3.2 Study 2: Nicotine and Copper Sulfate (30 min Tests)

Nicotine injection produced more emesis in male compared to female shrews [Fig. 9; $t(57)=3.8$, $p<0.0005$], but there were no sex differences in emesis after copper sulfate injection (Fig. 9).

Analysis revealed statistically significant behavioral patterns with a subset containing emesis. Figure 10 shows that the total number of t-patterns did not differ by stimulus, nicotine and copper sulfate. However, the number of emetic-related t-patterns was greater before and after emesis when animals were injected with nicotine compared to copper sulfate [$F(1,52)\geq 9.9$, $p\leq 0.005$, main effects of stimulus, before and after emesis].

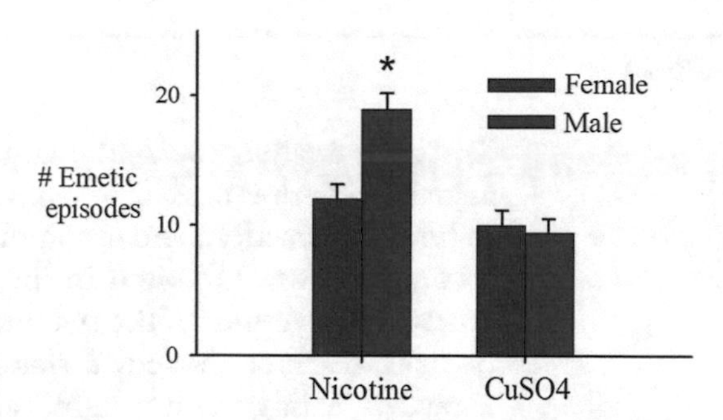

Fig. 9 Effects of nicotine (5 mg/kg, sc) and copper sulfate ($CuSO_4$, 120 mg/kg, po) on emesis in male and female shrews ($n \geq 27$, each group). Results represent the mean ± SEM. * = $p < 0.05$, two-tailed *t*-test

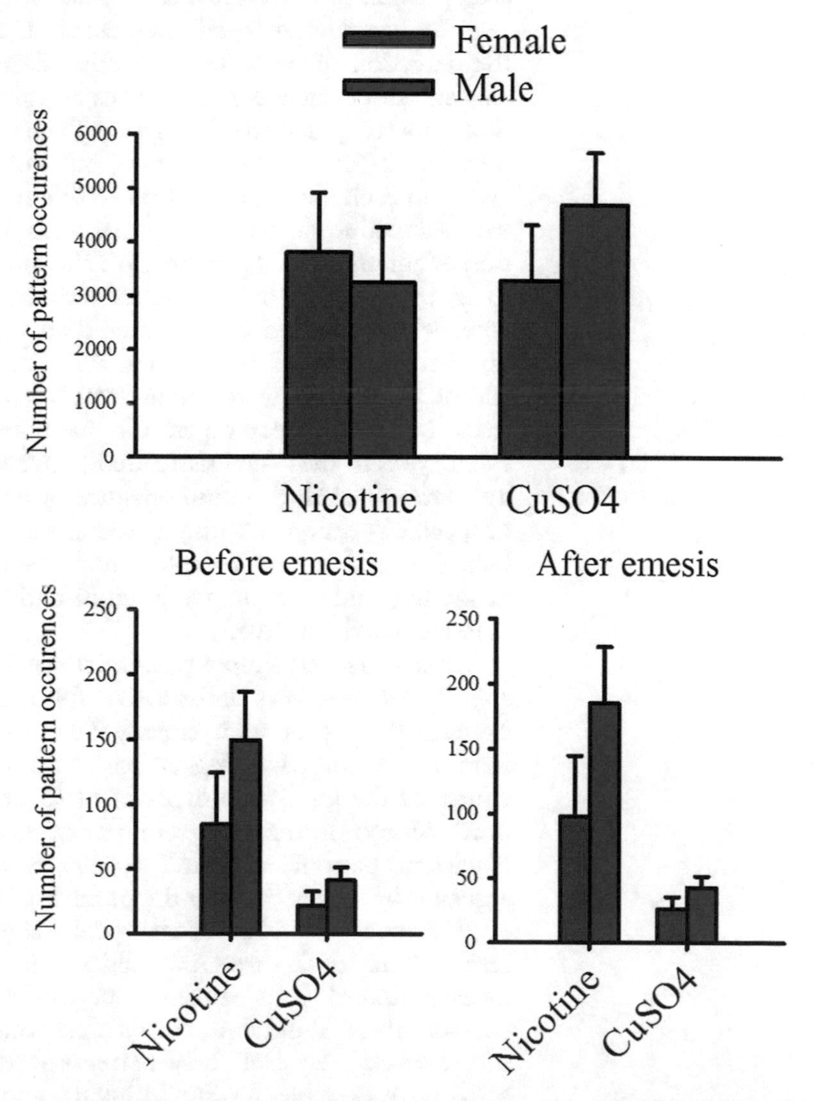

Fig. 10 Effects of nicotine (5 mg/kg, sc) and copper sulfate ($CuSO_4$, 120 mg/kg, po) on the total number and emesis-related T-patterns detected in male and female shrews ($n \geq 27$, each group). Results represent the mean ± SEM

4 Discussion

T-pattern analysis revealed a large number of patterns of behavior associated with emesis in musk shrews, including sniffing, changes in body contraction, and locomotion. There was little evidence that locomotion was inhibited in the 72 h study by the occurrence of emesis (using either of the two metrics; mc, discrete velocity cutoffs or dchi, distance moved, 2 standard deviations above the mean). However, eating was not significantly associated with emesis-related behavioral patterns. Eating and drinking, and other larger body movements, including rearing, grooming, and body rotation, were significantly less common events contained in emesis-related behavioral patterns in real versus randomized control data.

The current methods represent a balanced approach to the detection of behavioral patterns. Emesis is more difficult to analyze with t-pattern analysis because there are many fewer emesis events compared to other types of behavior. For example, there were 1000s of events of sniffing but only from 1 to 33 emetic events in each file. As a solution to this sparseness of emesis data, we used a minimum of 2 occurrences (in 2 animals) for the detection of emesis-related patterns. We also substantially narrowed the focus by selecting only patterns containing emesis as an event type, which resulted in a greater than tenfold reduction in the number of patterns in real and random detection of patterns. Although patterns were unique to some animals, it was obvious from the set of detected patterns that many animals share similar event types in these emesis-related patterns. We therefore looked more closely at these commonalities by analyzing the occurrence of specific event types within t-patterns in the 72 h data. Ingestive behavior and more complex movements, including rearing, grooming, and rotation, were significantly less common events in emesis-related patterns.

It appears that simple plots of single behaviors (Figs. 3, 4, 5, and 9) are not very informative for understanding behavioral changes that occur with emesis. There was little change in the number of eating, drinking, and other bouts of movement over the course of the long-term experiment (3 days after cisplatin injection). Moreover, the greater impact of nicotine on emesis-related behavioral patterns compared to copper sulfate treatment is not apparent by simply totaling the number of emetic responses.

We detected 1000s of statistically significant behavioral patterns in the musk shrew that did not include emesis. Although investigation of these patterns is beyond the scope of the current chapter, these should provide a rich source of information for future work. Many of these patterns were revealed as bursts of activity, for example, feeding, drinking, and locomotion. Although some of these patterns were not associated with emesis, their frequency might be affected by emesis and sickness. Reductions in non-emesis-associated behavioral patterns could provide insight into changes in the allocation of behavior during sickness. It is likely that animals will curtail other behavioral sequences as a result of sickness.

These studies represent the first use of a computational approach to understand emesis-related behavioral change and could provide a novel way to develop a behavioral index of nausea. Surrogate markers of nausea in animal studies have included conditioned flavor aversion, pica, and physiological measures. Many of these studies have relied on the use of rats or mice, which, as rodents, lack a vomiting response [31–33]. It has been suggested that conditioned gaping in the rat, produced by taste aversion learning, could be used as a marker for nausea [34]. However, injection of rats with emetic agents does not produce this response, and thus, gaping appears to be an emergent process of conditioning [35]. This indicates that the occurrence of gaping as an index of nausea is questionable. Laboratory rats also ingest kaolin clay (a pica response) when injected with toxins, such as cisplatin, and this response has been used as a marker of sickness [36, 37]. The amount of clay ingestion in the rat induced by different chemotherapies is related to the emetic potency of these agents in humans [38]. Like conditioned gaping, it is difficult to relate the pica response to nausea in humans. Physiological metrics of nausea, such as salivation [39], gastric dysrhythmia [7, 8], and systemic vasopressin release [3] occur in humans [40], but these can be difficult to measure and can require invasive procedures [7, 8, 41]. The current results suggest a useful, noninvasive, approach to measuring sickness or nausea in an animal model.

The current approach might eventually lead to a focus on patterns of behavioral change as appropriate targets for assessing the more global effects of potential anti-nausea drugs. Conversely, having a more thorough assessment of animal behavioral patterns might show potential limitations of antiemetic drugs to control nausea and/or visceral sickness (reduced food intake). These approaches could open a new door into nausea and emesis research to examine behavioral patterns associated with vomiting that are produced by a large number of drugs and diseases. Complex high dimensional behavioral analyses have become a growing force in the field of behavioral neuroscience, particularly in the areas of anxiety and pain research [13, 14, 30, 42, 43]. Researchers have been challenged to find reproducible behavioral metrics of psychological states and it is increasingly clear that single measures are unreliable. We believe that this also applies to work in the field of nausea and emesis. The current studies suggest a new approach that should foster the formation of computational models of behavioral change produced by emetic agents and facilitate drug development.

References

1. Olver I, Molassiotis A, Aapro M, Herrstedt J, Grunberg S, Morrow G (2010) Antiemetic research: future directions. Support Care Cancer 19(Suppl 1):S49–S55
2. Sanger GJ, Andrews PL (2006) Treatment of nausea and vomiting: gaps in our knowledge. Auton Neurosci 129:3–16
3. Billig I, Yates BJ, Rinaman L (2001) Plasma hormone levels and central c-Fos expression in ferrets after systemic administration of cholecystokinin. Am J Physiol Regul Integr Comp Physiol 281:R1243–R1255
4. De Jonghe BC, Horn CC (2008) Chemotherapy-induced pica and anorexia are reduced by common hepatic branch vagotomy in the rat. Am J Physiol Regul Integr Comp Physiol 294:R756–R765
5. Kanoski SE, Rupprecht LE, Fortin SM, De Jonghe BC, Hayes MR (2012) The role of nausea in food intake and body weight suppression by peripheral GLP-1 receptor agonists, exendin-4 and liraglutide. Neuropharmacology 62:1916–1927
6. Parker LA (2006) The role of nausea in taste avoidance learning in rats and shrews. Auton Neurosci 125:34–41
7. Percie du Sert N, Chu KM, Wai MK, Rudd JA, Andrews PL (2009) Reduced normogastric electrical activity associated with emesis: a telemetric study in ferrets. World J Gastroenterol 15:6034–6043
8. Percie du Sert N, Chu KM, Wai MK, Rudd JA, Andrews PL (2010) Telemetry in a motion-sickness model implicates the abdominal vagus in motion-induced gastric dysrhythmia. Exp Physiol 95:768–773
9. Kent S, Bluthe RM, Kelley KW, Dantzer R (1992) Sickness behavior as a new target for drug development. Trends Pharmacol Sci 13:24–28
10. Bermudez J, Boyle EA, Miner WD, Sanger GJ (1988) The anti-emetic potential of the 5-hydroxytryptamine3 receptor antagonist BRL 43694. Br J Cancer 58:644–650
11. Watson JW, Gonsalves SF, Fossa AA, McLean S, Seeger T, Obach S, Andrews PL (1995) The anti-emetic effects of CP-99,994 in the ferret and the dog: role of the NK1 receptor. Br J Pharmacol 115:84–94
12. Magnusson MS (2000) Discovering hidden time patterns in behavior: T-patterns and their detection. Behav Res Methods Instrum Comput 32:93–110
13. Bonasera SJ, Schenk AK, Luxenberg EJ, Tecott LH (2008) A novel method for automatic quantification of psychostimulant-evoked route-tracing stereotypy: application to Mus musculus. Psychopharmacology (Berl) 196:591–602
14. Casarrubea M, Sorbera F, Magnusson M, Crescimanno G (2010) Temporal patterns analysis of rat behavior in hole-board. Behav Brain Res 208:124–131
15. Chan SW, Rudd JA, Lin G, Li P (2007) Action of anti-tussive drugs on the emetic reflex of Suncus murinus (house musk shrew). Eur J Pharmacol 559:196–201
16. Horn CC, Henry S, Meyers K, Magnusson MS (2011) Behavioral patterns associated with chemotherapy-induced emesis: A potential signature for nausea in musk shrews. Front Neurosci 5:88
17. Huang D, Meyers K, Henry S, De la Torre F, Horn CC (2011) Computerized detection and analysis of cancer chemotherapy-induced emesis in a small animal model, musk shrew. J Neurosci Methods 197:249–258
18. Henningfield JE, Miyasato K, Jasinski DR (1985) Abuse liability and pharmacodynamic characteristics of intravenous and inhaled nicotine. J Pharmacol Exp Ther 234:1–12
19. Pizarro F, Olivares M, Uauy R, Contreras P, Rebelo A, Gidi V (1999) Acute gastrointestinal effects of graded levels of copper in drinking water. Environ Health Perspect 107:117–121
20. Warr DG, Grunberg SM, Gralla RJ, Hesketh PJ, Roila F, Wit R, Carides AD, Taylor A, Evans JK, Horgan KJ (2005) The oral NK(1) antagonist aprepitant for the prevention of acute and delayed chemotherapy-induced nausea and vomiting: Pooled data from 2 randomised, double-blind, placebo controlled trials. Eur J Cancer 41:1278–1285
21. Andrews PL, Okada F, Woods AJ, Hagiwara H, Kakaimoto S, Toyoda M, Matsuki N (2000) The emetic and anti-emetic effects of the capsaicin analogue resiniferatoxin in Suncus murinus, the house musk shrew. Br J Pharmacol 130:1247–1254
22. De Jonghe BC, Horn CC (2009) Chemotherapy agent cisplatin induces 48-h Fos expression in the brain of a vomiting species, the house musk shrew (Suncus murinus). Am J Physiol Regul Integr Comp Physiol 296:R902–R911
23. Horn CC, Still L, Fitzgerald C, Friedman MI (2010) Food restriction, refeeding, and gastric fill fail to affect emesis in musk shrews. Am J Physiol Gastrointest Liver Physiol 298: G25–G30

24. Sam TS, Cheng JT, Johnston KD, Kan KK, Ngan MP, Rudd JA, Wai MK, Yeung JH (2003) Action of 5-HT3 receptor antagonists and dexamethasone to modify cisplatin-induced emesis in Suncus murinus (house musk shrew). Eur J Pharmacol 472:135–145
25. Rissman EF, Silveira J, Bronson FH (1988) Patterns of sexual receptivity in the female musk shrew (Suncus murinus). Horm Behav 22:186–193
26. Castaner M, Torrents C, Anguera MT, Dinusova M, Jonsson GK (2009) Identifying and analyzing motor skill responses in body movement and dance. Behav Res Methods 41:857–867
27. Kemp AS, Fillmore PT, Lenjavi MR, Lyon M, Chicz-Demet A, Touchette PE, Sandman CA (2008) Temporal patterns of self-injurious behavior correlate with stress hormone levels in the developmentally disabled. Psychiatry Res 157:181–189
28. Kerepesi A, Jonsson GK, Miklosi A, Topal J, Csanyi V, Magnusson MS (2005) Detection of temporal patterns in dog-human interaction. Behav Processes 70:69–79
29. Lyon M, Kemp AS (2004) Increased temporal patterns in choice responding and altered cognitive processes in schizophrenia and mania. Psychopharmacology (Berl) 172:211–219
30. Casarrubea M, Sorbera F, Magnusson MS, Crescimanno G (2010) T-pattern analysis of diazepam-induced modifications on the temporal organization of rat behavioral response to anxiety in hole board. Psychopharmacology (Berl) 215:177–189
31. Andrews PL, Horn CC (2006) Signals for nausea and emesis: implications for models of upper gastrointestinal diseases. Auton Neurosci 125:100–115
32. Andrews PLR (1995) Why do some animals lack a vomiting reflex? Physiol Zool 68:61
33. Horn CC, Kimball BA, Gathright GR, Yates BJ, Andrews PLR (2010) Why don't rats and mice vomit? A behavioral and anatomical investigation. Appetite 54:650
34. Parker LA, Limebeer CL (2006) Conditioned gaping in rats: a selective measure of nausea. Auton Neurosci 129:36–41
35. Yamamoto K, Ngan MP, Takeda N, Yamatodani A, Rudd JA (2004) Differential activity of drugs to induce emesis and pica behavior in Suncus murinus (house musk shrew) and rats. Physiol Behav 83:151–156
36. Takeda N, Hasegawa S, Morita M, Horii A, Uno A, Yamatodani A, Matsunaga T (1995) Neuropharmacological mechanisms of emesis. II. Effects of antiemetic drugs on cisplatin-induced pica in rats. Methods Find Exp Clin Pharmacol 17:647–652
37. Takeda N, Hasegawa S, Morita M, Matsunaga T (1993) Pica in rats is analogous to emesis: an animal model in emesis research. Pharmacol Biochem Behav 45:817–821
38. Yamamoto K, Nakai M, Nohara K, Yamatodani A (2007) The anti-cancer drug-induced pica in rats is related to their clinical emetogenic potential. Eur J Pharmacol 554:34–39
39. Furukawa N, Fukuda H, Hatano M, Koga T, Shiroshita Y (1998) A neurokinin-1 receptor antagonist reduced hypersalivation and gastric contractility related to emesis in dogs. Am J Physiol 275:G1193–G1201
40. Koch KL (1997) A noxious trio: nausea, gastric dysrhythmias and vasopressin. Neurogastroenterol Motil 9:141–142
41. Lau AH, Rudd JA, Yew DT (2005) Action of ondansetron and CP-99,994 on cisplatin-induced emesis and locomotor activity in Suncus murinus (house musk shrew). Behav Pharmacol 16:605–612
42. Langford DJ, Bailey AL, Chanda ML, Clarke SE, Drummond TE, Echols S, Glick S, Ingrao J, Klassen-Ross T, Lacroix-Fralish ML, Matsumiya L, Sorge RE, Sotocinal SG, Tabaka JM, Wong D, van den Maagdenberg AM, Ferrari MD, Craig KD, Mogil JS (2010) Coding of facial expressions of pain in the laboratory mouse. Nat Methods 7:447–449
43. Ramos A, Pereira E, Martins GC, Wehrmeister TD, Izidio GS (2008) Integrating the open field, elevated plus maze and light/dark box to assess different types of emotional behaviors in one single trial. Behav Brain Res 193: 277–288

Chapter 14

Informative Value of Vocalizations during Multimodal Interactions in Red-Capped Mangabeys

Isabelle Baraud, Bertrand L. Deputte, Jean-Sébastien Pierre, and Catherine Blois-Heulin

Abstract

Social interactions can be viewed as "dialogs" during which messages between partners can be multimodal, consisting of several simultaneous or sequential signals from different sensory modalities. The nature of a message and the way it will be interpreted may depend on the signals, their order, but also the time intervals between them. This study analyzed the sequential organization of multimodal interactions with and without vocalizations to determine the influence of vocalizations on the behavioral structure of these interactions; to this purpose, we used Theme software which extracts behavioral structures ("patterns") within behavioral sequences. These behavioral structures are characterized by their temporal organization. Our results on a captive group of red-capped mangabeys (*Cercocebus torquatus torquatus*) showed that during sociosexual interactions, vocalizations were signals associated with decision-making process of the caller, to initiate contact with a partner some distance away. During grooming interactions, in contrast, vocalizations were not associated with the decision-making process of initiating an interaction. Patterns extracted from interactions with and without calls differed in the number, nature, and position of the decision points they included. So, information conveyed by vocalizations was neither fully original nor redundant with that conveyed by other sensory modalities. Calls seem to reduce complexity of interactions and to decrease the number of decision points partners used to mutually adjust their behaviors. Calls also play a role in structuring interactions as they may occur in patterns either at the beginning or in the "middle" of the interactions but never at the end of them.

Key words Cercocebus torquatus torquatus, Multimodality, Red-capped mangabeys, Sequential analysis, Social interactions, Vocalizations

1 Introduction

According to the seminal study of Weaver and Shannon [1], communication is an exchange of information between a sender and a receiver, by one or several transmission channels. During social interactions, the sender and the receiver may use signals in several sensory modalities [2–4]. During a social interaction, potentially most of behaviors, whatever their complexity and diversity, could be considered as communicative events [5]. As pointed out by

Magnus S. Magnusson et al. (eds.), *Discovering Hidden Temporal Patterns in Behavior and Interaction: T-Pattern Detection and Analysis with THEME™*, Neuromethods, vol. 111, DOI 10.1007/978-1-4939-3249-8_14,

Marler [2], the meaning of a combination of signals affects the meaning of each signal taken separately ("metacommunication," 6). This multimodality has already been highlighted in some birds [7–11], in primates (*Macaca fascicularis*, [12]; *Macaca arctoides*, [13], *Macaca mulatta*, [14, 15]) and in some other mammals (black-tailed prairie dog, [16]; ground squirrels, [17]). In birds, courtship and territorial displays typically involve simultaneous and/or sequential bimodal (vocal and visual) communicative processes. In nonhuman primates, simultaneous olfactory and visual signals are used in sexual communication (*Macaca mulatta*, 14).

A dyadic interaction is a behavioral flow between two individuals, characterized by a certain time structure [18]. During a dyadic interaction, some behaviors can be viewed as "decision points" made by the partners. For example,

Pitcairn and Schleidt [19] have analyzed how these decisions structure courtship dances in the New Guinean Medlpa, leading to some behavioral regularities ("patterns"). Generally speaking, in an interaction behaviors might be used by both partners to make their own decisions. Such sequential contingencies have been represented as a chain of events [20, 21]. The output of the interaction thus depends on the precise order in which the various signals are emitted and perceived [22].

The decision one individual can make may be based not on the immediately previous event but on one having occurred some time previously [23, 24]. This emphasizes the importance of the temporal structure of the interaction and the time intervals between signals [18]. When an individual's behavioral flow is divided into a succession of behavioral units, these units (monads) show a certain amount of repetition. When these units are grouped in pairs (dyads) or higher-order sequences, the repetition rate rapidly decreases. Therefore, complex sequences are likely to be unique, making any comparative process difficult. Hence, to compare sequences of behaviors from interacting partners, we have to find a means to preserve the sequence as it was observed, without using any rule to transform it into a shorter sequence with a certain repetition rate [25].

This study analyzes sequences to account for the multiple decisions made during multimodal interactions occurring with and without vocalizations, in red-capped mangabeys, highlighting the temporal and behavioral structures of these interactions in order to determine the influence of vocalizations on the behavioral structure of these interactions. We applied the THEME method developed by Magnusson [26], as an adequate tool to identify structures of interactions and to make possible the comparisons of complex sequences of behavior. We therefore (1) analyze the sequential organization of multimodal interactions while preserving their behavioral richness, and (2) compare the structure of interactions involving vocalizations with those lacking vocalizations in order to illustrate the role of multimodal signals in shaping social interactions.

We hypothesize that the emission of vocalizations will add information and thus reduce the complexity of other behavioral acts observed.

2 Methods

2.1 Subjects

We carried out this work on a family group of four red-capped mangabeys (*Cercocebus torquatus torquatus*) living at the primate colony of the Biological Station in Paimpont (France): one adult male (B, 19 years), one adult female (A, 12 years), and their two sexually mature offspring (F, an 8-year-old male; C, a 6-year-old female). The group lived in a cage including both indoor (around 8.30 $m^2 \times 2.8$ m) and outdoor (21 $m^2 \times 4$ m) sections. The two sections were linked via a sliding door, and both were equipped with perches and climbing structures. Animals were fed twice a day with fresh fruits in the morning and pellets (SAFE®) in the afternoon. Water was available ad libitum.

2.2 Data Collection

A system of "scheduling sample sessions" was used [27]. This involved sampling sequential dyadic interactions. The onset of an interaction was defined as the occurrence of a behavior such as a gaze directed at a partner, an addressed vocalization (i.e., calling while looking at a partner), an approach, etc. which leads to a behavioral response from a congener (adapted from Altmann's [27] "behavior dependent" "scheduling of session onsets"). The ensuing interaction implied a contact or at least a close proximity between partners. The sampling session ended either when an individual left its partner (interindividual distance greater than one arm length) or initiated an interaction with another partner (adapted from Altmann's [27] "scheduling of session terminations").

All interactions were videotaped—with a camcorder Sony DCR-TRV 330E—(45 min in the morning and 45 min in the afternoon, before mealtimes) for a total of 12 h of video footage, during 2 months. Observations were made simultaneously with the recordings.

In groups of primates where social relations are stable, as the one we studied, there often are not enough interactions to get a large sample of interactions for each dyad of individuals. The individuals we studied were closely related, but even if communication was idiosyncratic because of these relationships, we do not think that this would generate modes of communication impossible for the species. So this study should be viewed as a case study; we will be cautious in the interpretation of the results and we cannot generalize to other groups.

We used an open and detailed repertoire of 191 behaviors, adapted from Deputte's [22] repertoire for grey-cheeked mangabeys. These behaviors were associated with four sensory modalities:

visual behaviors (78 motor-gestural behaviors, corresponding to the body movements of the animal, and 23 gazes), six olfactory behaviors, 74 tactile behaviors, and 10 vocal types (Bovet, pers. comm.). See examples of behaviors from the behavioral catalogue in the Appendix.

2.3 Sequences Sorting

We compared behavioral sequences with and without vocalizations. From an initial sample of 224 interactions, we extracted, according to a "target behavior," grooming and sociosexual interactions, 62 interactions with calls and 34 without calls. The "target behaviors" permit us to define the interactions. We considered two types of interactions: (1) For sociosexual interactions, the target behavior was "individual X presents its genitals to individual Y" and was not followed by copulation [28] or grooming. (2) For grooming interactions, the target behavior was "individual X grooms individual Y." A target behavior never signals the beginning of an interaction.

2.4 Definitions of Terms

Here are the definitions of terms used in this study.

Act: a behavioral event = a behavior starting or ending; for example: starting walking towards, or ending looking at.

Behavioral sequence: a succession of different acts.

Interaction: a behavioral sequence including behaviors from two communicating individuals.

Decision point: "a moment of high information content from which the rest of the behaviour, being redundant, can, potentially be reconstructed. A decision then is an event which leads to a sudden decrease in the uncertainty of future behaviour" [29].

Pattern: a cluster of decision points, i.e., a combination of events characterized by fixed event order and significantly similar time distances between the consecutive parts over occurrences of the cluster. As this cluster is stable, once the first act of a pattern has occurred, the other acts constituting this pattern will appear in a predictable order. So a pattern is included in a behavioral sequence (i.e., in an interaction in our study). Even if a sequence contains vocalizations, a pattern extracted from this sequence does not necessarily contain these vocalizations. For more details, see Data Analysis below.

Based on our results, we distinguished, a posteriori, four types of patterns (Fig. 5):

- A "starting pattern" is a pattern containing only acts uttered towards the beginning of the interactions, but the act marking the onset of an interaction does not necessarily appear in such a pattern.
- "Ending pattern" is a pattern containing only acts uttered towards the end of the interactions, but the act marking the end of an interaction does not necessarily appear in such a pattern.
- "Middle pattern" is a pattern containing only acts uttered during the course of the interactions. It may or may not contain the target behavior.

– "Complete pattern" is a pattern containing acts that occur towards the beginning, the middle, and the end of the interactions.

2.5 Data Analysis

We analyzed the sequential ordering of signals in the four sensory modalities from both partners in the interactions, as well as the time intervals between the behaviors. As previously mentioned two complex sequences are almost never identical and are therefore not directly comparable. However software implementing a new approach to analyzing sequences is now available, called THEME [26, 30, 31]. THEME was primarily developed for analyzing verbal and/or nonverbal interactions in humans [32, 33] and has been successfully applied to the study of various types of behavior in mice [34], hen chicks [35], humans [36], Drosophila [37], and to the study of neuronal interactions, DNA and proteins, strings, traces, and control in cells and cities [38], etc. Instead of calculating the transition matrix for two successive acts within sequences [39], THEME detects temporal and behavioral structures, or "patterns."

For the sake of pattern detection with THEME, interactions for a given dyad were concatenated into a single sequence. Within this new sequence, all interactions were "separated" by setting a temporal gap between them. Hence, if a pattern was detected, it would be restricted to a single interaction and not bridge two successive interactions.

In Magnusson's THEME method, a T-pattern is a repetitive structure with a dual feature:

1. Sequential organization. The elements of the pattern are ordered in time, but not in direct succession. In the period separating two successive elements of the pattern, several other elements may occur randomly.
2. Hierarchical organization. T-patterns are represented as hierarchical trees, where the length of the branches is proportional to the time lag between the linked elements (i.e., leaves).

To summarize, T-patterns may be viewed as classification trees subject to a temporal restriction between leaves which must be ordered on an increasing time scale. The building of such trees proceeds from the bottom up. Critical time intervals are detected between pairs of elementary acts (e.g., behaviors, vocalizations). A critical interval is detected when the time lag between two acts deviates strongly from random (according to an exponential model) and is statistically restricted between a lower and an upper bound. Those critical intervals are detected on the basis of a risk of first kind, α defined by the program user. New critical intervals are built up recursively, by linking lower-level T-patterns together. A number of tuning parameters may be used in the process of pattern detection, the two most important being α (defined above) and Nmin, the minimum number of times a pattern must occur. There is no available theory giving the probability of detecting by chance a completely random pattern for a given α and Nmin. A statistical validation of the type of pattern obtained is therefore provided by a random permutation test of the acts followed by the pattern-detection process achieved with the same set of parameters as for the true succession. As an example, if act Q is detected as occurring in a given critical interval after P, and similarly S after R, then two T-patterns (PQ) and (RS) are detected. If, furthermore, a critical interval exists between P and R or P and S, or between Q and R or Q and S, then (PQ) and (RS) would be grouped in the T-pattern ((PQ), (RS)). In the next step, a single act T may be incorporated to give ((PQ), (RS), T), and so on.

This method is very powerful in detecting complex temporal structures where random acts may be inserted between other acts that systematically succeed one another. After having analyzed the data, we distinguished two kinds of acts: random acts which reflect the uncertainties of the individual (i.e., unpredictable acts) and nonrandom acts which correspond to a decision made by the individual. This decision results in the expression of "decision points" [29]. These decision points are linked by a temporal structure (i.e., predictable acts). Figure 1 summarizes our sampling decisions, showing how the THEME software has allowed us to compare a large sample of sequences and how interactions are structured (incorporating patterns, decision points, and target behaviors).

For the comparison between silent interactions and those with vocalizations, among the entire sample of dyads, we only retained the dyads with the most interactions with and without calls and which were represented in the two types of sequences. We focused especially on the longest patterns (because the shorter patterns are included within the longest ones) which included the target behavior and/or calls. In this way, six types of patterns could be defined: In interactions with calls, four patterns contained as decision points either (1) the target behavior and vocalization(s), either (2) the target behavior but no vocalization, either (3) vocalization(s) but not the target behavior, or (4) no vocalization and no target behavior. And in interactions without vocalization, two patterns contained (5) the target behavior or (6) not. We note that even when there are vocalizations within the interactions, these vocalizations are not necessarily in patterns. The risk of first kind was set at $\alpha = 0.005$ (because given all the simultaneous tests made by THEME, we needed to set a threshold of risk less than 0.05) and we required at least two occurrences for each pattern ($N \geq 2$). We determined the number of different longest patterns of each type, the number of acts they included, and the number of times they occurred. Finally, we compared the number, nature, and position of the acts present in the longest patterns from interactions with and

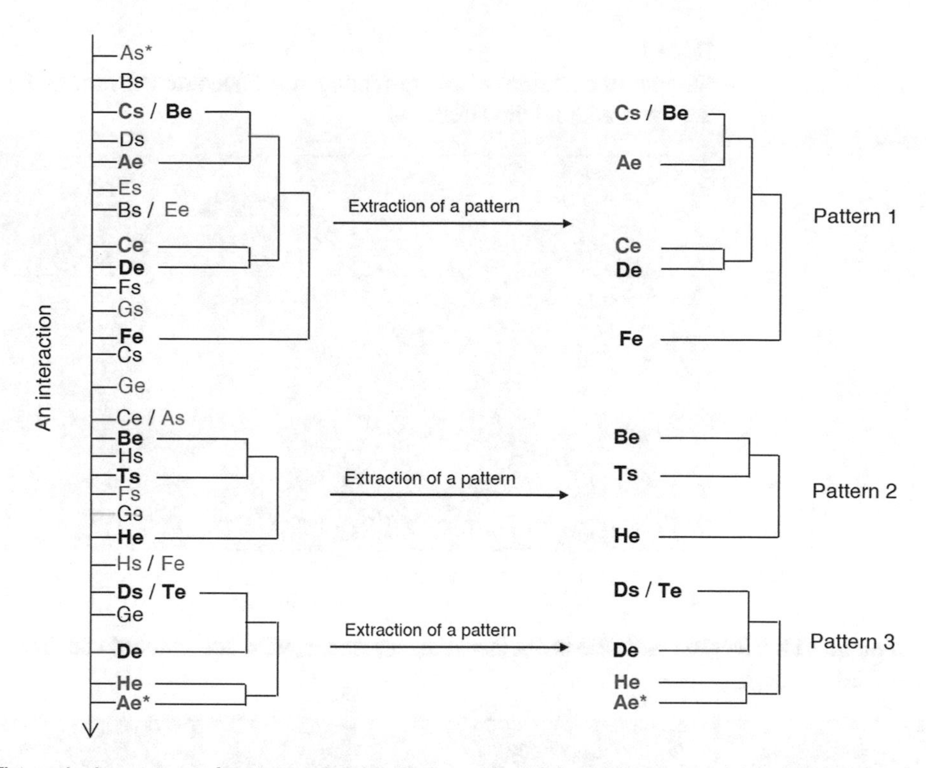

Fig. 1 Theoretical structure of an interaction. *Letters* = acts displayed by two social partners. T letter = target behavior. *Letters in grey* = acts emitted by an individual. *Letters in black* = acts emitted by another individual. *Bold letters* = acts as decision points; they are participating to three different patterns. Asterisk (*) indicates onset and ending acts of the interaction; see "Data collection" for details. : time of occurrence of an act. Slash (/) separates 2 simultaneous acts. *s*: onset of an act. *e*: ending of an act

without vocalizations (nonparametric tests: Mann–Whitney, $\alpha = 0.05$). We also observed the position of the patterns within the interactions (Chi-square tests, $\alpha = 0.05$).

3 Results

3.1 Sociosexual Sequences

The dyad BC, composed of the male B and the female C, is the one for which we sampled the most sociosexual interactions (Table 1). The 11 sociosexual sequences out of 26 total sequences for this dyad were used for extraction of patterns with THEME (Table 1).

In this dyad BC, four types of patterns emerged within the interactions with vocalizations ("vocal" interactions, Table 2). The number of acts within these patterns varied from 6 to 14 (Table 2), some of which were "Hon" calls. These patterns with "Hon" calls are 6 or 8 act long. It is important to note that a vocalization could have been recorded during an interaction but not be included in the patterns extracted by THEME on the basis of temporal regularities

Table 1
Number of sociosexual and grooming sequences for the different dyads with and without vocalizations

	DYADS					
Types of interactions	**BC**	**BA**	**FC**	**AF**	**AC**	**BF**
Sociosexual sequences						
With vocalizations	6	2	3	0	1	0
Without vocalization	5	3	2	1	2	1
Grooming sequences						
With vocalizations	8	2	5	9	24	0
Without vocalization	3	1	2	3	11	0

Table 2
Characterization of the longest patterns in sociosexual sequences, with and without vocalizations, for the dyad BC

	Variables		
Types of patterns	**Number of different patterns**	**Number of occurrences of each pattern**	**Number maximal of acts in a pattern**
Sequences with vocalizations			
Pattern with Presenting (no call)	1	2	14
Pattern with "Hon" call (no presenting)	1	2	6
Pattern with presenting and "Hon" call	1	2	8
Patterns without presenting and without call	1	2	7
Sequences without vocalization			
Pattern with presenting	2	2	7
Pattern without Presenting	2	3	5

(expression of acts with a statistically constant time interval between some acts). The pattern including "presenting" but no vocalization was the longest one (i.e., containing the most acts; Table 2). In silent interactions, two types of patterns, five and seven acts long, were detected (Table 2). These patterns had almost the same length as those including a "Hon" call (Table 2).

The patterns extracted from vocal interactions and including "presenting" and/or vocalizations began in two different ways:

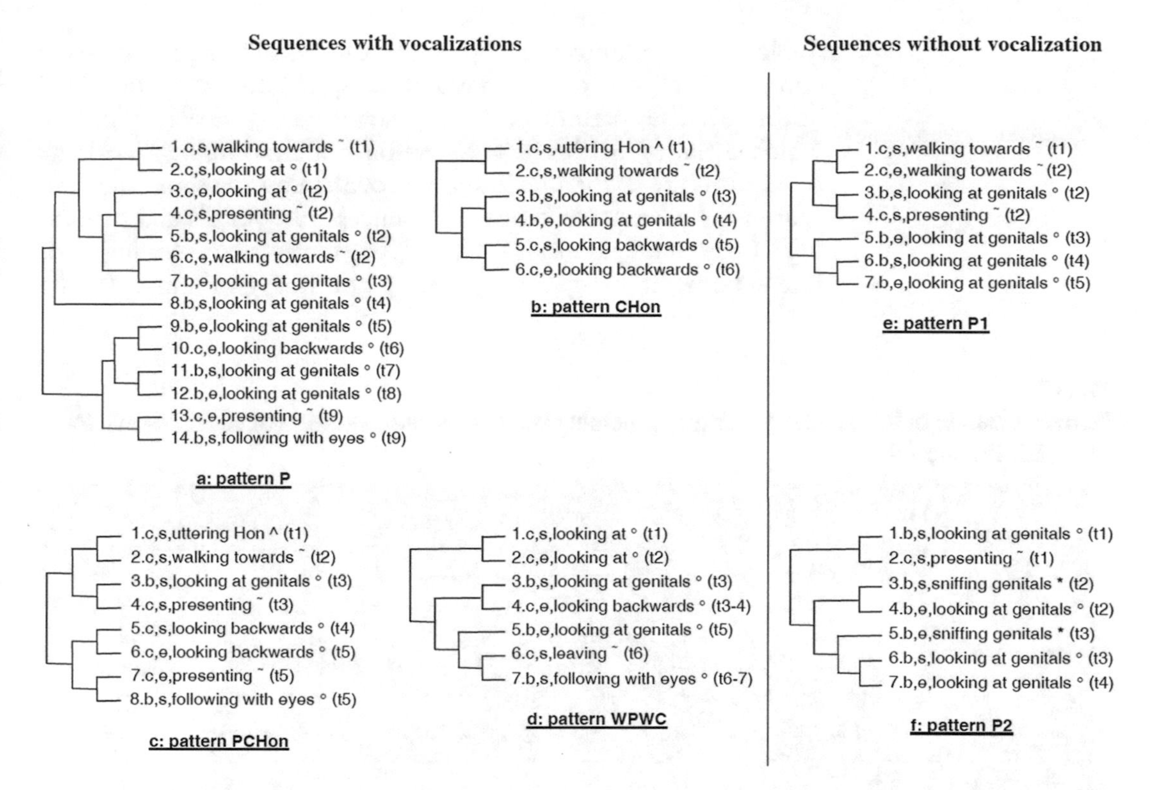

Fig. 2 The different types of the longest patterns in male–female sociosexual interactions with and without vocalizations. *c*: the female; *b*: the male; *s*: start; *e*: end; *tx*: the act appears at the moment *x*, in a chronological order. ~ = motor-gestural acts; ° = gazes; ^ = vocal acts; # = tactile acts; * = olfactive acts. *P*: pattern with presentation (without call); CHon: pattern with "Hon" call (without presentation); PCHon: pattern with presentation and "Hon" call; WPWC: pattern without presentation and without call; WP: pattern without presentation (without call). For more details about the construction of the trees, see paragraph "Data Analysis"

either with an addressed approach ("X walks towards Y" associated with a gaze; Fig. 2a) or with a "Hon" call, uttered by the female, just prior to the approach ("starting call," Fig. 2b, c). The first nonvocal act of patterns from silent interactions is the same as those after the "starting call" in vocal patterns (Fig. 2e vs. b–c).

When interactions were fully structured ("complete pattern"), or partially structured ("starting pattern" or "ending pattern"), starting and/or ending acts of the patterns were similar ("Hon" call and/or approach as the starting act, and "ended presenting" and/or leaving as the ending act; Fig. 2a–e). When patterns were detected in the course of interactions ("middle pattern"), they started with the target behavior (Fig. 2f).

3.2 Grooming Sequences

The dyads BC and AC emitted the most grooming interactions we sampled (Table 1). In the male–female dyad (BC), the number of patterns detected with THEME software in "vocal" sequences was greater than that in silent sequences (10 versus 1 respectively;

Table 3). In addition, the patterns of "vocal" sequences were composed of 6 or 10 nonvocal acts, including "grooming" (Table 3). One pattern of "vocal" sequences (WGWC; Fig. 4e), which included neither a vocalization nor "grooming," had the same structure as the previous ones, containing the same acts present in the patterns of "vocal" sequences, except calls and grooming. Only one pattern (WG; Fig. 4g) was detected in silent interactions. It comprised 11 acts, but not grooming. In the

Table 3
Characterization of the longest patterns in grooming sequences with and without vocalizations, for the dyads BC and AC

Types of patterns	Variables					
	Dyad BC			Dyad AC		
	Number of different patterns	Number of occurrences of each pattern	Number maximal of acts in a pattern	Number of different patterns	Number of occurrences of each pattern	Number maximal of acts in a pattern
Sequences with vocalizations						
Pattern with grooming (no call)	2	5	10	2	8 or 10	5
Pattern with "Hon" call (no grooming)	2	5	7	2	10	5
Pattern with "Tipiak" call (no grooming)	1	5	7	–	–	–
Pattern with grooming and "Hon" call	1	5	6	1	12	3
Pattern with grooming and "Tipiak" call	2	5	6	–	–	–
Pattern without grooming and without call	2	5	7	2	10	6
Sequences without vocalization						
Pattern with grooming	–	–	–	1	4	8
Pattern without grooming	1	2	11	9	4	8

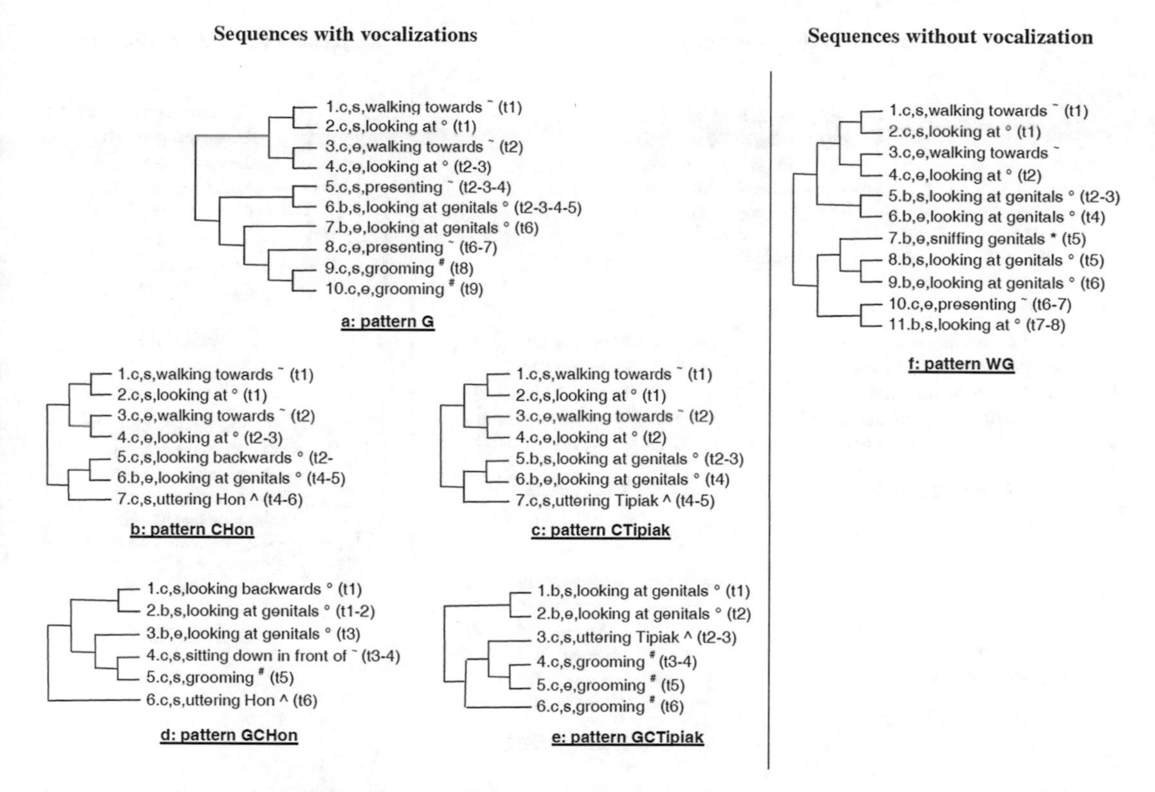

Fig. 3 The different types of the longest patterns in male–female grooming interactions with and without vocalizations. For sequences with calls, only one tree of each type was presented. *G*: pattern with grooming (without call); *CHon*: pattern with "Hon" call (without grooming); *CTipiak*: pattern with "Tipiak" call (without grooming); *GCHon*: pattern with grooming and "Hon" call; *GCTipiak*: pattern with grooming and "Tipiak" call; *WGWC*: pattern without grooming and without call; *WG*: pattern without grooming (without call). Other legends: see Fig. 2

female–female dyad (AC), four and two types of patterns were detected in "vocal" and silent interactions, respectively (Table 3). In this dyad, the most complex pattern in "vocal" sequences comprised six acts, without grooming and vocalization. The patterns containing calls were three or five acts long (Table 3). In silent interactions, the detected patterns involved eight acts (Table 3). Hence, for both dyads, "vocal" patterns, those including vocalizations, were shorter, i.e., containing fewer nonvocal acts, than those present in silent interactions (Table 3; Mann–Whitney tests: $Z=2.37$, $N1=2$, $N2=30$, $P=0.0088$ for the dyad BC; $Z=7.25$, $N1=32$, $N2=40$, $P=0.0001$ for the dyad AC).

All interactions considered from the very beginning started with the same silent approach (Figs 3a–c, f and 4a–c, f, g). All "middle patterns" included the target behavior (Figs. 3d, e and 4d) whereas only three out nine "starting patterns" included that behavior (Figs. 3a and 4a, f).

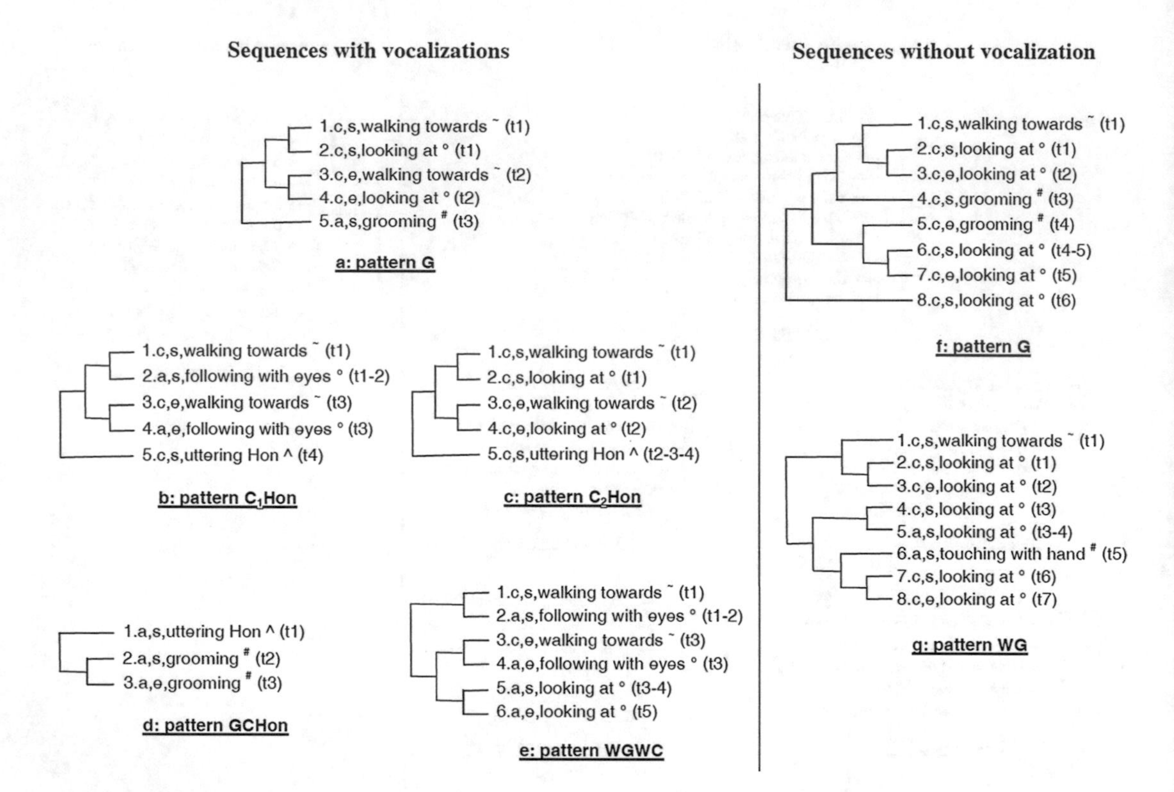

Fig. 4 The different types of the longest patterns in female–female grooming interactions with and without vocalizations. Only one tree for each type of patterns G, WGWC from sequences with calls, and WG from sequences without call are presented. c: the female; a: the other female. Other legends: see Figs. 2 and 3

In both BC and AC dyads, calls appearing in the patterns ("Hon" and "Tipiak") were usually preceded by visual acts and followed by tactile ones. Indeed, when a call was preceded by an act in the patterns (which is the case in six patterns: Figs. 3b, e and 4b, c), it was a visual act (gaze) in five out of six cases. In the sixth case, it was a tactile act of grooming (cf dyad BC; Fig. 3d). When a call was followed by an act in the patterns (which was the case in two patterns: Figs. 3e and 4d), it was a tactile act (grooming) in two of two cases. But, these calls, followed by a tactile act, were not the same, depending on the dyad: a "Tipiak" call in the male–female dyad BC and a "Hon" call in the female–female dyad AC. So in the dyad (BC), the female, C, uttered a "Tipiak" call before or when she started grooming her father (Fig. 3e), and uttered a "Hon" call after she started grooming her mother, A (Fig. 3d).

Grooming patterns never ended with an act that also constituted an ending act (leaving) for a grooming interaction. Patterns

Table 4
Rates at which different patterns were observed according to their temporal position during an interaction (in reference to Fig. 5)

	Types of patterns				
Types of interactions	No pattern	"Starting patterns"	"Ending patterns"	"Middle patterns"	"Complete patterns"
Sociosexual interactions					
Dyad BC (out of 6 cases)	0	3	1	0	2
Grooming interactions					
Dyad BC (out of 6 cases)	0	4	0	2	0
Dyad AC (out of 7 cases)	0	6	0	1	0

ended with a variety of acts (grooming, vocalizations, gazes beginning or ending; Figs. 3 and 4).

The rates at which the different patterns were observed indicated that the majority of patterns detected in grooming interactions are "starting patterns" (Table 4; Chi-square test for both dyads: $\chi^2 = 3.77$, $P < 0.052$).

Though the sample size was small, it is worth noting two differences between the male–female and the female–female dyads. (1) "Tipiak" call was noted only in the male–female dyad. As already mentioned (2) patterns detected in male–female interactions were significantly longer than female–female ones ($T+ = 15$, $N = 5$, $P = 0.03$).

4 Discussion

Most social interactions are analogous to dialogs where behaviors from each partner form a chain of events [20, 21]. A social interaction implies several processes: motivation, intention, perception, information processing, decision-making process, expression of emotion, memory. Most animal behavior implies intentionality. Dennett's [40] description of intentionality refers to purposive behavior. He argued that behavior is most fruitfully described at a level of integration reflecting its goal-directed nature. In our study, the goal of the individuals who interacted may be what we called a "target" behavior (i.e., a consummatory act [20], a "grooming act" for grooming interactions, and a "genitals presenting" for sociosexual interactions). This consummatory act is the outcome of a certain motivation. The motivation of the individuals thus corresponds to an appetitive phase of the sequence. Buck [41] considered motivation as being responsible for the control of behavior, that is, the activation and direction of behavior. Motivation is thus defined as a potential that is inherent in the structure of systems that control behavior and that is manifested in emotion [41]. In our study, if an individual, when moving towards a partner, did not have a particular intention, e.g., to get involved in a grooming rather than another kind of interaction for example, then the patterns from grooming interactions would start in the same way as patterns from sociosexual interactions (same acts in same number and in same order). And yet, this was not the case. So we might assume that individuals interacted in an intentional way. Hinde [42] considered a social interaction to be a negotiation between emotional states of the two interacting partners. The number of decisions that partners have to make during the course of an interaction implies a large flexibility in processing of information [43]. This flexibility arises also from the fact that the meaning of signals could be partially context-dependent [44, 45], i.e., individuals may process not only combinations of signals from their partners but also information from a wider context that they perceive simultaneously.

Considering sociosexual interactions, we showed that vocalizations seemed to play a role in the decision-making process to start an interaction, by shortening the interaction and making it less complex. This conclusion is based on the comparison of complete patterns. As this kind of patterns reflects the unfolding of the interaction from its beginning to its end, they are more interesting than the other kinds for studying the complexity of an interaction. Vocalizations occurred at the very beginning of the patterns in interactions with vocalizations. Such a "starting vocalization" constitutes a "decision point" in reference to the concept of "decision structure" introduced by Dawkins and Dawkins [29]. These authors argue that "behaviour can be described in terms of its changing uncertainty or decision structure over time" [29]. The "starting vocalization" indicates the intention of the initiator to start an interaction [40]. This intention could be perceived from a distance by any potential receiver whose attention has been aroused [2, 46]. Like the "Coo" call of macaques [47], the "Hon" call uttered by mangabeys is easily localized [48] and increases the probability of a subsequent interaction. In interactions without a vocalization, the first decision point, a nonvocal act, could be the same as the one found after a "starting vocalization." However, in the nonvocal case, the intention of the initiator is not so easily perceived by intended receivers as a vocalization from a distance.

In patterns detected within grooming interactions, we observed only two types of vocalizations, "Hon" and "Tipiak." In the patterns, the "Hon" grunt calls always occurred simultaneously or preceding or even succeeding with either visual or tactile signals from the caller. This association yields multimodal signals addressed to a specific partner and thus represents a "one-to-one" communication [49]. Conversely, the high pitched "Tipiak" calls were never observed in association with visual behaviors within patterns. According to Morton's [50] "motivation-structural rules," high pitched vocalizations are associated with fearfulness. In many cases touching a partner is likely to be riskier than just approaching or looking at them [47, 51], and this risk may induce "Tipiak" calls that are higher pitched and louder than the "Hon" call. This is supported by the fact that the adult male was involved in the only interactions with Tipiak calls. In several instances, interactions contained both patterns with "Tipiak" and others with "Hon." As the "Hon" patterns followed the "Tipiak" ones, it is likely that the emotional level (in terms of risk-taking) of the caller was decreasing as grooming proceeded. Actually, in grooming interactions, we may observe patterns where vocalizations follow visual acts and precede tactile ones. According to the "negotiation" concept and to the emotional load of touching a partner, we propose a gradient of successive decisions leading to the consummatory act of touching [20]. The "Hon" calls seem to play a role as a low emotional signal whereas "Tipiak" call seems to broadcast the expression of a high emotional level.

When interactions contained no pattern at all (Fig. 5a), we may assume that the behaviors of the interacting partners are opportunistic, resulting in apparently random sequences. When patterns can be detected, they constitute regularities in the course of different interactions between different partners, revealing a structure to the negotiation between the partners. The patterns may include vocalizations or not. When vocalizations are part of the pattern, they appear as "decision points." Consequently, the role of vocalizations in making the communication process more efficient is increased. This is what we observed in grooming interactions where interactions with patterns including vocalizations were less complex than those without vocalizations, assuming that complexity is related to the number of "decision points" (length of patterns). The structure of interactions was inferred from the presence of patterns at different times: a sequence might include patterns only at its beginning (Fig. 5b), only at its end (Fig. 5c), or at several locations along the sequence (Fig. 5d), or a sequence might be thoroughly structured constituting a single pattern from beginning to end

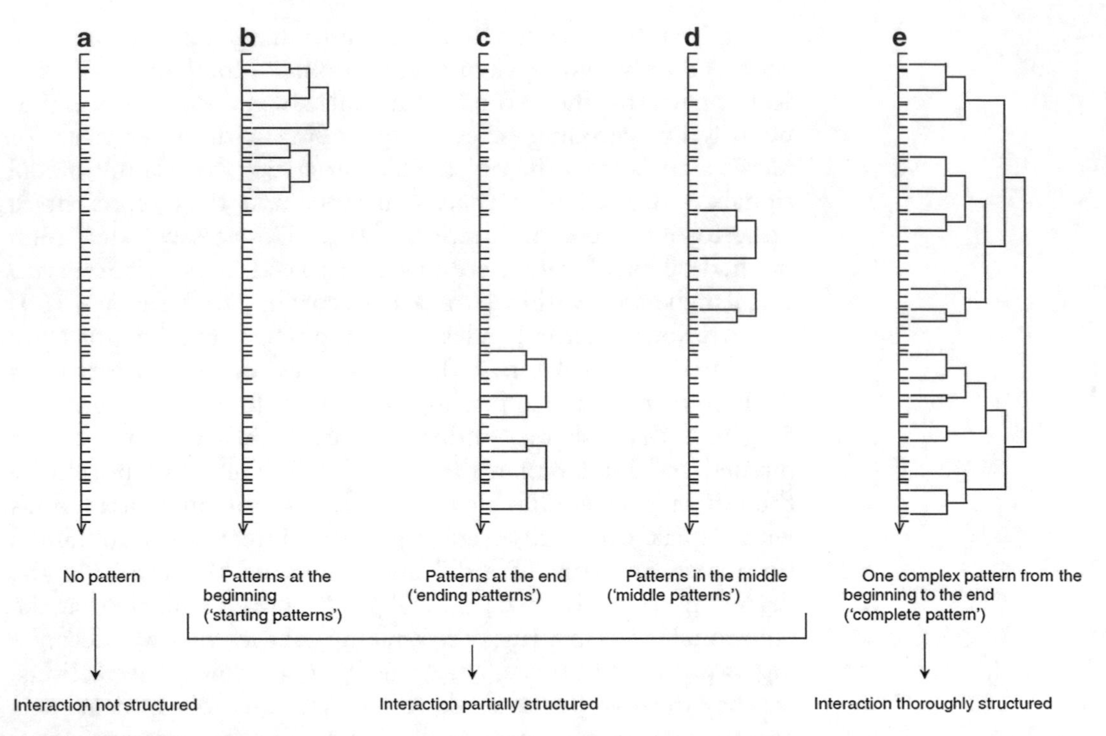

Fig. 5 Structure of an interaction inferred from the presence of patterns at different times. Time runs from *top* to *bottom*

(Fig. 5e). This latter case was only observed in sociosexual interactions involving vocalizations. Sociosexual interactions without vocalizations were only partially structured: patterns occurred at the beginning, middle, or end of the interaction (Fig. 5b–d). Therefore, vocalizations helped to convey clearer messages. In contrast, grooming interactions with or without vocalizations were always partially structured (Fig. 5b, d). When interactions were initially structured (Fig. 5b), vocalizations were either present or not. Therefore, in these cases, vocalizations did not seem to play a major role in structuring grooming interactions. However, when structure emerged in interactions after some exchange of behaviors (Fig. 5d), the patterns included both vocalizations and grooming ("target behavior"). This suggests that vocalizations were tightly linked (immediately prior, during, or after) to the process of grooming.

Although our sample size was small, and its further reduction due to analytical constraints prevents us from being conclusive, it is worth mentioning that, on the one hand, only vocalizations from females were included in the patterns detected within the interactions. Therefore, only female vocalizations seem to have a predictive value on the development of the interactions. On the other hand, some existing differences between female–female and male–female dyads would suggest that the role of vocalizations in interactions differs according to the sex, the age, the social status of the partners involved, and maybe to the body part that is groomed. This point would need to be investigated on a larger sample. A young female's relationships with an adult male may be different than with a female that it is her father or a mother or an individual with which she has another kinship relationship.

In conclusion, the THEME method developed by Magnusson [26] seems to be a valuable tool to analyze sequences in accounting for the multiple decisions made during multimodal interactions, and the temporal and behavioral structures of these interactions. The patterning of interactions is a result of the capacity of animals to perceive and process immediate information in reference to past experiences. The richness of the information these monkeys process is a consequence of the multimodal nature of communication which results in multimodal signals evoking responses not elicited by unimodal signals [3]. From our results on this captive group of four individuals, we conclude that in sociosexual interactions, the "Hon" call plays a role in the decision-making process to start an interaction. In grooming interactions, vocalizations do not play a role in the decision-making process to start an interaction but seem to be tightly linked to the process of grooming. The "Hon" call seems to play a role as a low emotional signal, whereas the "Tipiak" call seems to broadcast the expression of a higher emotional level and was used only in interactions with the adult male.

In sociosexual and grooming interactions, vocalizations are involved in a process of increasing the efficiency of interactions and in the decision-making process, in participating to the construction of a less complex message. So, the informative value of vocalizations uttered during these two kinds of interactions should be complementary to those from other sensory signals. Multimodal signals may facilitate the addition of a referential feature to an emotional expression, as it might be the case in visual-vocal bimodal signal [46]. Patterns detectable in social interactions might function to reduce the complexity of negotiation.

We have showed that vocalizations play a role in structuring interactions as they may occur in patterns either at the beginning or in the "middle" of the interactions but never at the end of interactions. In most nonhuman primate species, vocalizations are emotional signals. The fact that a vocalization is embedded within an interaction likely provides an additional informative value to guide an interaction rather than just the mere expression of an emotion.

Acknowledgements

We thank OHLL program for its financial support, Mr Philippe Bec for his help during the handling of the animals. All of the experiments complied with the current laws of the country in which they were conducted (France).

Appendix 1: Examples of Behaviors from the Behavioral Catalogue of Cercocebus T. Torquatus Used for the Study

Except when otherwise mentioned, only subsets of behaviors used to describe mangabeys' interactions are presented here.

Visual behaviors, including locomotor behaviors, facial expressions, and gaze behaviors:

Locomotor behaviors:

- To avoid a partner (with a movement of the whole body)
- To chase a partner in a playful context (pursuit)
- To chase a partner while displaying aggressive signals
- To present to a partner = Present, while having the tail up putting the ischial callosities and the callosities against the face of a partner
- To turn towards a partner: the subject was assuming a sitting posture and ended facing his partner. In addition the subject was within an arm's reach of his partner
- To walk away from a partner (the partner was close or in contact with the subject) = to leave
- To walk towards a partner

Facial expressions:

- To threaten a partner
- To lip smack towards a partner from a distance
- To display a "play face" towards a partner from a distance

Gazes:

- To follow (≥2 s) a partner with one's eyes (the partner is moving)
- To glance (<2 s) at a partner (other body parts than the genitals, callosity, tail, mouth; see below)

- To glance at what a partner is doing (e.g., looking at fingers of a partner who is eating or grooming or handling an object, etc.)
- To glance at what a partner is holding (implies that the partner is not doing anything with the held item)
- To glance backward at a partner
- To look at a partner (who is motionless) (≥2 s)
- To look at a partner's genitals
- To look at what a partner is doing (looking at fingers of a partner who is eating or grooming or handling an object, etc.)
- To look at what a partner is holding (see above)
- To look backward at a partner (who is motionless)

Olfactory behaviors (all the olfactory behaviors used in the study are reported here):

- To sniff at a partner (on a body part other than mouth, genitals, and tail)
- To sniff at what a partner holds in his hands
- To sniff the callosity of a partner
- To sniff the genitals of a partner
- To sniff the mouth of a partner
- To sniff the tail of a partner

Tactile behaviors:

- To "wrestle play" with a partner
- To bite a partner
- To climb up onto a partner
- To do pelvic movements while mounting a partner (with or without intromission in case of male to female mounting)
- To embrace a partner
- To grasp a partner (a body part other than genitals)
- To groom a partner (on a body part other than genitals or callosity)
- To hit a partner
- To lick a partner
- To touch a partner with hand (somewhere on the partner's body other than mouth, genitals, callosity, and head)

Appendix 2: Vocal Behaviors (All the Basic Vocalizations of the Species-Specific Repertoire are Reported Here). The Names given to the Vocalizations Correspond to Onomatopoeia (Figs. 6 and 7)

Most of the other behaviors correspond to variants of the ones cited here, i.e., precisions of the speed of the locomotor movements (walking, running, jumping), or the position in relation to the partner (back/face/aside), or the body parts of the partner (leg, arm, foot, hand, callosity, genitals, mouth, tail), or the general context of the interaction, i.e., the nature of all the behaviors expressed within a given interaction (playing, aggressive) or the distance between the individual and his partner (from a distance or nearby).

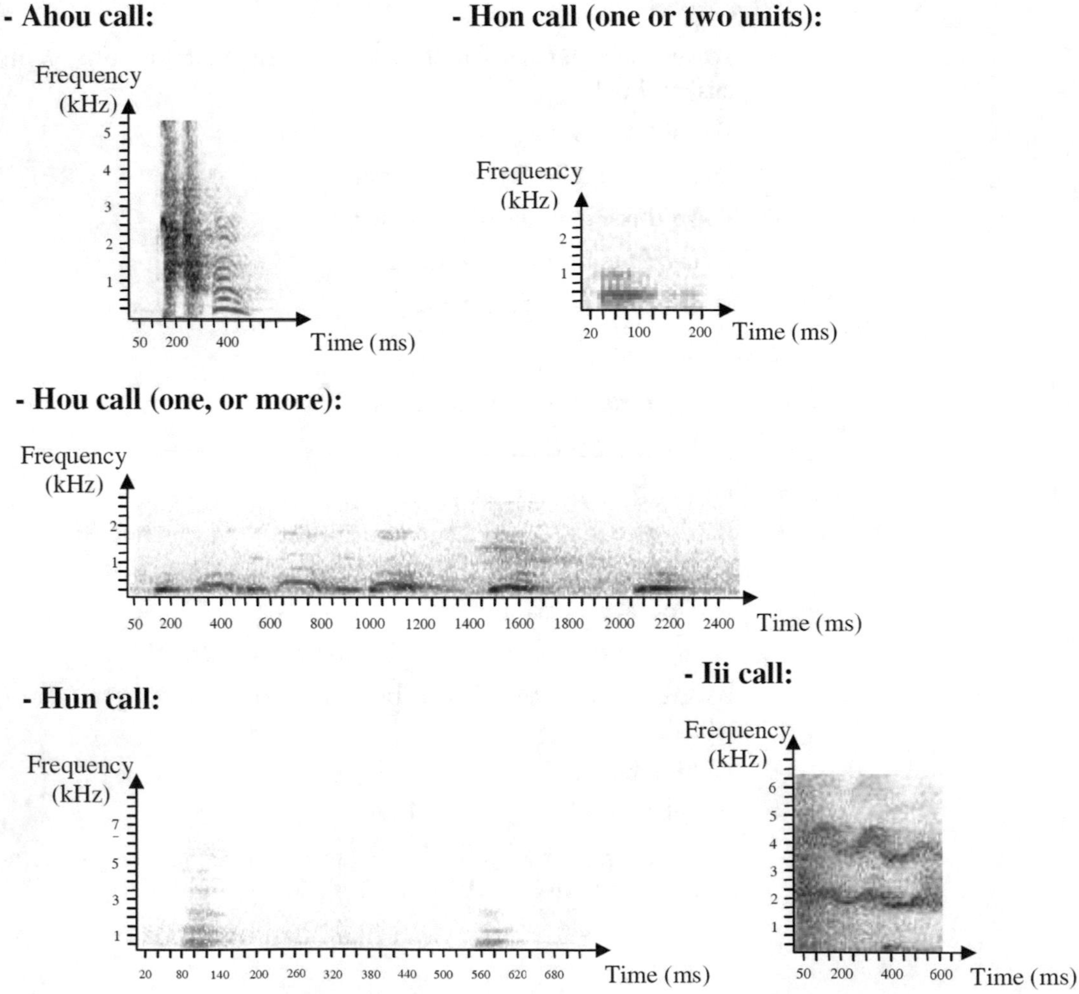

- Ahou call:

- Hon call (one or two units):

- Hou call (one, or more):

- Hun call:

- Iii call:

Fig. 6 Sonograms (frequency/time display) of 5 common calls from the vocal repertoire of the redcapped mangabey (*Cercocebus torquatus torquatus*)

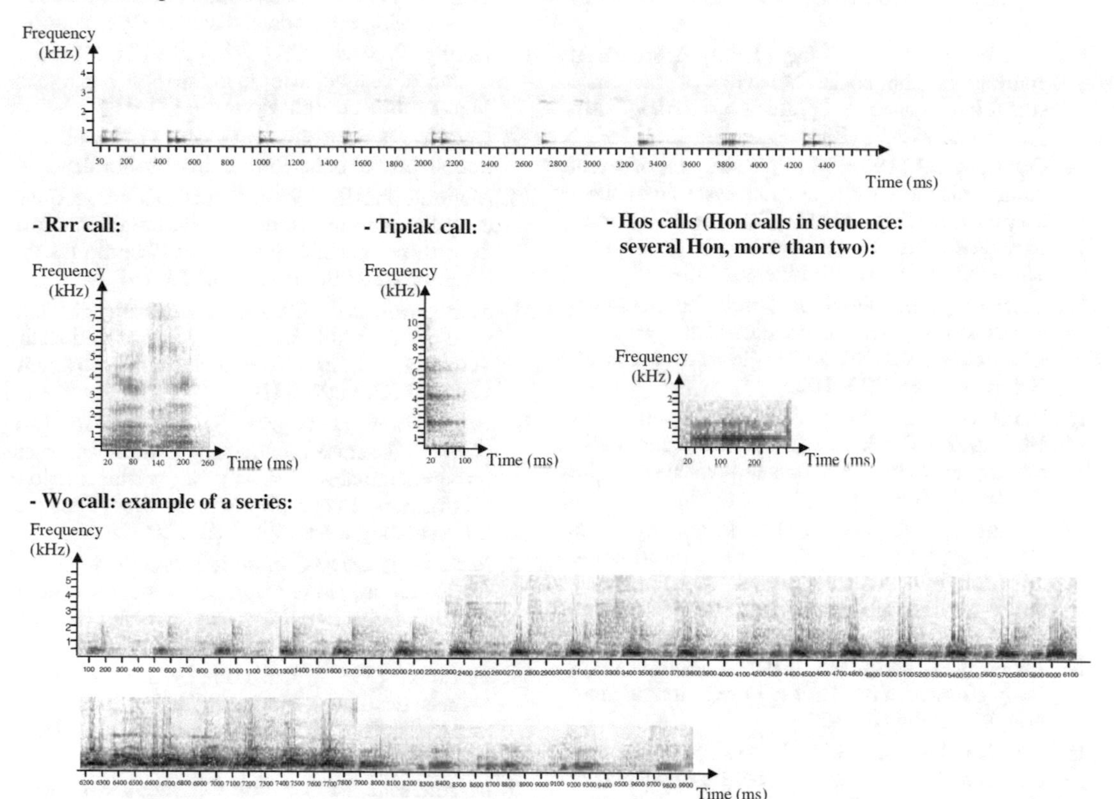

Fig. 7 Sonograms of 5 additional calls from the vocal repertoire of the red-capped mangabey (*Cercocebus torquatus torquatus*). The Wo call is the red-capped mangabey's male loud call

References

1. Weaver W, Shannon CE (1975) Contributions récentes à la théorie mathématique de la communication. In: Retz CE (ed) Théorie mathématique de la communication. C.E.P.L, Paris, pp 29–56
2. Marler P (1965) Communication in monkeys and apes. In: DeVore I (ed) Primate behavior, field studies of monkeys and apes. Holt, Rinehart and Winston, New York, pp 544–584
3. Partan S, Marler P (1999) Communication goes multimodal. Science 283:1272–1273
4. Rowe C, Skelhorn J (2004) Avian psychology and communication. Royal Soc Rev 1435–1442
5. Burling R (1993) Primate calls, human language, and non verbal communication. Curr Anthropol 34:25–53
6. Altmann SA (1962) Social behavior of anthropoid primates: analysis of recent concepts. In: Bliss EL (ed) Roots of behavior. Harper and Brothers, New York, pp 277–285
7. Friedman MB (1972) Auditory influences on the reproductive system of the ring dove. Unpublished doctoral dissertation. Rutgers–The State University
8. Friedman MB (1977) Interactions between visual and vocal courtship stimuli in the neuroendocrine response of female doves. J Comp Physiol Psychol 91(6):1408–1416
9. Fusani L, Hutchison RE, Hutchison JB (1997) Vocal-postural co-ordination of sexually dimorphic display in a monomorphic species: the barbary dove. Behaviour 134:321–335
10. Gilliard ET (1969) Birds of paradise and bower birds. Weidenfeld, Nicolson, London
11. Peek FW (1972) An experimental study of the territorial function of vocal and visual display in the male red-winged blackbird (Agelaius phoeniceus). Anim Behav 20:112–118
12. Deputte BL, Goustard M (1980) Copulatory vocalizations of female macaques (Macaca fas-

cicularis): variability factors analysis. Primates 21:83–99

13. Adams DB, Schoel WM (1982) A statistical analysis of the social behavior of the male stumptail macaque (Macaca arctoides). Am J Primatol 2:249–273
14. Goldfoot DA (1982) Multiple channels of sexual communication in rhesus monkeys: role of olfactory cues. In: Snowdon CT, Brown CH, Petersen MR (eds) Primate communication. Cambridge University Press, Cambridge, pp 413–428
15. Partan S (2002) Single and multichannel signal composition: facial expressions and vocalizations of rhesus macaques (Macaca mulatta). Behaviour 139:993–1027
16. Smith WJ, Smith SL, DeVilla JG, Oppenheimer EC (1976) The jump-yip display of the black-tailed prairie dog Cynomys ludovicianus. Anim Behav 24:609–621
17. Hennessy DF, Owings DH, Rowe MP, Coss RG, Leger DW (1981) The information afforded by a variable signal: constraints on snake-elicited tail flagging by California ground squirrels. Behaviour 78:188–226
18. Dienske H, Metz HAJ (1977) Mother-infant body contact in macaques: a time interval analysis. Biol Behav 2:3–37
19. Pitcairn TK, Schleidt M (1976) Dance and decision, an analysis of a courtship dance of the Medlpa, New Guinea. Behaviour 58:298–316
20. Hinde RA (1966) Animal behaviour: a synthesis of ethology and comparative psychology, 1st edn. McGraw-Hill, London
21. Tinbergen N (1951) The study of instinct. Oxford University Press, Oxford
22. Deputte BL (1986) Ontogenèse du Cercocèbe à joues blanches, en captivité, Lophocebus albigena, Développement des comportements de communication et des relations sociales. Thèse d'état, Université de Rennes I
23. Altmann SA (1962) A field study of the sociobiology of rhesus monkeys, Macaca mulatta. Ann N Y Acad Sci 102:338–435
24. Altmann SA (1965) Sociobiology of rhesus monkeys. II: Stochastics of social communication. J Theoretical Biol 8:490–522
25. Deputte BL (1994) Ethological study of yawning in primates. I. Quantitative analysis and study of causation in two species of Old World monkeys (Cercocebus albigena and Macaca fascicularis). Ethology 98:221–245
26. Magnusson MS (1996) Hidden real-time patterns in intra- and interindividual behavior: description and detection. Eur J Psychol Assess 12:112–123
27. Altmann J (1974) Observational study of behavior: sampling methods. Behaviour 49:227–266
28. Hanby JP, Brown CE (1974) The development of sociosexual behaviours in Japanese macaques Macaca fuscata. Behaviour 49:152–196
29. Dawkins R, Dawkins M (1973) Decisions and uncertainty of behaviour. Behaviour 45:83–103
30. Magnusson MS (1989) Structures syntaxiques et rythmes comportementaux: Sur la détection de rythmes cachés. Sciences et Techniques de l'Animal du Laboratoire 14:143–147
31. Magnusson MS (2000) Discovering hidden time patterns in behavior: T-patterns and their detection. Behav Res Methods Instrum Comput 32(1):93–110
32. Beaudichon J, Legros S, Magnusson MS (1991) Organisation des régulations inter et intrapersonnelles dans la transmission d'informations complexes organisées. Bulletin de Psychologie 44(399):110–120
33. Bernicot J, Caron-Pargue J, Trognon A (1997) Conversation, interaction et fonctionnement cognitif. Collection 'Processus discursifs, langage et cognition'. Presses Universitaires de Nancy
34. Féron C (1992) Les comportements sociosexuels des souris staggerer mâles: caractéristiques et effets de l'expérience sociale. Thèse d'état, Université Paris XIII
35. Martaresche M, Le Fur C, Magnusson MS, Faure JM, Picard M (2000) Time structure of behavioral patterns related to feed pecking in chicks. Physiol Behav 70:443–451
36. Sakaguchi K, Jonsson GK, Hasegawa T (2005) Initial interpersonal attraction between mixed-sex dyad and movement synchrony. In: Anolli L, Jr Duncan S, Magnusson MS, Riva G (eds) The hidden structure of interaction: from neurons to culture patterns. Ios Press, Amsterdam, pp 107–120
37. Arthur BI Jr, Magnusson MS (2005) Microanalysis of Drosophila courtship behaviour. In: Anolli L, Jr Duncan S, Magnusson MS, Riva G (eds) The hidden structure of interaction: from neurons to culture patterns. Ios Press, Amsterdam, pp 3–22
38. Magnusson MS (2005) Understanding social interaction: discovering hidden structure with models and algorithms. In: Anolli L, Jr Duncan S, Magnusson MS, Riva G (eds) The hidden structure of interaction: from neurons to culture patterns. Ios Press, Amsterdam, pp 3–22
39. Maurus M, Pruscha H (1973) Classification of social signals in squirrel monkeys by means of cluster analysis. Behaviour 47:106–128
40. Dennett DC (1983) Intentional systems in cognitive ethology: The 'Panglossian paradigm' defended. Behav Brain Sci 6:343–390

41. Buck R (1986) The psychology of emotion. In: LeDoux JE, Hirst W (eds) Mind and brain: dialogues in cognitive neuroscience. Cambridge University Press, Cambridge, pp 275–300
42. Hinde RA (1985) Was 'The expression of the emotions' a misleading phrase? Anim Behav 33:985–992
43. Tomasello M, Call J (1997) Primate cognition. Oxford University Press, New York
44. Deputte BL (1993) Ethologie et cognition: le cas des primates. Intellectica 16:21–44
45. Tavolga WN (1974) Application of the concept of levels of organization to the study of animal communication. In: Krames L, Pliner P, Alloway T (eds) Advances in the study of communication and affect, Vol. 1: Nonverbal communication. Plenum Press, New York, pp 51–76
46. Seyfarth RM, Cheney DL (2003) Meaning and emotion in animal vocalizations. Ann N Y Acad Sci 1000:32–55
47. Bauers KA, de Waal FBM (1991) "Coo" vocalizations in stumptailed macaques: a controlled functional analysis. Behaviour 119:143–160
48. Brown CH (1986) The perception of vocal signals by blue monkeys and grey-cheeked mangabeys. Exp Biol 45:145–165
49. Altmann SA (1967) The structure of primate social communication. In: Altmann SA (ed) Social communication among primates. The University of Chicago Press, Chicago, pp 325–362
50. Morton ES (1977) On the occurrence and significance of motivation-structural rules in some bird and mammal sounds. Am Nat 111:855–869
51. Weary DM, Fraser D (1995) Signalling need: costly signals and animal welfare assessment. Appl Anim Behav Sci 44:159–169

Chapter 15

Identification and Description of Behaviours and Domination Patterns in Captive Vervet Monkeys (*Cercophitecus Aethiops Pygerythrus*) During Feeding Time

Gerardo Ortiz, Gudberg K. Jonsson, and Ana Lilia del Toro

Abstract

In the current study we explore dominant behaviours and interactive patterns during feeding of a group of captive vervet monkeys (Cercophitecus aethiops pygerythrus). Observations were carried out in a group including an adult male, an adult female, two immature individuals, and an infant, living in the Guadalajara Zoo. They were located in a confined area that was divided in six zones according to the proximity or distance to where the spectators are located. For the collection of behavioural data we used the Observer 5.0 and Theme 5.0 for detection and analysis of behavioural patterns. Three recordings for every day were made (i.e. Wednesday to Sunday), during feeding time, at approximately 17:30 h and each lasting 1½ h, beginning 30 min before food delivery. We registered general activities (i.e. eating, resting), as well as domination behaviours (i.e. allogrooming, agonism, direct and indirect access to food). Results indicate significant differences in the use of space/proximity between female and male, as well as the proportion of general activities, although resting was the main registered behaviour. Differences by sex (i.e. male, female) and by location of the place of food delivery (i.e. inside, outside) were observed in expressed behaviours and interactive patterns. In wildlife conditions, female hierarchy usually dominates over male hierarchy, possibly because of social networks established between female relatives. However, in captivity conditions the group structure changes, modifying its function (i.e. male predominant over female). The current type of research might provide us with clues on how to improve the design of the facilities for captive animals. Future objectives of the project concern comparing the findings to data collected in the wild and implementing the results into a new design of the facilities for captive vervet monkeys.

Key words Vervet monkey, T-patterns, Theme

1 Introduction

Because recreational functions are the main objective of most zoos, the enclosures are generally designed so that the visitors can easily see the animals, and thus they are not built in accordance with the animal's natural habitat or the behavioural characteristics of the diverse species on exhibition (e.g. 1). For this reason, emitted

Magnus S. Magnusson et al. (eds.), *Discovering Hidden Temporal Patterns in Behavior and Interaction: T-Pattern Detection and Analysis with THEME™*, Neuromethods, vol. 111, DOI 10.1007/978-1-4939-3249-8_15,

behaviours will vary in respect to those emitted in wildlife, developing specific types of interaction in captivity (i.e. "artificial" environments), situations created and modified by humans, as in the case of the social structure of the captive animals. For example, Manning [2] suggests, in relation to domination aspects, that unlike what happens in freedom, fights are more frequent in zoos, usually because spaces are reduced and there is not sufficient space for the subordinates to stay far away from the dominant individuals.

Vervet monkeys (*Cercophitecus aethiops pygerythrus*) are old-world primates; they are of a greenish-brownish colour with the chest and contour of the face in white colour. Face, hands, and legs are of black colour [3–5]. The male has, around its genitals, a shining blue colour, contrasting with its penis, which is of red colour [3, 5, 6]. As far as weight, males weigh approximately 6 kg with height 1 m, and females are smaller and usually weigh 4 kg [7].

The natural habitat of vervet monkeys is the semi-desert zones in the African savannah of the Sahara, where there are a great amount of trees; this species is considered semi-terrestrial, because usually they are transferred between zones by ground [5]. Although they feed basically on fruit, flowers, and insects (i.e. leaves, seeds, nuts, different kinds of grass, fruits, berries, eggs), their diet varies according to available resources (e.g. 5, 6, 8). Their predators in wildlife are lions, leopards, cheetahs, and other types of felines, in addition to hyenas and baboons, among others [5, 9, 10].

This species has a matriarchal linear hierarchy, separate from the male hierarchy; in this type of hierarchy, the alpha females (beta, gamma) are supported by their close relatives forming coalitions in order to maintain their hierarchic position. Daughters usually inherit the rank of their mothers and mothers still support their female offspring after the daughters have reached maturity [11]. Females of higher hierarchy and relatives (i.e. mothers, sisters, and daughters), are those who can look for and select more and better quality food [7, 12, 13], thus they have more advantage in comparison with lower hierarchical level individuals, better nutritional reserves, better health, and more opportunities for successful reproduction. Dominant individuals determine access to the food, places of rest, and access to the females (e.g. 2, 13). Also, the dominant one usually receives more allogrooming than other members of the group; this activity, presented frequently in primates, serves them as a pacifier gesture.

As a way to maintain domination in the group, and to communicate the status that the individual has in the hierarchy rank, males usually present ritual displays to show their dominant position, a cheaper energetic way than physical fights; one of the basic domination displays is to exhibit their scrotum that works as a symbol of their power in the troop [7]. There are two different forms of scrotum exhibition, the first known as *splay-legged*, in which the male is sited down on a tree branch or on the floor with its knees opened, leaving the genitals exposed. This kind of display appears as an aggression sign when a vigilant male detects another male near the limits of his territory, signalling that the intruder not approach either the territory or the female group [14].

A second form of domination display is known as the *red, white, and blue display*, and consists of the approach of the dominant male to a subordinate, walking in circles around him, raising his tail and exposing the genital–anus area to the minor rank male [15, 16]. Other forms to express domination within the group are face gestures or expressions, such as: (a) the exhibition of the eyelid, because the skin of this area is of a shining colour, contrasting with the rest of the face, which is black colored; and (b) retracting the forehead while the dominant individual is fixedly watching the dominated individual [2, 15]. At the same time, to stand up in a bipedal position in the face of the subordinate or to shake its body facing another individual, can work as threat expressions. It has been observed that if the subordinate responds to these expressions, it bends over or crouches, making some grunts.

Another important characteristic of the vervet monkey, which allows them to organize as a group, are their vocalizations, because they have an alarm system that announces to the other members of the group that a predator is close, changing the type of sound according to the identified predator (e.g. 17–19). According to Cheney and Seyfarth [20] the vervet monkey has a great ability to recognize signals emitted by other monkeys and predators in their wild habitat, although they do not seem to have the capacity to generalize in other contexts, as in the case of captivity environments.

From an interbehavioural approach (e.g. 21, 22), Ortiz et al. [1] and Ortiz [27] suggest that when an individual behaves in a specific situation, it is responding to elements and particular events that compose that situation. Such elements have both a physical–chemical nature (i.e. the environmental temperature or colours and scents of certain types of plants and animals), and a quasi-conventional nature (i.e. formation and dynamics of the social group in which it is immersed). In this sense, each species, and each individual of that species, due to its psychological characteristics, keeps a singular relation with the environment and other species. Thus, each species (anatomically and physiologically adapted, as well as behaviourally adjusted to its surroundings), develops specific behavioural modifications both for its survival [23] as well as for the sustainable maintenance of the dynamics of the ecosystem of which it is part.

In this sense, we can identify some elements that can allow us to describe the ecological milieu and analyse its function-related animal behaviour. Ortiz et al. [1] and Ortiz [27] propose that we can, and must, identify geophysical elements (i.e. scents, shapes, flavours, colours, weather, seasonality), geoecological elements (i.e. type of land or enclosure, location and size of feeding, location and distribution of resting and protection zones), interactive elements (i.e. presence or absence of the same or other species or individuals), as well as the relationship between behaviour and adjustment criteria in a specific situation (i.e. behaviour specific to a situation, related to a situation, functional in a situation). Under this scope, if domination behaviours are psychological ones, they can differ from a captive group to those emitted in wild conditions, because in captivity the context often differs with respect to the wildlife conditions on population density, space in which the group lives, the fact that they do not have to search for food, as well as the composition of the group (i.e. ratio gender).

The present study is focused on identifying and describing domination-related behaviours during feeding circumstances. Although any other circumstance could have been selected, we chose feeding; it was considered pertinent because in captivity there are few circumstances in which animals can be immersed and because it seems to be part of the most relevant events in their day. Also, during feeding time in wildlife, this species emitted behaviours that make reference to domination aspects, where females of higher hierarchy have food access primacy (e.g. 7, 9, 13).

2 Methods

2.1 Subjects and Enclosure

The observations were carried out in a captive group ($n=5$) of vervet monkeys (*Cercophitecus aethiops pygerythrus*) that lives in the Guadalajara Zoo. The group contained an adult male, adult female, two immature individuals, and an infant. This troop was located in an enclosed area 6 m wide long by 6 m and 1.70 m height, with an unevenness of 3°; the area is delimited by three cement walls and one metallic grate that separates the animals from the spectators; also, the ceiling is constituted of the same material as the grate (see Fig. 1). The floor of the back part of the confinement (2 m^2) is covered with cement, whereas the floor of the rest of the confinement is compacted earth (4 m^2). In the centre bottom of the confinement, there is access to the dormitory area.

For analysis aims, the confinement was divided in six zones (Fig. 1): far left (FL), far central (FC), far right (FR), close left (CL), close central (CC), and close right (CR) (see Fig. 1).

2.2 Materials

A camcorder Sony 8 CCD-TR413 was used to register the behaviour of the subjects for approximately 1½ h during feeding time. In order to make the registration and analysis of the observations, the Observer version 5.0 program was used. Also, for the analysis of sequences of interactions we used the Theme 5.0.

2.3 Procedure and Data Analysis

Recordings were carried out from Wednesday to Sunday, covering three recordings per day in three different weeks and Wednesday and Thursday of the fourth week, during March, 2006, obtaining

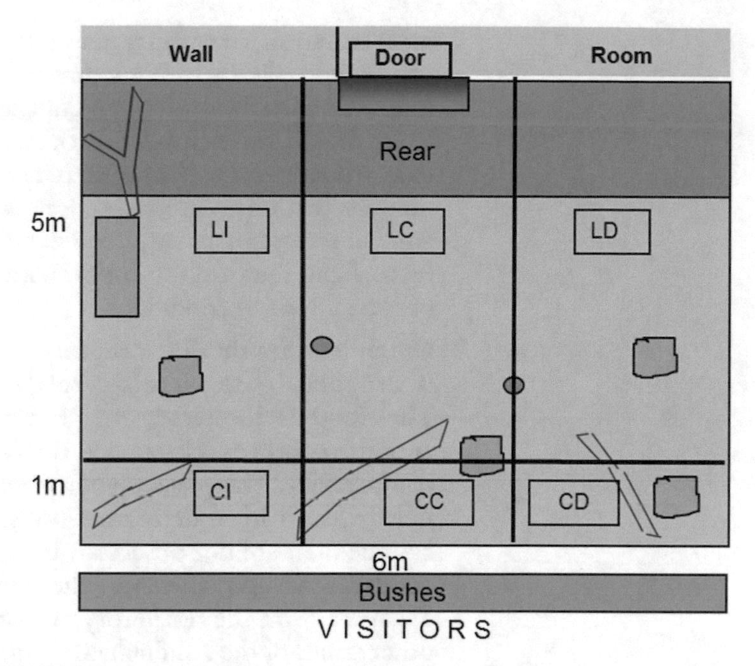

Fig. 1 The vervet monkeys enclosed area, divided into six zones

a total of 17 videos. The recordings were made when the caretakers fed the subjects, approximately at 17:30 h; each recording lasted 1½ h, initiating 30 min before the food delivery (17:00 h). In order to make the recordings, we placed the camcorder on a tripod, at a distance that allowed us to observe all the confinement areas and all five subjects.

We used an animal focal continuous method of registering [24] based on the behaviour catalogue, using the Observer 5.0. Once the observed behaviours were registered, the obtained data were analysed with Theme version 5.0 obtaining information on temporal sequences of frequent interactions.

Behavioural sequences were analysed as displayed patterns when the food was inside or outside the dormitory area, thus having two conditions of study, with the purpose of making comparisons of the displayed pattern by the subjects in both conditions, considering the change an important element during the feeding circumstance (i.e. the place where feeding behaviour occurs).

Next, we describe both conditions (i.e. inside and outside) in terms of some elements of the proposed methodology by Ortiz et al. [1] and Ortiz [27], such as the ecological milieu, some geophysical and geoecological aspects, and intra- or interindividual interactions:

1. Condition "outside the dormitory": Condition when the food is located in the far central zone (LC) just outside the dormitory

main entrance. The food is put on the ground. To the right side of the food pile there is a stone and on the left side a cement wall. When the food is located outside, all individuals have access to the food, inasmuch as it is in open condition; that is to say, the food is put almost at the centre of the confinement and there are no walls that obstruct access. Because the food is in the central zone of the confinement, it is a zone that has less humidity and more light than in the inside condition. Seven videos were obtained in these conditions.

2. Condition "inside the dormitory": Condition when the food is dropped by the zoo's caretakers inside the dormitory. The food is located at the dormitory's centre, the main entrance of which is located in the far central zone (LC) of the confinement. This place is approximately 1 m × 2 m, of rectangular form, with four cement walls. More objects within the confinement are not observed. In order to access the food the animal must go through the entrance of approximately 1.5 m × 1.5 m; the dormitory is more humid and darker than other zones of the confinement. Eight videos were obtained in these conditions.

Also, within these two conditions (inside and outside), it was considered pertinent to select more consistent sequences of interaction, recovering and separating those patterns or interaction sequences that were more general, due to the lack of relationship with domination behaviours (i.e. information on the zones with the subjects and with those who interact).

2.4 Behavioural Catalogue

General activities

- Eating: to take food and to ingest it, registering the zone where this behaviour happened.
- Resting:
  - Sitting: with legs doubled and half of the body on the ground or on a treetrunk.
  - Hanging: when the subject was hanging with four or two legs on the gratings of the confinement, or some of the trunks that are in the confinement.
  - Lying down: when the subject had its back or belly on the ground or the trunks.
  - Domination-related behaviours.
- Allogrooming: to groom the body of other individual.
  - To give.
  - To receive.
- Agonistic behaviours: set of behaviours that constitute the proximal mechanism of competition. We identify the individual

that initiated or presented/displayed the conduct and towards which individual it was directed (i.e. who received the behaviour).

- Displacement: the emitter moves towards the place that occupies the receiver causing the latter to leave the space, and the emitter occupies it in its place.

  To give.

  To receive.

- Hitting with hands*:* the emitter strikes another member of the group with the legs or hands.

  To give.

  To receive.

- Pursuing*:* a member of the group runs behind another one, by more than a meter of distance; otherwise it is considered as play.

  To give.

  To receive.

- Snatching*:* the emitter takes the food of another member of the group, or it stands up opposite the individual that brings food, in order that the receiver leaves the food on the ground.

  To give.

  To receive.

- Showing teeth: the emitter directs a glance towards another member of the group while it opens the mouth showing the teeth.

  To give.

  To receive.

• Food Access

- Direct: when a member of the group has direct access to the food source, manipulates the food, selects, eating or not what it has selected.

- Indirect: when a member of the group is eating (which implies having the food in the hand or the ground and putting it to the mouth to swallow it), and suddenly this individual stops eating, or throws part of its food, and another member of the group has access to that food left by the emitter. Taking food from another one must occur almost immediately to consider it as indirect access. We register who left or threw the food and who took it.

  To leave food.

  To take food.

3 Results

Figure 2 displays the proportion of male and female activity during observation time. White bars show data when food is located inside the enclosure, the gray bars show data when food is located outside the enclosure, and the black bars shows the activity's total frequency. The left graphs show results of male behaviours, and right-sided graphs show female data. The upper graphs display the result of resting activities (i.e. sitting, hanging, lying down); the upper middle graphs show data related to given domination activities (i.e. displacement, pursuing, snatching); whereas the lower middle graphs show data on received domination behaviours. Finally, the lower graphs display the proportion of indirect grabbing of food (i.e. given to or taken from).

In both subjects almost all the activities are related to resting behaviours; in the male the total frequency of resting behaviours is greater when food is located outside, whereas in the female the relation is the opposite. Sitting is the most frequent behaviour emitted in both conditions, followed by hanging. There are no important differences between food conditions.

There are observable differences in domination behaviours between male and female and between food locations. The male shows more dominant behaviours when food is located outside, whereas the female shows it more frequently when food is located inside the enclosure, emitting more dominant behaviours than the male in this condition. In the male, the most frequent dominant behaviour is displacement when food is located inside, and hitting with hands and displacement when food is outside. In the female, the most frequent dominant behaviours emitted are the same (i.e. displacement, hitting with hands) but the relationship of emitted behaviours and food location is the opposite of the male.

The male does not receive any domination behaviour when food is located inside the enclosure, but receives hitting with hands, pursuing, and snatching when food is located outside. The female receives a greater amount of dominant behaviours when food is inside (i.e. displacement, hitting with hands, and pursuing), than when food is located outside (i.e. displacement, pursuing, and hitting with hands).

Finally, results show that the male emits a greater frequency of leaving food when food is outside; of the food left for the male, the juvenile grabs that food more frequently, followed by the female when food is inside or the infant when food is outside. The male only grabbed food indirectly 10 times from the juvenile and female when the food is inside (i.e. two times each), and from the female when the food is outside (i.e. six times). The female leaves food more frequently when food is inside than when food is outside; the juvenile grabs the

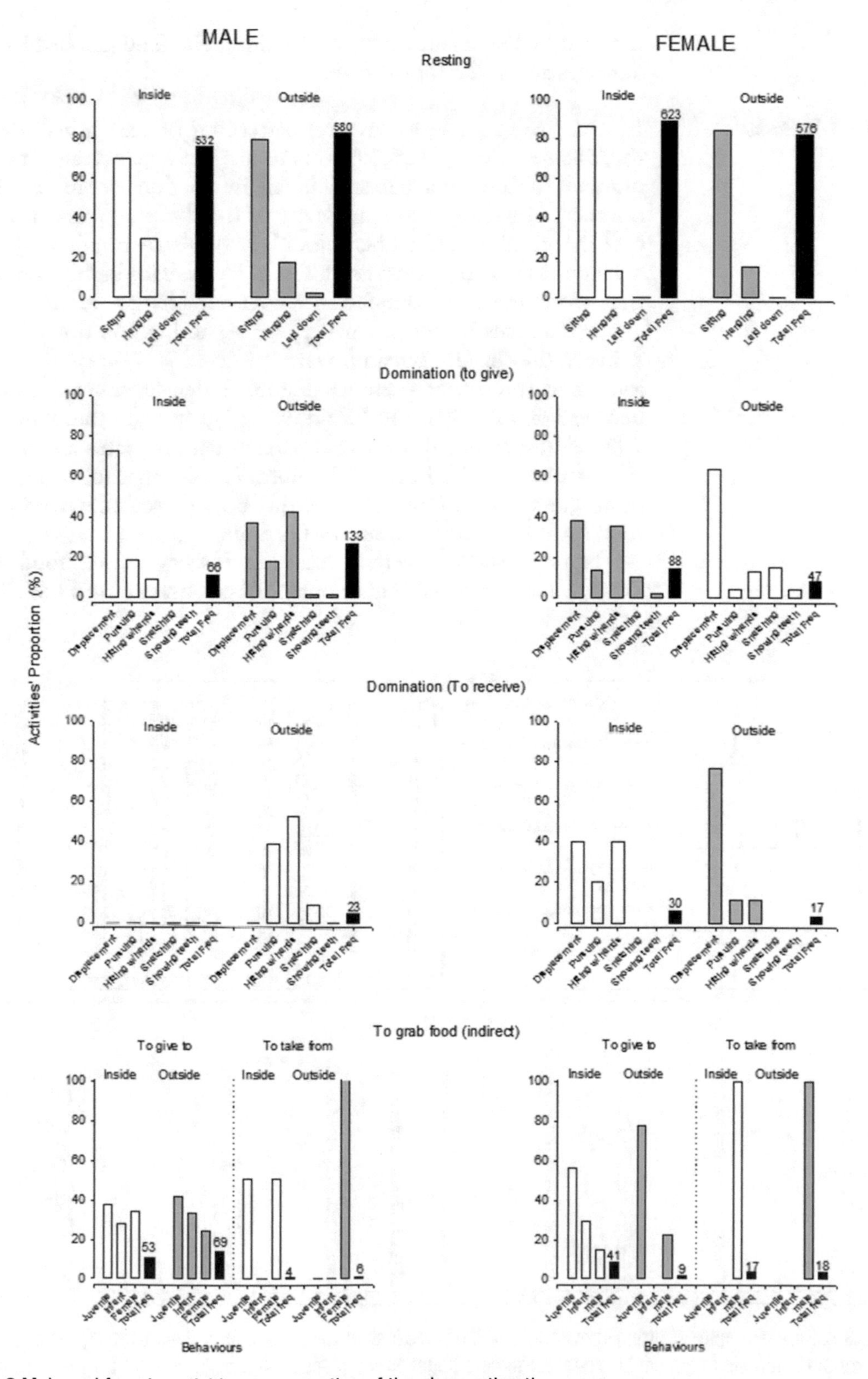

Fig. 2 Male and female activities as proportion of the observation time

food left by the female almost all the time. The food grabbed by the female is always left for the male.

3.1 T-Patterns

In order to examine a temporal patterning of transcribed events the Theme software [25, 26] was used. The data revealed a high number of temporal patterns in all interaction situations. The number, frequency, and complexity of the detected patterns indicate that the transcribed behaviour was very structured. This synchrony was found to exist on different levels, with highly complex time structures that extended over considerable time spans.

Differences by sex (i.e. male, female) and by location of food delivery (i.e. inside, outside) were observed in expressed behaviours and interactive patterns. Figure 3 demonstrates a pattern detected exclusively in the "food inside" group and found in over 90% of observational files. This pattern demonstrates an alternation pattern to food access (hembra/female, macho/male) as a nondomination pattern, as well as that both subjects eat nearby the food source (comerlejoscentro = LC zone).

Figure 4 shows a pattern found exclusively in the "food outside" group and was found in over 90% of observational files. This

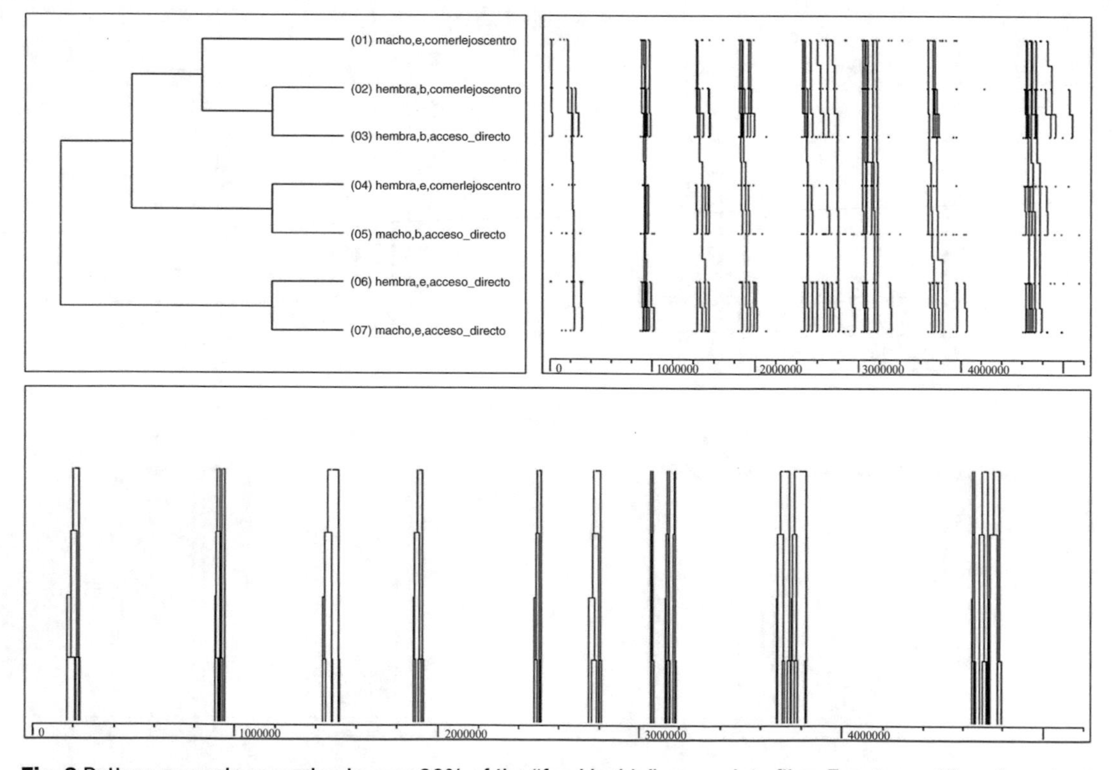

Fig. 3 Pattern example occurring in over 90% of the "food inside" group data files. Events are (1) male end eat LC zone; (2) female begin eat LC zone; (3) female begin direct access; (4) female end eat LC zone; (5) male begin direct access; (6) female end direct access; and (7) male end direct access

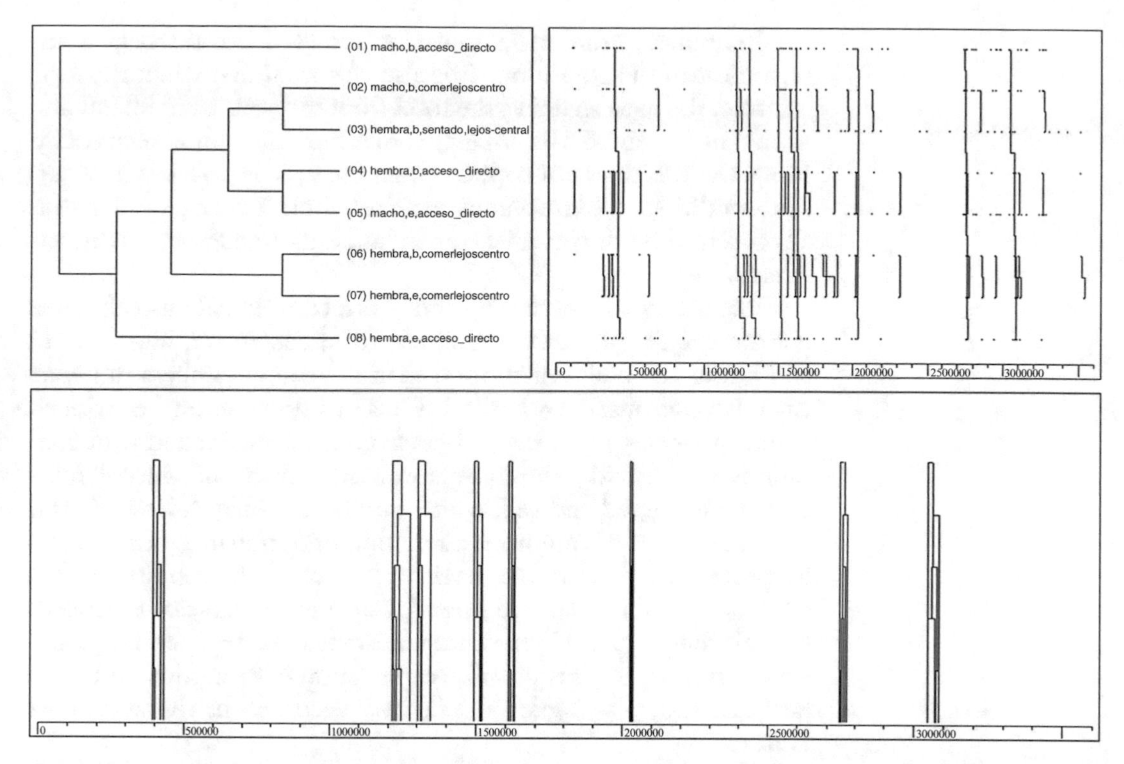

Fig. 4 Pattern example occurring in over 90% of the "food inside" group. Events are (1) male begin direct access; (2) male begin eat LC zone; (3) female begin sit far central; (4) female begin direct access; (5) male end direct access; (6) female begin eat LC zone; (7) female end eat LC zone; and (8) female end direct access

pattern demonstrates the same alternation pattern to direct food access, although the male (macho) seems to begin this food pattern showing a possible domination pattern.

4 Discussion and Conclusion

The results suggest that the females show consistency in patterns related to domination, regardless of where food was placed (i.e. outside-inside). Unlike some data in wild conditions, first access to food is slightly smaller in the female than the male; it appears that the male dominates the female, who seems to dominate the rest of the group suggesting that the conditions of confinement (i.e. group composition) may be a relevant factor for the expression of such patterns. Similarly, the female is consistent with respect to the areas where usually rest (i.e. close to dormitory area) and eating (i.e. far center and right areas) take place. However, we can observe changes related to domination (i.e. agonistic) behaviours associated with the condition. These results may relate to the fact that, in the outside condition, the food is more dispersed and does not have to be fought for.

In general, the male typically shows agonistic behaviours regardless of the condition. It is also observed that when the food is inside, the male snatches the food from the youth and infant, and shows his teeth to the infant, sometimes obtaining reciprocity. However, this does not happen when the food is placed in the outside condition. Allogrooming remains without change and is usually received by the juvenile, the infant, and, to a lesser extent, the female.

Regardless of where the food is placed, the male usually takes it from the female, and sometimes from the youth, and in both conditions the male leaves food that all group members can take. Also, we can observe a limited display of domination behaviours, mainly agonistic ones, that differs from data obtained in wild conditions (i.e. eyelid exhibition, forehead retraction, bipedal position, splay-legged and red, white, and blue displays). Possibly this is because in wild conditions a function of dominance behaviour is to protect or delimit the territory; in captivity conditions (i.e. zoos), the confinement is a bounded area where it is not functional to emit these types of behaviours, inasmuch as there are no predators or intruders. This possible explanation is strengthened by the fact that the splay-legged display was recorded in the areas near visitors.

Meanwhile, female behaviour is usually consistent in receiving domination, displacements, and slapping only by the male. However, the male does not snatch her food nor show his teeth; we did not register dominance behaviours received from members of the group other than the male. These data seem to suggest that the female is really just dominated by the male and manages to dominate the rest of the group because nobody displays dominance behaviours to her, gaining first access to food and eating more than the other members of the group, though slightly less than the male. The female usually displaces and chases the youth in both conditions, generally associated with movement that pushes the youth away from the food source. She does not eat food left by the other members of the group. In wild conditions, the female is usually the dominant one, however, in captivity seems to be dominated by the male. Thus, it appears that the number of females in a group can be an important variable for the development of female dominance behaviours. Data showed a great variety of agonistic behaviour in the female that could lead us to assume a higher hierarchy, allowing her first access to food; however, first access is made by the male.

Ortiz et al. [1] and Ortiz [27] suggest that the classification of this situational adjustment should be made based on the relationships between the emitted behaviours and the situation in which they are presented. This situation should be seen as an array composed of a group of elements, factors, and/or variables that keep a peculiar relationship between them, structuring a network of interactions. In this way, we can identify at least four types of functional relationships between the behavioural patterns and the situation in which they are presented: (a) interaction or situation-specific behaviour, (b) interaction or behaviour required by the situation, (c) interaction or functional behaviour in the situation, and (d) interaction or irrelevant behaviour to the situation. In this sense, data suggest that this kind of behaviour is required but not quite functional in the situation; and in this sense, it seems necessary to be careful with taking into account only the morphology of behaviour, without identifying the organism adjustment.

Our data suggest the difficulty of sustaining the emergence of fixed patterns of behaviours, especially when conditions in captivity are so different from the wild. We observed changes when altering or changing an element of the situation in the enclosure (i.e. placing the food inside or outside the enclosure), which seems to suggest that modifying an element of a situation can mean that the individual has to change in order to adjust to that change. It is generally considered that the vervet monkey is a species that is able to adapt to conditions of captivity because their eating habits are not very complex and can be adapted to feed on what they have in confinement. In addition, they are considered to be behaviourally flexible, in the sense that when the ecological milieus change, their behaviour changes rapidly. Such as in the case of the female, with no other females in her group, adjusts her behaviours "surrendering" to the male, due to it being stronger, but dominates the rest of the group resulting, in any way, in first access to food, along with the male, selecting the better food. Meanwhile, the male may not display other dominance behaviours (such as those given in wild conditions) because it would not be functional in captivity, as there are no predators or intruders. In this sense, what they do, although it is different from what the species do in wild conditions, seems to be functional and allows adjustment to the condition they are in in captivity, while facilitating adaptation (i.e. reproductive success).

The individuals analysed are adjusting to very particular environmental conditions (i.e. captivity in the Guadalajara Zoo); based on the idea that captivity does not have the same stimuli as that compounding wild conditions, from this perspective so-called psychological well-being would be considered related to deployment of behaviours related to the required criteria by the contingency array or situation (i.e. functional, required, specifics). That is, we need to identify whether the organism is adjusted according to what the situation demands. It is possible that the individuals of a particular troop do not display the fixed action patterns in captivity, and this does not necessarily mean a lack of psychological well-being or that the organism behaviour is "bad" or "inadequate" per se, because the subject does not do what it is supposed to do, according to its species and by what it does in the wild. However, as the captivity condition is different, the animal will behave differentially given the condition of the particular confinement, adjusting in a different way than in wild conditions. But it is an adjustment and to lower the risk of the animal failing to adapt and reproduce, we must identify what factors contribute to the animal adjusting and adapting to a given confinement. Furthermore, the psychological well-being can be related to the identification of variability, differentiation, modification, integration, and delayed inhibition of its reactions [21], displaying not only behavioural variability, but adjustment to the situation that demands it.

In regard to the temporal patterns detected, the synchrony was found to exist on different levels, with highly complex time structures that extended over considerable time spans, as well as less complex patterns with a shorter time span. The results show that pattern analysis can be used to track elements in the social hierarchy during feeding time (i.e. aggression and dominant behaviour, team structure) in a novel way, indicating that pattern analysis is useful in enhancing existing methods used in animal research. Moreover, some answers have already suggested questions, such as: Are there certain patterns that are related to more aggressive behaviour and place of food delivery? What responses seem to be evoked by certain actions or sequences of actions?

Researchers could use this kind of structural information to increase their understanding of the subject being studied and zoos might benefit from such information when in the process of designing facilities for animals.

References

1. Ortiz G, Correa L, Gallardo MF (2006) Proposal for a methodology of classification, identification and manipulation of relevant factors to adaptation. ABMA Annual Conference Proceedings, 128–131
2. Manning A (1985) Introducción a la conducta animal. Alianza Editorial, España
3. Rowe N (1996) The pictorial guide to the living primates. Pogonias Press, East Hampton, NY
4. Groves C (2001) Primate taxonomy. Smithsonian Institute Press, Washington, DC
5. Cawthon-Lang KA (2006) Primate Factsheets: Vervet (Chlorocebus) Behavior. <http://pin.primate.wisc.edu/factsheets/entry/vervet/behav>. Accessed 24 May 2013
6. Fedigan L, Fedigan LM (1988) Cercopithecus aethiops: a review of field studies. In: Gautier-Hion A, Bourlière F, Gautier JP, Kingdon J (eds) A primate radiation: evolutionary biology of the African guenons. Cambridge University Press, Cambridge, pp 389–411
7. Cosme J (2004) Fauna y flora de Botswa-na y Namibia. Recuperado en abril de 2006. www.shelios.com/sh2004/divu/charlas/cosmez.pdf
8. Harrison MJS (1984) Optimal foraging strategies in the diet of the green monkey, Cercopithecus sabaeus, at Mt. Assirik, Senegal. Int J Primatol 5(5):435–471
9. Isbell LA, Enstam KL (2002) Predator (in) sensitive foraging in sympatric female vervets (Cercopithecus aethiops) and patas monkeys (Erythrocebus patas): a test of ecological models of group dispersion. In: Miller LE (ed) Eat or be eaten: predator sensitive foraging among primates. Cambridge University Press, Cambridge, pp 154–168
10. Zinner D, Peláez F, Torkler F (2002) Distribution and habitat of grivet monkeys (Cercopithecus aethiops aethiops) in eastern and central Eritrea. Afr J Ecol 40:151–158
11. Fairbanks LA, McGuire MT (1986) Age, reproductive value, and dominance-related behaviour in vervet monkey females: cross-generational influences on social relationships and reproduction. Anim Behav 34(6):1710–1721
12. Isbell LA (1995) Seasonal and social correlates of changes in hair, skin, and scrotal condition in vervet monkeys (Cercopithecus aethiops) of Amboseli National Park, Kenya. Am J Primatol 36(1):61–70

13. Slater PJB (1991) Introducción a la Etología. Crítica, México
14. Henzi SP (1985) Genital signaling and the coexistence of male vervet monkeys (Cercopithecus aethiops pygerythrus). Folia Primatol 45(3-4):129–147
15. Vervet Monkey Foundation (S/F). Understanding vervets. http://www.vervet.za.org/vervets/vervet_home.asp. Accessed 23 May 2013
16. Gerald MS (2001) Primate colour predicts social status and aggressive outcome. Anim Behav 61(3):559–566
17. Seyfarth RM, Cheney DL, Marler P (1980) Vervet monkey alarm calls: semantic communication in a free-ranging primate. Anim Behav 28(4):1070–1094
18. Seyfarth RM, Cheney DL, Marler P (1980) Monkey responses to three different alarm calls: evidence of predator classification and semantic communication. Science 210(4471):801–803
19. Enstam KL, Isbell LA (2002) Comparison of responses to alarm calls by patas (Erythrocebus patas) and vervet (Cercopithecus aethiops) monkeys in relation to habitat structure. Am J Phys Anthropol 119:3–14
20. Cheney DL, Seyfarth RM (1986) The recognition of social alliances among vervet monkeys. Anim Behav 34:1722–1731
21. Kantor JR (1924) Principles of psychology vol. 1. Knopf, Chicago
22. Ribes E, López F (1985) Teoría de la Conducta. Trillas, México
23. Maier R (1998) Comparative animal behavior: an evolutionary and ecological approach. Allyn and Bacon, Boston
24. Altmann J (1974) Observational study of behavior: sampling methods. Behaviour 49:227–266
25. Magnusson MS (1996) Hidden real-time patterns in intra- and inter-individual behavior: description and detection. Eur J Psychol Assess 12(2):112–123
26. Magnusson MS (2000) Discovering hidden time patterns in behavior: T-patterns and their detection. Behav Res Methods Instrum Comput 32(1):93–110 (View/Download PDF version)
27. Ortiz G (2014) Classification, identification, and manipulation of relevant factors for adaptation and behavioural adjustment from a psychological point of view. Psychology 5:1517–1526. doi:10.4236/psych.2014.513162

Chapter 16

Tidal Location of Atlantic Cod in Icelandic Waters and Identification of Vertical and Horizontal Movement Patterns in Cod Behavior

Gudberg K. Jonsson, Vilhjalmur Thorsteinsson, and Gunnar G. Tomasson

Abstract

The Atlantic cod is one of the most important commercial species known. The behavior of this species, important for fisheries, research, and stock assessment, is in many ways masked by extensive horizontal and vertical dispersion in its habitat. A tidal location model has been developed to predict location and movement of adult cod in Icelandic waters from signals from data storage tags. The time series were prepared for T-pattern analysis, including detection and delimitation of tidal influence in the data and event basing raw data according to predefined events. A high number of temporal patterns were detected, patterns of repeated vertical and horizontal movements and speed and acceleration changes. Number of specific temporal patterns was also identified across vertical and horizontal movements of individual cod.

Key words Cod, *Gadus morhua*, Behavior, DST tags, Tidal wave model, T-patterns

1 Introduction

Seasonal variations in the environment, and consequently variations in food supplies, continuously induce annual migrations of most species in birds, mammals, or fishes over short or vast distances. Due to the development of promising tracking methods, migration studies have been rapidly increasing in recent decades, with particular emphasis on multiple year patterns [1–6]. In this context, data storage tags (DSTs) allow the investigation of thermobathymetric migration patterns of fish, and the variability of individual migrations, using relatively long-term storage DSTs [7].

The Atlantic cod (*Gadus morhua* L.) is historically one of the most important commercial species known [8]. The behavior of this species, important for fisheries, research, and stock assessment, is in many ways masked by extensive horizontal and vertical

Magnus S. Magnusson et al. (eds.), *Discovering Hidden Temporal Patterns in Behavior and Interaction: T-Pattern Detection and Analysis with THEME™*, Neuromethods, vol. 111, DOI 10.1007/978-1-4939-3249-8_16,

dispersion in its habitat [9]. Its individual behavior related to feeding migrations has been shown to vary within a single spawning ground. Two behavior types have been observed, i.e., a coastal, relatively stationary type in shallow waters, and a migratory frontal (offshore) type, breeding in shallow waters but migrating to deeper waters to commonly feed near thermal fronts [7, 9–11]. Both behavior types have been studied in Norwegian [10, 11] and Icelandic waters [9, 12], and have been shown to segregate by depth both during spawning and feeding time [10, 13]. Recent studies have also shown genotype differences at the pantophysin locus (Pan I, see [12, 14–16]. The pantophysin Pan I^B allele was predominantly observed in northeast Arctic cod and in the Icelandic frontal type, while the Pan I^A allele was predominantly observed in Norwegian coastal cod and in the Icelandic coastal type. However, despite intensive research on these two behavior types, no studies have dealt with year-to-year repeatability of individual migrations, or the timing and routes of migrations. This information can be retrieved from DSTs [17, 18] and may shed new light on feeding migration processes (i.e., whether individuals move in groups/shoals) and on the consistency of the migration habits of these behaviors.

The current pilot study uses a new approach to analyze behavior such as horizontal and vertical movements of tagged cod in Icelandic waters. The approach, known as T-pattern detection [19, 20], has successfully been used within other research fields but never before in this particular field.

1.1 What Theme Does

Theme (see www.patternvision.com) [19, 20] looks for relationships between events. It takes into account the order and relative timing (critical interval relationship). If the critical interval is less than would be expected by chance, it defines a pattern called T-pattern. Theme starts with simple patterns and gradually adds them together to form more complex patterns. Less complete patterns do not survive, longer chains are "fitter"; as the patterns recombine and grow, only the fittest survive.

2 Method

The data was collected using the Data Storage Tags (DST centi series) developed by Star-Oddi (see www.star-oddi.com). The DST centi is a small underwater data logger, available with sensors for underwater temperature and depth logging. The cod was tagged with Data Storage Tags (DSTs) with memory capacity of up to 260,000 records measuring temperature and depth at 10 min intervals. All measurements are time related, utilizing a real time clock inside the DST.

During the years 2002–2005, 1104 cod were tagged and released with DSTs in Icelandic waters; of these, 347 (31.4 %) were recaptured up to 1807 days later. Tagging localities at the southwest (SW), southeast (SE), west (W), and northeast coasts (NE) of Iceland were selected for the study (Fig. 1). Spawning cod, 65–107 cm in length, were captured using gillnets off the southern and western coasts and by Danish seine off the NE coast.

2.1 Seasonality and Timing of Migration

In total, 41 DST-tagged fish were recaptured at least 18 months after release and were used for the analysis of temporal stability and repetitive behavior. The classification of behavior types as coastal versus frontal was performed according to previous studies [9, 12, 13]. In general, coastal and frontal behaviors were defined according to the annual temperature and depth history of the tagged individuals. Coastal types (C) spend at least 70 % of their time in shallow waters showing an annual rise in temperature to a maximum in September/October and decline in temperature to a minimum in February/March. The frontal types (F) share the depth range of the coastal types during spawning migrations, but during feeding migrations, they move to deeper waters (250–600 m). The temperature history showed visits to thermal fronts and frequent vertical migrations moving between extremes in temperatures typically found at such locations (<0 °C and >7 °C).

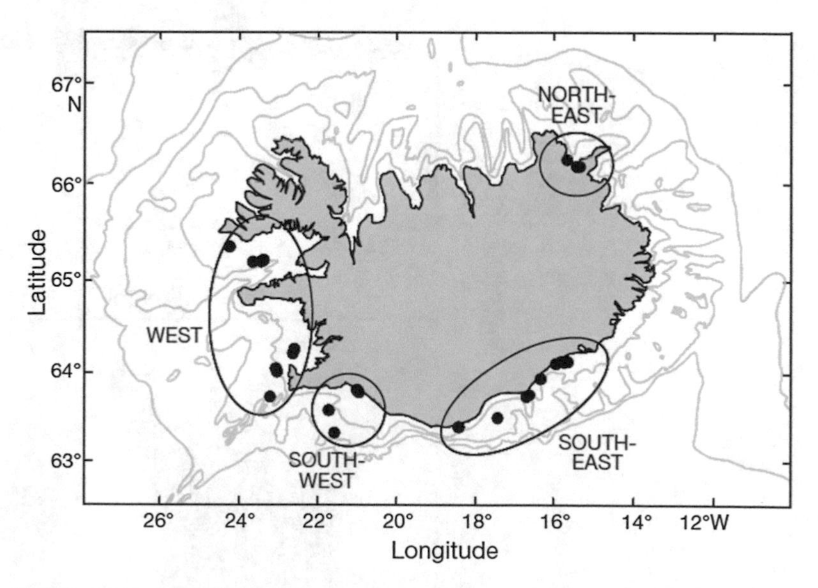

Fig. 1 *Gadus morhua*. Release areas of adult cod with data storage tags on spawning grounds in 2003–2005 that were recaptured after at least 18 month. Depth contours at 100, 200, 500 m

The temporal stability in depth and temperature profiles between years of each individual was estimated with Pearson correlation between monthly mean depth and temperature of a particular month in 2 consecutive years. Between-year consistency in timing of arrival at, and departure from, a spawning ground was tested with analysis of covariance (ANCOVA), with time of migration in the latter year as the dependent variable, time of migration in the former year as a covariate and behavior type as a random factor. Timing of arrival and departure was estimated from the behavior pattern as day of the year. Individual cod are found at smaller depths on the spawning grounds than during the feeding migrations. Therefore, patterns are easily seen in the depth profiles (Fig. 2), and the time of arrival on the spawning ground can be estimated from the vertical activity. Periods of low vertical activity, when cod are found at the same depth for more than 2 days at the end of the migratory period, are characteristic of individuals arriving at the spawning ground. During the following period of low activity, the cod remain at the spawning ground. When the behavior pattern reverts to increased vertical activity, and the cod moves into deeper waters, the fish has left the spawn-

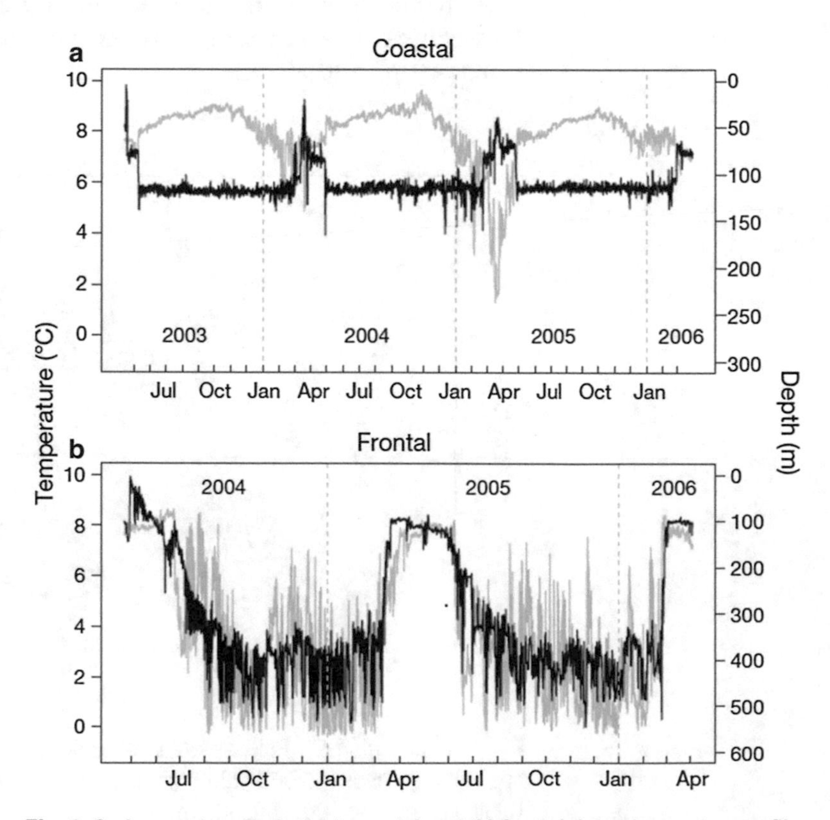

Fig. 2 *Gadus morhua*. Typical (**a**) coastal and (**b**) frontal data storage tag profiles over more than 2 years. Depth = *black line*, temperature = *grey line*

ing ground. Eight comparisons of each behavior type were included in the model, as migration from and to the spawning ground could only be compared using 8 coastal and 8 frontal individuals. Behavior type was not significant and was removed from the final model for simplification.

2.2 Level, Speed, and Acceleration Changes

From the DST tags original data events were defined by start time and end time, and also a label indicating the meaning of the event. For example, an event may indicate that the tag was located at a certain depth level, or that temperature increased over a certain time interval. The three types of events used in the current pilot study are:

- Level-based events that are based on the range of the measurement data.
- Speed-based events, which indicate changes in the data.
- Acceleration-based events, which indicate changes in speed over time.

2.3 Estimating Location from a Tidal Model

A numerical model has been developed and set up to predict sea level variations and tidal currents along the Icelandic coastline, taking into account both the astronomical and meteorological forcing. The model is run on an operational basis at the Icelandic Maritime Administration to predict sea level and tidal currents in Icelandic coastal waters using a weather forecast from the European Center for Medium Range Weather Forecasts. The model is based on the two-dimensional part of the Princeton Ocean Model (POM) [21], solving the nonlinear shallow water equations numerically using a staggered finite difference scheme.

The tidal model location was applied to the 41 DSTs retrieved more than 18 months after release. In order to save memory space on the DSTs the measurement frequency was set at 6 h for a part of the measurement period. These measurements were omitted from the tidal location studies due to insufficient temporal resolution to detect the tidal oscillation, and the limit was therefore set at 10 min for the measurement frequency. Furthermore, the fish must stay at the seafloor for several hours to detect a long enough tidal signal for the tidal location model to be applied. A data base of amplitude and phase of 7 tidal harmonic constituents on a 0.25° longitude and 0.10° latitude grid was used as a basis for the tidal location of the fish. The data originated from a numerical model of the tides in the North Atlantic Ocean around Iceland, developed by Tómasson and Káradóttir [22]. The model extends over an area of 5.7×10^6 km^2 with 10×10 km resolution over the whole model domain but 2×2 km resolution on the Icelandic shelf. It is based on the seven most important tidal harmonic constituents and takes boundary conditions at open ocean boundaries from a larger global model. The model has been calibrated and verified extensively with data from harbors and mooring stations around Iceland and elsewhere. Its accuracy on the Icelandic shelf is within a few centimeters in amplitude and a few minutes in phase.

The tidal location method used here is an extension of that presented by Pedersen et al. [23, 24]. The first step is to search the entire data series for possible tidal patterns. A tidal pattern is identified where fitting to a sine curve over a sliding window of 10 h in length satisfies three criteria, i.e., root mean square error (RMSE) <0.42 m, coefficient of determination $R^2 > 0.85$ and amplitude of the fitted sine wave $A > 0.3$ m. The second step, applied when a fit to a sine curve is identified, consists of finding the most probable location of the fish by comparison of the amplitude and phase of the fitted sine curve with calculated tidal signals within the model domain. These steps are identical to those of Pedersen et al. [23] (except for the criteria on minimum amplitude, which were twice as small in this study to adapt to tidal conditions in the sea around Iceland) resulting in an automatically proposed location of the fish. However, those criteria alone were found to propose too many false or unreliable tidal signals and locations when applied to data from Icelandic cod. Therefore, a third step was added based on visual comparison of observed tidal DST-pattern of the cod and the modeled tidal oscillation at the predicted location. Based on a comparative evaluation of the agreement between the 2, and an inspection of other possible tidal patterns in nearby time intervals, taking into account cod swimming speed and distance from the last accepted location, a final decision was made on acceptance or rejection of the tidal location. With this additional step, around one-third of the originally proposed tidal patterns and locations were rejected. The accuracy and reliability of the predicted tidal locations was verified by applying the method to time series from stationary tags (tags moored at fixed locations). The results showed that if a signal was correctly identified as a tidal signal, the predicted location was reliable with an accuracy of 10–20 km in most cases.

3 Results

3.1 Seasonality

Data retrieved from the 41 individual DSTs revealed a high degree of interannual regularity in temperature and depth patterns, with a typical pattern of coastal cod inhabiting the same depths and temperatures during 3 years (Fig. 2) and a typical pattern of frontal cod visiting the same depths over a period of 2 year at low and fluctuating temperatures, and returning to shallower water at a fairly constant time of the year (Fig. 2).

The two behavior types clearly displayed distinct seasonal patterns of temperature and depth (Figs. 2 and 3), and individuals of

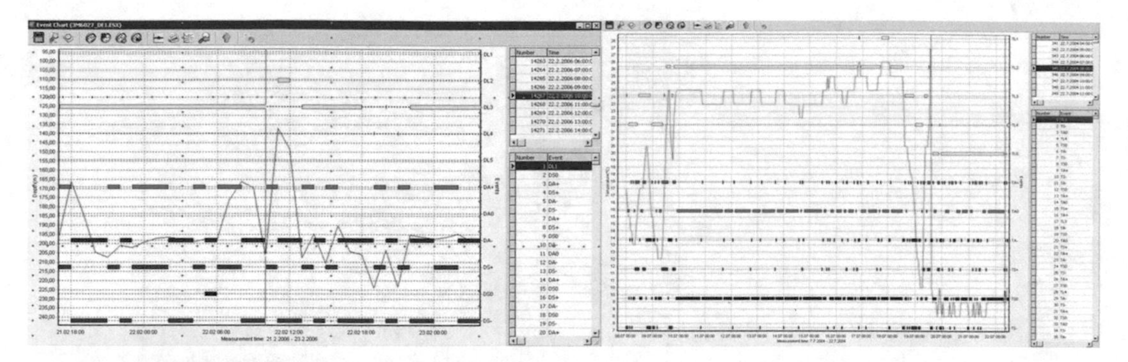

Fig. 3 Event basing depth data involved partitioning time-series into defined periods/events

both types retained their behavior characteristics through the observation time, i.e., coastal behavior types did not demonstrate frontal behavior patterns or vice versa.

The depth patterns of coastal cod showed largely similar trends across areas with shallowest distributions in April at the time of peak spawning, although the depth pattern in the NE area was shallower. The observed depth distribution patterns of frontal cod were more highly pronounced than for the coastal type, showing clear migrations into shallower waters during the spawning season in March and April, followed by migrations into deeper waters, associated with increased variability in depth, and reaching maximum depths in autumn.

In addition, the comparison of migration patterns of both behaviors showed that individuals consistently occupied similar depth/temperature niches during the same month in consecutive years.

3.2 Timing of Migration

Individuals of both behavior types arrived at the spawning grounds around mid-March and left for the feeding grounds at various times from the end of May to the end of June. Although based on a relatively small number of samples, the arrival on the spawning ground in the second year was significantly related to that of the first year (ANCOVA; $F=5.274$; $p=0.038$). No significant differences could be observed between dates of arrival at spawning grounds of the behavior types. Likewise, the departure from the spawning ground in the second year was significantly related to departure in the first year (ANCOVA; $F=6.312$; $p=0.0249$), and no differences were observed between behavior types.

3.3 Geo-location and Shoal Migration

Geographic positions of two coastal individuals tagged at similar locations (Fig. 4) showed that they stayed within the same bay during the 2 years time series. The tidal model furthermore showed that coastal individuals were repeatedly found close to each other (within 25 km in distance and 15 days in time) during 3 seasons

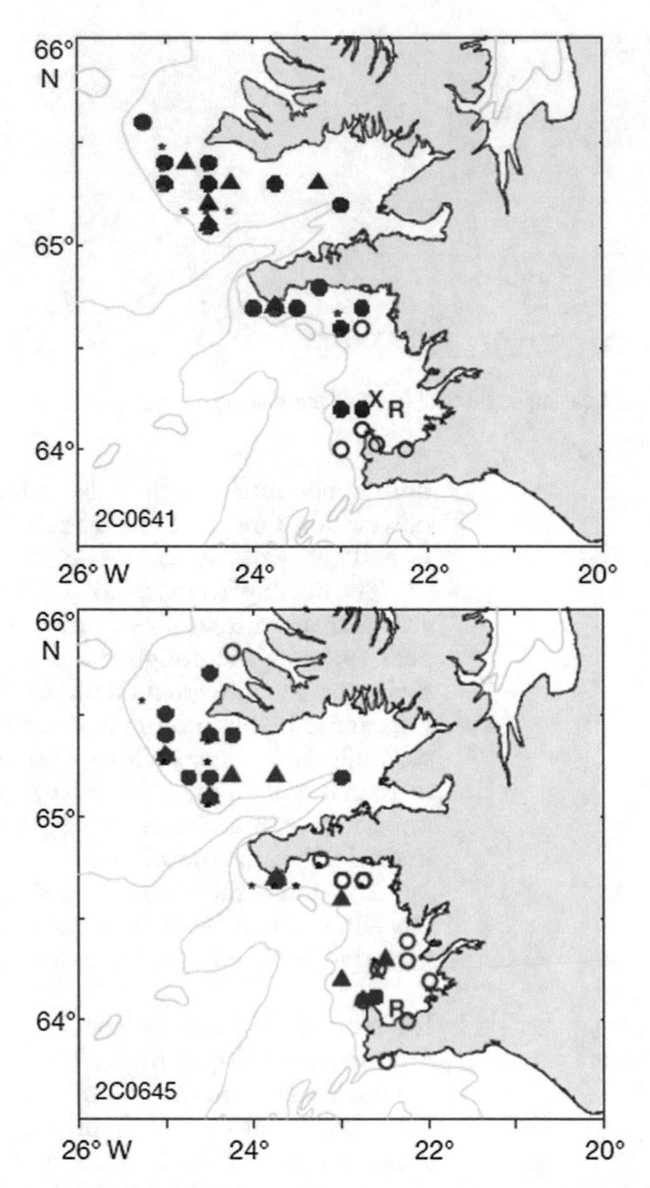

Fig. 4 *Gadus morhua*. Example of geo-location of two coastal individuals (2C0641, *black*; 2C0645, *red*) which were repeatedly located together during a 2-year period, indicating that the two individuals stayed together. *Symbols* show locations in February to May (*open circles*), June to September (*filled triangles*), October to January (*filled circles*); simultaneous (within 25 km and 15 days) location of other individual (*asterisks*), release (*R*) and recapture (*x*) locations

(Fig. 5), which tends to confirm that they moved together in a concurrent way and exhibited fidelity to migrating routes, areas and groups or shoals. The tidal location data including all recovered DSTs (Fig. 5) also showed that during spawning time and spawning migrations (February to May), individuals of each behavior type were distributed in the same area, but frontal cod occupied

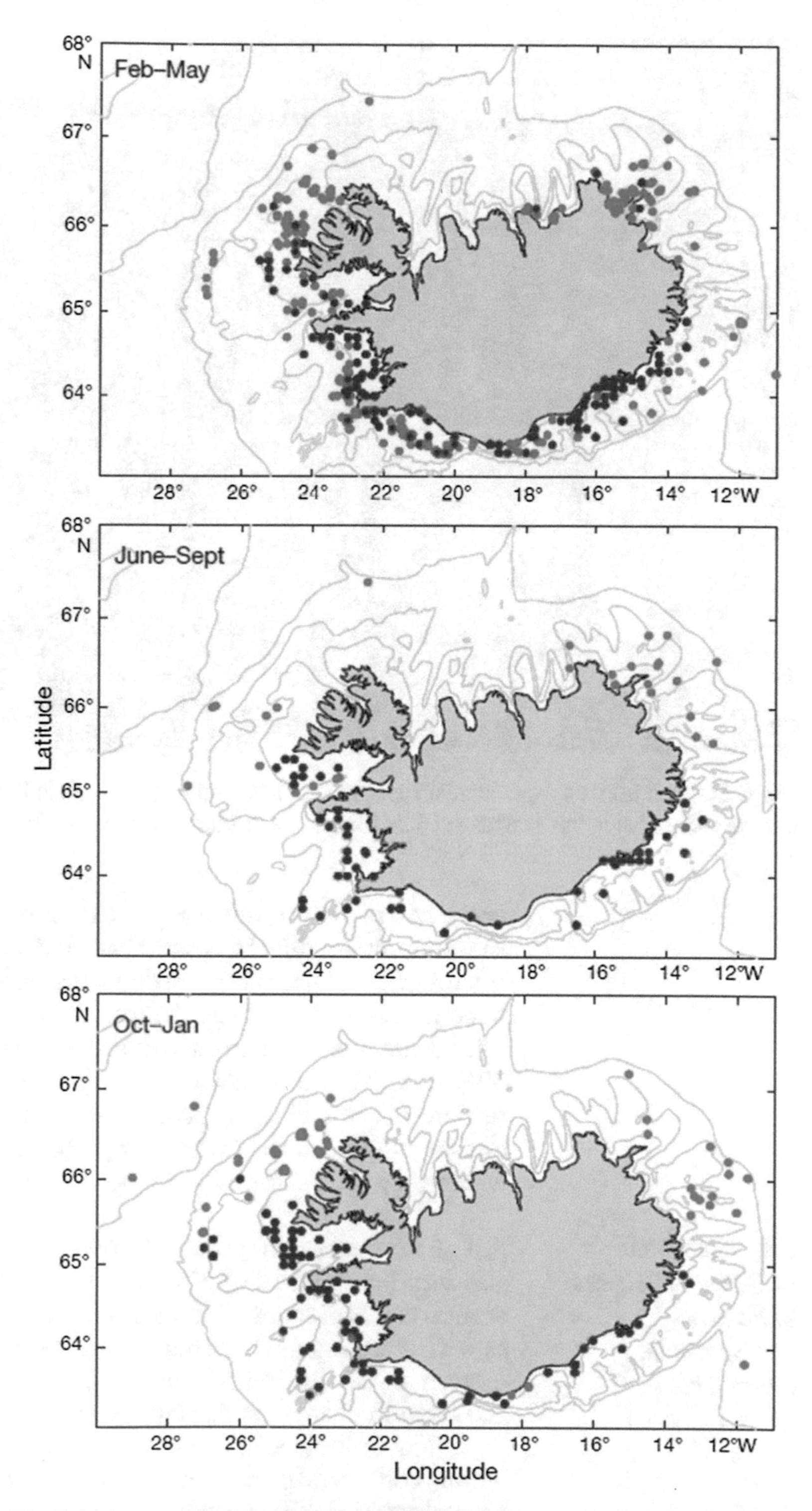

Fig. 5 *Gadus morhua*. Geo-location of individual cod (coastal = *red*; frontal = *blue*) based on an improved model of Pedersen et al. [23, 24]. All individuals (41) are presented by regions and seasons

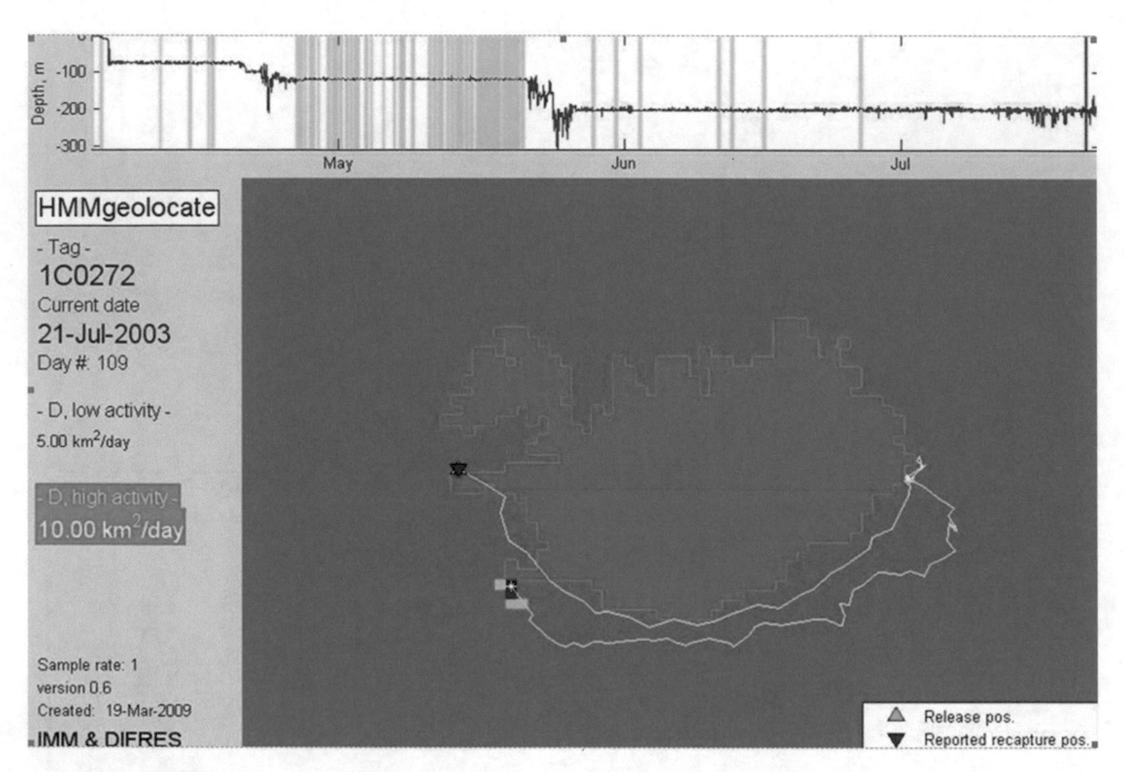

Fig. 6 Example from an experimental application of this method tracing an individual cod tagged in Icelandic coastal waters using the database of tidal components generated by the tidal model

deeper parts of the spawning grounds compared to the coastal cod. In the summer (June to September) and winter feeding migrations (October to January) the distributions were uneven, as coastal cod were more prominent in shallow waters in the south, SW, and SE. Frontal cod were mostly found in deeper waters in the north, NW, and NE, where they were likely to en counter polar temperature fronts. In addition, most tagged individuals exhibited spawning site fidelity, as they were recaptured at the same GPS location where they were released (data not shown) (Fig. 6).

3.4 Examples of Temporal Patterns of Vertical and Horizontal Movements

A high number of temporal patterns were detected in the cod DST and location data using Theme. These patterns were of repeated vertical and horizontal movements, speed, and acceleration changes as well as resting at the same defined vertical level. An example of pattern of an individual cod's vertical movements, speed, and acceleration changes is displayed in Fig. 7.

Number of specific temporal patterns were also identified within cod vertical movements (see examples in Figs. 8 and 9). Several multiple year temporal patterns were also identified across individual Costal or Frontal cod vertical movements. No multiple year vertical temporal patterns were discovered between Costal and Frontal cod individuals analyzed.

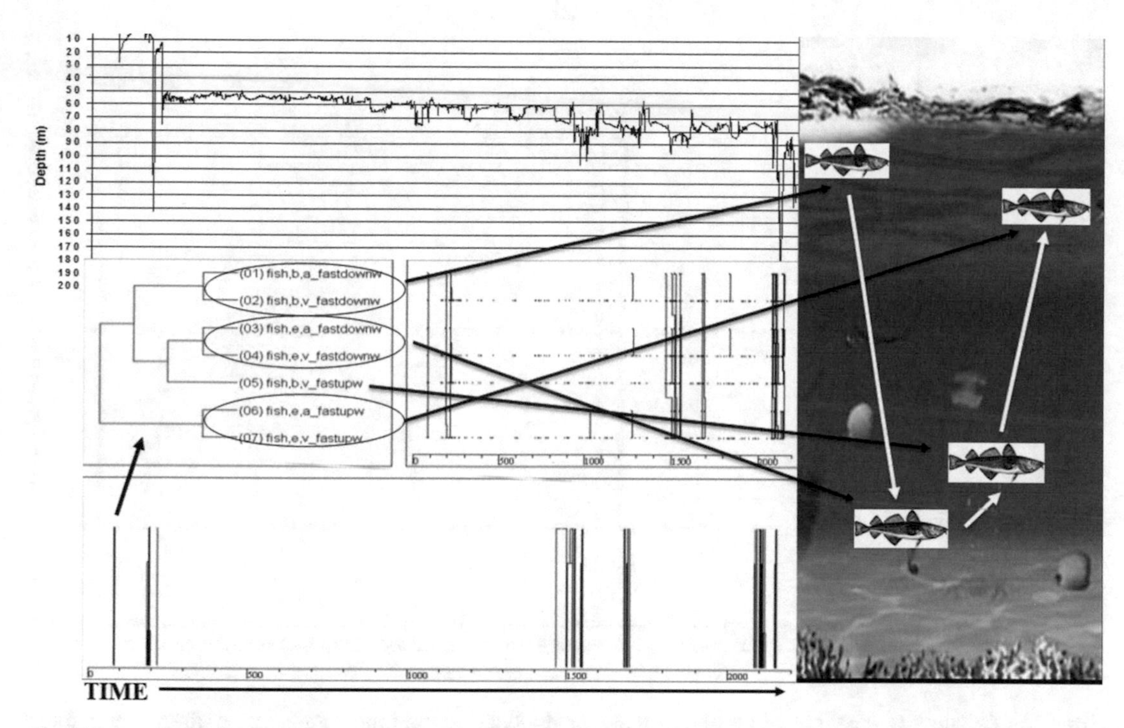

Fig. 7 An example of one pattern of vertical movements found in the time series. The pattern occurs 12 times during the recording period with same order and approximately the same time interval between event types (depth, speed, and acceleration change)

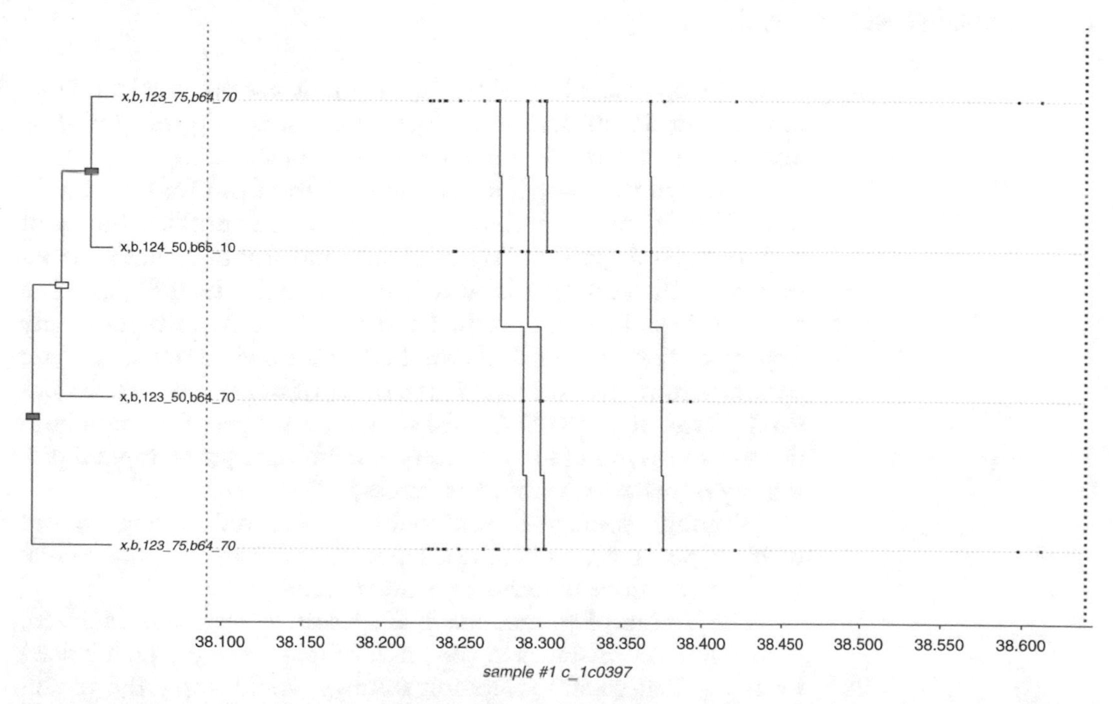

Fig. 8 An example of a pattern of horizontal movements found in the time series. The pattern occurs three times during the recording period with the cod moving between four different locations with same order and approximately the same time interval between event types (locations)

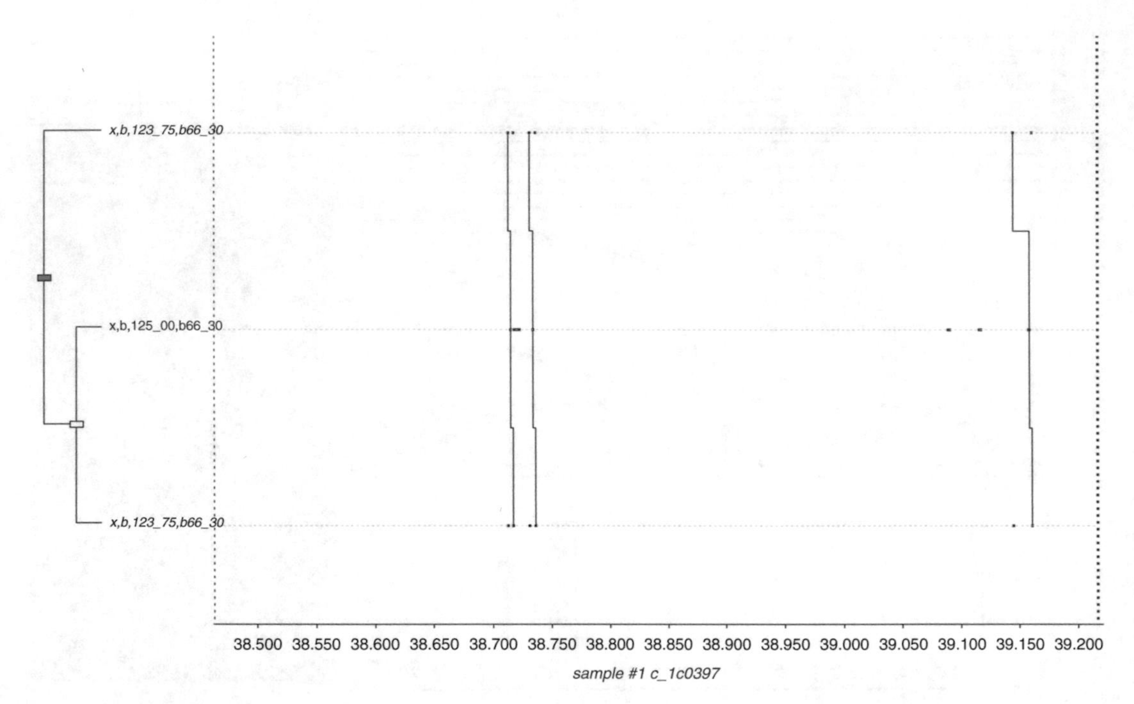

Fig. 9 An example of a pattern of horizontal movements found in the time series. The pattern occurs three times during the recording period with the cod moving between three different locations with same order and approximately the same time interval between event types (locations)

4 Discussion and Conclusion

The development of new tracking methods has provided a unique opportunity to study the spatiotemporal variability of individual migration patterns over more than one annual cycle.

In this study, using information retrieved from DSTs placed in individual Atlantic cod, the consistency in DSTs profiles, timing of migration, and geo-location of coastal and frontal behavior types over more than 18 months were investigated for the first time. The results revealed that (1) coastal and frontal behavior types clearly display distinct seasonal thermobathymetric patterns which are constant from year to year, (2) the onset of migration is consistent from year to year, (3) DST tidal signatures suggest feeding migrations in shoals, and (4) the presence of multiple year temporal patterning of behavior within and across individuals.

Although some variation could be observed among regions within types, the two behavior types demonstrated highly consistent and repetitive patterns over the period investigated.

The timing of migrations in 2 successive years was analyzed, suggesting a consistency in the onset of migration for each behavior type. A successful migration strategy would imply the retainment of migratory behaviors for the benefit of reduced food competition [25]. The consistency observed here might reflect successful, concurrent strategies to avoid resource competition within a large, coastal population.

One of the most striking observations during the present study was that the tidal location model suggested that the feeding migrations of cod in Icelandic waters were undertaken in groups or shoals during the whole year. Shoaling is commonly described as groups of fishes that remain together for social reasons with no implication of structure or function [26, 27]. One possible explanation for the observed results is that social transmission might enable the rapid transfer of beneficial behavioral traits (feeding migration, homing, and timing of the migration) from older spawning cod to new recruits (see "adopted migrant" hypothesis of McQuinn [28]).

The results of the preliminary analysis of temporal patterns in the event based and location data indicate that a Theme analysis can make a significant contribution to the analysis of cod behavior, offering an increased advantage to view and understand hidden and complex patterns within a large number of data points. A high number of temporal patterns were detected, patterns of repeated vertical and horizontal movements and speed and acceleration changes. Number of specific temporal patterns were also identified across individual cod vertical and horizontal movements.

Future objective is to further search for patterns of vertical movement in relation to environmental parameters emphasizing (a) behavioral patterns that are tidal-wave related, (b) behavioral patterns related to temperature and depth data, (c) behavioral patterns related to observed patterns of wind speed and direction, tidal stage, lunar stage, day length, and other local oceanographic features, and (d) behavioral patterns related to location of the cod in the Icelandic waters.

References

1. Egevang C, Stenhouse IJ, Phillips RA, Petersen A, Fox JW, Silk JRD (2010) Tracking of Arctic terns Sterna paradisaea reveals longest animal migration. Proc Natl Acad Sci U S A 107: 2078–2081
2. Klaassen RHG, Alerstam T, Carlsson P, Fox JW, Lindstrom A (2011) Great flights by great snipes: long and fast nonstop migration over benign habitats. Biol Lett 7:833–835
3. Matthews CJD, Luque SP, Petersen SD, Andrews RD, Ferguson SH (2011) Satellite tracking of a killer whale (Orcinus orca) in the eastern Canadian Arctic documents ice avoidance and rapid, long-distance movement into the North Atlantic. Polar Biol 34: 1091–1096
4. Olifiers N, Loretto D, Rademaker V, Cerqueira R (2011) Comparing the effectiveness of tracking methods for medium to large-sized mammals of Pantanal. Zoologia (Curitiba) 28:207–213
5. Vardanis Y, Klaassen RHG, Strandberg R, Alerstam T (2011) Individuality in bird migration: routes and timing. Biol Lett 7:502–505
6. Forsythe PS, Crossman JA, Bello NM, Baker EA, Scribner KT (2012) Individual-based analyses reveal high repeatability in timing and location of reproduction in lake sturgeon (Acipenser fulvescens). Can J Fish Aquat Sci 69:60–72
7. Neat FC, Wright PJ, Zuur AF, Gibb IM et al (2006) Residency and depth movements of a coastal group of Atlantic cod (Gadus morhua L.). Mar Biol 148:643–654
8. Kurlansky M (1997) Cod: a biography of the fish that changed the world. Walker and Company, New York, p 294

9. Pálsson ÓK, Thorsteinsson V (2003) Migration patterns, ambient temperature, and growth of Icelandic cod (Gadus morhua): evidence from storage tag data. Can J Fish Aquat Sci 60:1409–1423
10. Nordeide JT (1998) Coastal cod and north-east Arctic cod—do they mingle at the spawning grounds in Lofoten? Sarsia 83:373–379
11. Godo OR, Michalsen K (2000) Migratory behaviour of northeast Arctic cod, studied by use of data storage tags. Fish Res 48:127–140
12. Pampoulie C, Jakobsdóttir KB, Marteinsdóttir G, Thorsteinsson V (2008) Are vertical behaviour patterns related to the Pantophysin locus in the Atlantic cod (Gadus morhua L.)? Behav Genet 38:76–81
13. Grabowski TB, Thorsteinsson V, McAdam BJ, Marteinsdóttir G (2011) Evidence of segregated spawning in a single marine fish stock: sympatric divergence in Icelandic cod? PLoS ONE 6, e17528
14. Fevolden SE, Pogson GH (1997) Genetic divergence at the synaptophysin (Syp I) locus among Norwegian coastal and north-east Arctic populations of Atlantic cod. J Fish Biol 51:895–908
15. Sarvas TH, Fevolden SE (2005) Pantophysin (Pan I) locus divergence between inshore v. offshore and northern v. southern populations of Atlantic cod in the north-east Atlantic. J Fish Biol 67:444–469
16. Skarstein TH, Westgaard JI, Fevolden SE (2007) Comparing microsatellite variation in north-east Atlantic cod (Gadus morhua L.) to genetic structuring as revealed by the pantophysin (Pan I) locus. J Fish Biol 70: 271–290
17. Hunter E, Metcalfe JD, Arnold GP, Reynolds JD (2004) Impacts of migratory behaviour on population structure in North Sea plaice. J Anim Ecol 73:377–385
18. Metcalfe JD (2006) Fish population structuring in the North Sea: understanding processes and mechanisms from studies of the movements of adults. J Fish Biol 69:48–65
19. Magnusson MS (2000) Discovering hidden time patterns in behavior: T-patterns and their detection. Behav Res Methods Instrum Comput 32(1):93–110 (View/Download PDF version)
20. Magnusson MS (1996) Hidden real-time patterns in intra- and inter-individual behavior: description and detection. Eur J Psychol Assess 12(2):112–123
21. Blumberg AF, Mellor GL (1987) A description of a three-dimensional coastal ocean circulation model. In: Heaps NS (ed) Three-dimensional coastal ocean models. American Geophysical Union, Washington, DC, pp 1–16
22. Tómasson GG, Káradóttir ÓR (2005). A two dimensional numerical model of astronomical tide and storm surge in the North Atlantic Ocean. In: Viggosson G (ed) Second International Coastal Symposium in Iceland. Höfn, Hornafjörður, 5-8 June, 2005, Icelandic Maritime Administration, Abstract volume, p 266–267
23. Pedersen MW, Righton D, Thygesen UH, Andersen KH, Madsen H (2008) Geolocation of North Sea cod (Gadus morhua) using hidden Markov models and behavioural switching. Can J Fish Aquat Sci 65:2367–2377
24. Thorsteinsson V, Pálsson OK, Tómasson GG, Jónsdóttir IG, Pampoulie C (2012) Consistency in the behaviour types of the Atlantic cod: repeatability, timing of migration and geolocation. Mar Ecol Prog Ser 462:251–260
25. Brodersen J, Nilsson PA, Chapman BB, Skov C, Hansson LA, Brönmark C (2012) Variable individual consistency in timing and destination of winter migrating fish. Biol Lett 8:21–23
26. McQuinn IH (1997) Metapopulations and the Atlantic herring. Rev Fish Biol Fish 7:297–329
27. Pitcher TE (1983) Heuristic definitions of shoaling behaviour. Anim Behav 31:611–613
28. Pitcher TJ, Parrish JK (1993) Functions of shoaling behaviour in teleosts. In: Pitcher TJ (ed) Behaviour of teleost fishes. Chapman & Hall, London, pp 363–440

Chapter 17

Complex Spike Patterns in Olfactory Bulb Neuronal Networks

Alister U. Nicol, Anne Segonds-Pichon, and Magnus S. Magnusson

Abstract

Using T-pattern analysis, a procedure developed for detecting a particular kind of nonrandomly recurring hierarchical and multi-ordinal real-time sequential patterns (T-patterns), we have inquired whether such patterns of action potentials (spikes) can be extracted from extracellular activity sampled simultaneously from many neurons across the mitral cell layer of the olfactory bulb (OB). Spikes were sampled from urethane-anesthetized rats over a 6 h recording session, or a period lasting as long as permitted by the physiological condition of the animal. Breathing was recorded as markers of peak inhalation and exhalation. Complex t-patterns of up to ~20 elements were identified with functional connections often spanning the full extent of the array. A considerable proportion of these sequences were related to breathing. By comparing sequence detection in our real data with that in the same data when randomized (using either of two procedures, one preserving the interval structure of each spike train, and so the more conservative), we find that the incidence of sequences is very much greater in the real than in the random data. Further, in cases where recordings were terminated before completion of the full recording session, the difference between pattern detection in real data and that of randomized data strongly correlated with the physiological condition of the animal—in recordings leading to the preparation becoming physiologically unstable, the number of patterns detected in real data approached that in the randomized data. We conclude that such sequences are an important physiological property of the neural system studied, and suggest that they may form a basis for encoding sensory information.

Key words T-pattern analysis, Spike patterns, Neuronal network, Olfactory bulb, Rat

1 Introduction

Much of the computational power of the brain undoubtedly resides in the activities of cooperating and competing networks of neurons. Functional coupling between simultaneously sampled neurons in an in vivo preparation was first reported 1963 by Griffith and Horn [1] and has long been recognized as theoretically important in neuroscience. Synchronized activation between neurons has been linked to perceptual cognition, namely "the binding problem" [2], whereby the combined features of a complex

Magnus S. Magnusson et al. (eds.), *Discovering Hidden Temporal Patterns in Behavior and Interaction: T-Pattern Detection and Analysis with THEME™*, Neuromethods, vol. 111, DOI 10.1007/978-1-4939-3249-8_17,

stimulus are associated by the synchronization of the activities of neurons responding to one or more of those features. The most widely accepted theory of the physiology of memory formation [3], is also based upon the occurrence of such interactions in memory systems—"When an axon of cell A is near enough to excite B and repeatedly or persistently takes part in firing it… A's efficiency, as one of the cells firing B, is increased."

Experimental models have provided evidence that memory formation may adhere to so-called "Hebbian" principles. Long-term potentiation (LTP) is a physiological process that has been studied extensively since its first description in 1974 by Bliss and Lomo [4]. In LTP, presynaptic and postsynaptic elements in a neural pathway are simultaneously activated by repeated electrical stimulation of the presynaptic elements, thereby fulfilling the first of Hebb's principles [3]—neuron A repeatedly activates neuron B through the synaptic connection between the two elements. Subsequently, the efficiency of neuron A in activating neuron B is increased, and so a simple hebbian assembly is formed. In this paradigm, a large number of ***neuron A's*** activate a large number of ***neuron B's*** (i.e., there is little noise in the system), and the enhanced efficiency of transmission from one to the other is evident in the increased amplitude of the field potential generated when a single pulse is delivered to the presynaptic elements. However, evidence for such a process occurring across a single synapse in a functioning system remains elusive.

Attempts have been made to discover spike patterns within populations of neurons, but so far these have not produced the desired kinds of results. Thus, Abeles [5] proposed a search algorithm for the detection of multi-neuron patterns called "synfire." While numerous patterns were detected, doubt remains regarding the statistical significance of the findings [6, 7]. In this chapter, a more flexible pattern model, called a T-pattern, is applied (see Chapter 1 in this volume). T-pattern detection uses an evolution algorithm for the detection of the repeated hierarchical and multi-ordinal real-time patterns in data sets consisting of a number of time point series all occurring within the same observation period [8–10]. The large number of T-patterns detected frequently far exceeds those found in randomized data thus the complex patterns discovered through T-pattern analysis provide a dynamic view of neuronal interaction which may be invaluable in understanding the mechanisms of neuronal networks and the way they encode sensory information.

Olfactory encoding is of specific interest as behavioral paradigms underpinning studies of the neurobiology of olfactory learning and memory are considered particularly robust [11] and considerable progress has been made in establishing the neural substrates and pathways involved [12–14]. Much of the encoding takes place at the level of the olfactory bulb (OB), the primary cortical projection area for olfactory input, and an area that is entirely committed to processing this information. Structurally, the OB is widely conserved across vertebrate taxa. The area has been confirmed as playing an important role in olfactory memory formation. Thus, understanding the processes involved in encoding olfactory information is of great importance to understanding the fundamental neuronal mechanisms of learning and memory. Olfactory receptor neurons in the olfactory epithelium in the nasal cavity project mitral cells in glomeruli in the OB [15]. Optical imaging studies demonstrate that different odorants elicit spatially defined spatial patterns of glomerular activity in the olfactory bulb [16, 17]. The quality of an olfactory stimulus is thus encoded by the combined specific activation of glomeruli by a given odorant. Gaining access to the olfactory bulb with a microelectrode array (MEA) allows in vivo electrophysiological sampling of neuronal activity over a relatively large area of cortex (>2 mm^2). We have applied t-pattern analysis to spike data collected simultaneously across many OB neurons, using microelectrode arrays [18], and established that recurring complex sequences of spikes can be detected in the activity sampled across the mitral cell layer of the OB. These patterns have been characterized in light of a putative role in processing sensory information.

2 Methods

The present data were collected from the olfactory bulb of anesthetized rats (see Fig. 1 for further details). Throughout surgical and experimental procedures, humidified air was supplied through a mask over the nose, and breathing was monitored and recorded using a thermistor in the mask. Data were sampled in 10s periods (trials) at 5 min intervals through a recording session of 6 h, or for as long as the animal remained physiologically stable as judged by its breathing. Some animals reached a point when irregular breathing clearly indicated degrading physiological condition. In such cases recordings were terminated, and degrading condition noted as an experimental variable. Here we compare one such animal with another which remained physiologically viable throughout the full sampling period.

Microelectrode arrays of sharpened tungsten electrodes, arranged in a 6×5 array with 350 μ spacing, were advanced laterally into the OB (for one animal a 6×8 electrode array, with 250 μ spacing was used). Action potentials (spikes) were sampled from mitral layer OB neurons across an area of ~2.2 mm^2 using a 100 channel laboratory interface (Bionic Technologies Inc./ Cyberkinetics Inc., USA). Spikes sampled in the mitral cell layer are assumed to be generated by mitral cells, as the other cell type in this layer, the granule cell, does not possess an axon [19]. After completion of recordings, offline discrimination of spikes from

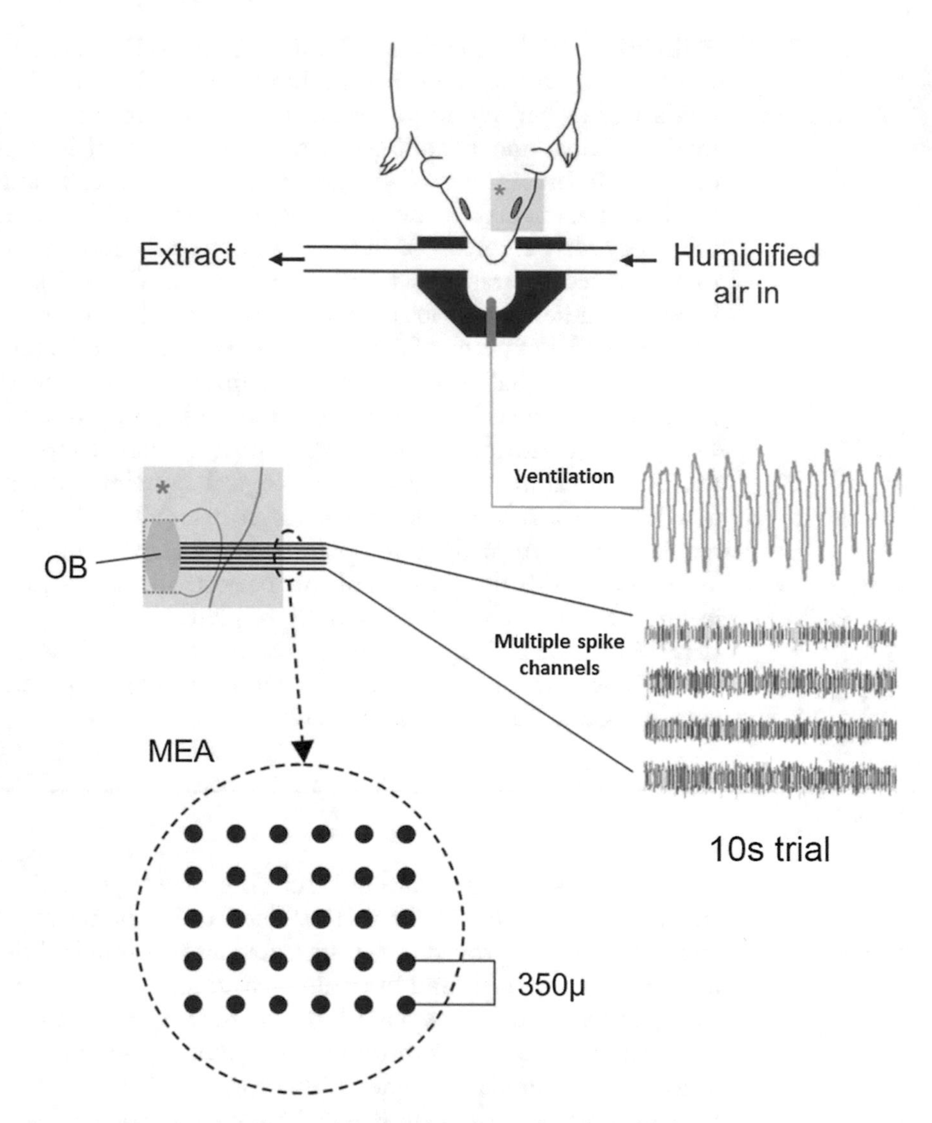

Fig. 1 Action potentials (spikes) from individual neurons were sampled from the olfactory bulb of urethane-anesthetized rats using a microelectrode array positioned laterally in the mitral cell layer of the olfactory bulb of urethane-anesthetized rats. Each electrode in the array could sample spikes from multiple neurons. Here, individual neurons are coded in varied greyscale in the four traces displayed. Humidified air was supplied through a mask over the nose, and breathing was monitored and recorded via a thermistor in the mask

individual neurons was performed using a spike sorting procedure based upon machine learning algorithms to combine the features acquired using principle components analysis (PCA) with features describing the geometric shapes (curvature) within the spike waveform. This procedure was developed specifically to process these data and allows discrimination of activity from multiple neurons at each active electrode [20]. Typically, spikes were sampled simultaneously from ≥100 neurons across the array. Times of occurrence of spikes generated by individual neurons were stored as events coded with the identity of the neuron and its location on the

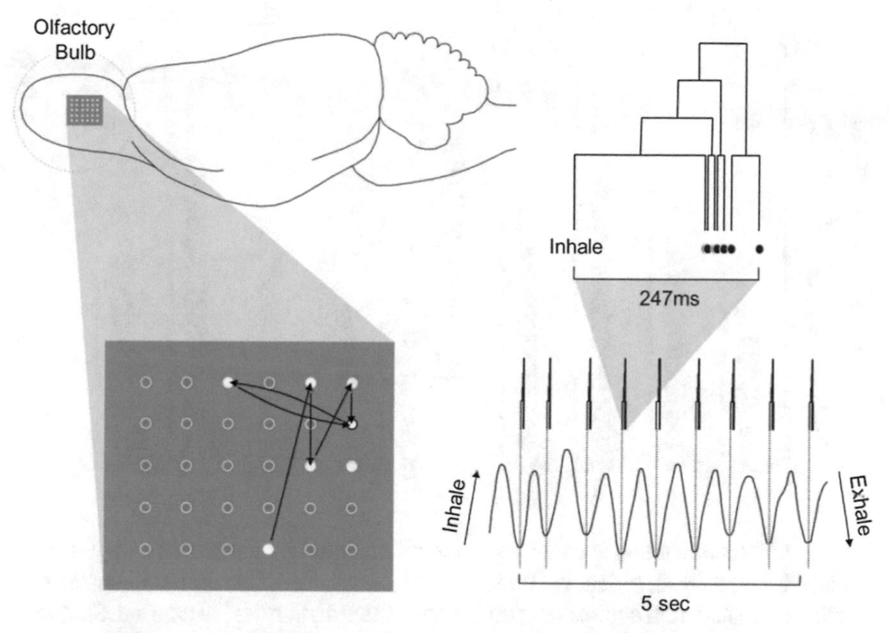

Fig. 2 An example breathing-related sequence of spikes across the olfactory bulb. This pattern spanned the full vertical extent of the 6 × 5 microelectrode array, incorporating 6 neurons, each sampled on a separate electrode in the array. One neuron, sampled at the electrode (*outlined in black* on the array), was activated twice in the sequence, and provided the final spike in the pattern. The first event in the t-pattern, not shown on the array, was the onset of inhalation—the sequence of spikes occurred regularly as the animal breathed in

MEA. Also stored were events marking onset of expiration and onset of inspiration in the breathing cycle to allow this data to be related to patterns identified in the neural data. The data stored in this way are suitable for t-pattern analysis using Theme™ software (PatternVision, www.patternvision.com, Iceland). In each trial, data were collected during five consecutive complete breathing cycles, each cycle beginning with the onset of inhalation, and ending with the offset of exhalation (see Fig. 2). Event types entered into t-pattern analyses were times of occurrence of spikes from individual neurons, and times of onset of inhalation and exhalation in the breathing cycle.

3 Results

In these data, the relationship between pattern lengths and firing rates amongst the neurons was not simply determined by a stochastic process. It might be assumed that neurons with higher firing rates would by chance have a greater opportunity to participate in longer patterns. We categorized the neurons as slow, medium, or fast, according to firing rate, based approximately around modal firing rates determined across the entire sample. Slow neurons were sufficiently slow that they contributed to no patterns of any length. Amongst the fast neurons, the *longest* patterns to which these

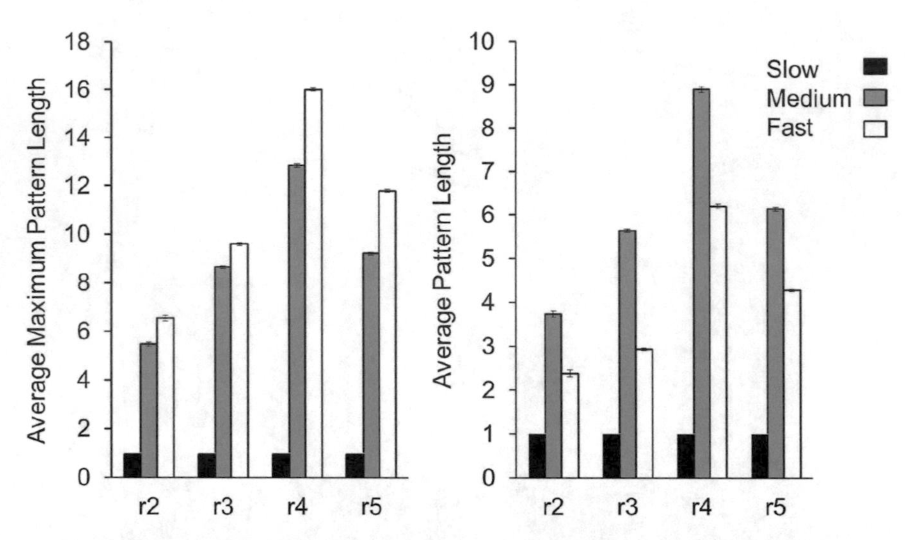

Fig. 3 Distribution of pattern lengths according to neuron firing rate. Neurons were categorized as slow (0–7 spikes), medium (8–100 spikes), or fast (>100 spikes) according the total number of spikes per trial. These firing rates were determined from approximate modal firing rates across the full data set. Slow neurons did not contribute to patterns, and so occur only independently (i.e., pattern length = 1). The remaining fast and medium neurons each, on average, participated in multiple patterns. The average maximum length of pattern that neurons participated in, increased with increasing firing rate. Thus, the average maximum pattern length was longer for fast neurons than for medium neurons. However, this distribution was reversed when considering the average length of patterns in which these neurons participated; the average length of patterns for fast neurons was shorter than that for medium neurons

neurons contributed were indeed longer than those amongst the medium neurons (Fig. 3a). However, conversely to what might have been expected by chance, the *average* length of patterns to which the medium neurons contributed was longer than that for the fast neurons (Fig. 3b). In this way, pattern length is not a simple function of the firing rate of the underlying neurons.

3.1 Statistical Validation of T-Patterns

The detection of critical intervals is based on a null hypothesis that is tested a large number of times when seeking patterns in a single data set. Accordingly, many significant patterns are expected to be found even in random data. A crucial issue, therefore, is whether significantly more t-patterns are detected in the initial data than in the same data after they have been randomized. For this the Theme™ software allows randomizing and reanalyzing the data an optional number of times using the same detection parameters as for the real data and then comparing the findings. For this two types of randomization are provided: shuffling and rotation. In both cases the number of series and the number of points in each, remain unchanged.

- *Randomization by shuffling:* Here the time points in each series in the real data are randomly redistributed (shuffled) over the observation period.

– *Randomization by rotation:* This can be imagined thus: all the series in the data set are wrapped around the same cylinder, each series thus forming a circle that is independently rotated by a random number of degrees (0–360°). This procedure, the more conservative of the two, leaves the structure of each series practically unchanged while randomizing the temporal relationship between the series.

By repeatedly randomizing and then searching for patterns in the same data set with the same search parameters as for the real data, an occurrence distribution with mean and a standard deviation is obtained for each pattern length. This allows statistical comparison of the number of patterns of each length detected in the real data as compared to the randomized data and differences can be expressed in terms of the number of standard deviations between the two.

Comparing the lengths and incidence of t-patterns extracted from this data, t-patterns were similarly extracted from the same data when randomized either by shuffling or by rotation. Using either procedure, significantly fewer sequences were detected in the randomized data than in the original electrophysiological data (e.g., see Figs. 4, 5, 6, 7, 8, and 9). The incidence of t-patterns in

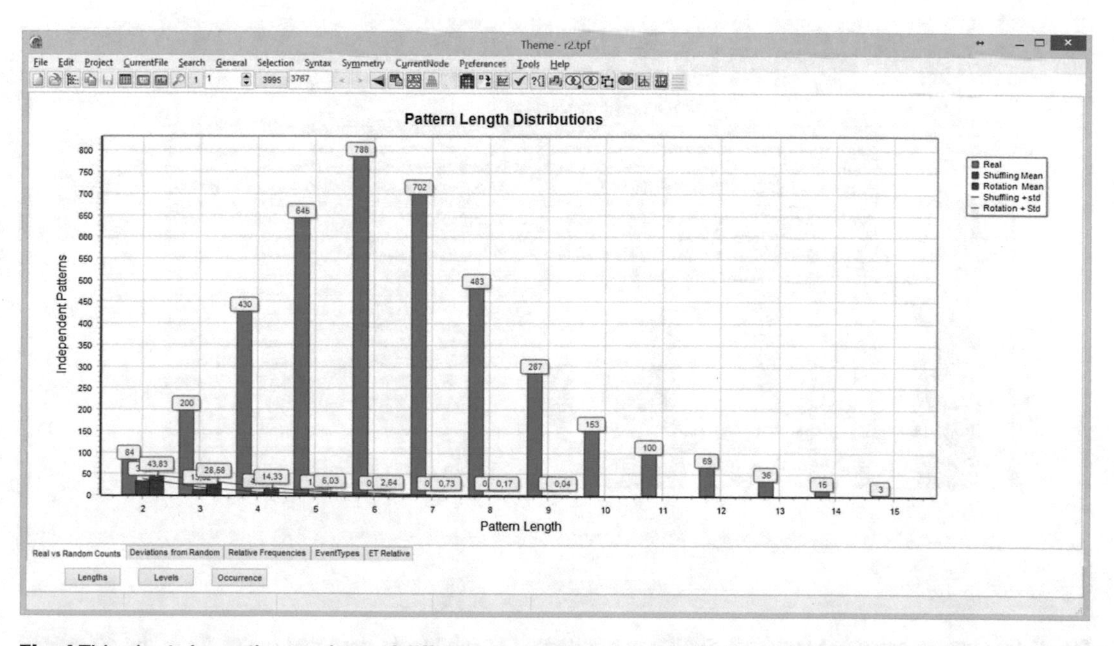

Fig. 4 This chart shows the numbers of different patterns of each length detected in first of the 12 control trials of animal r2, which became physiologically instable and finally had to be terminated after the 12th trial. It also shows the number of different patterns of each length detected on average over 100 repetitions of randomization and search for each of two types of randomization, shuffling and rotation. Here the number of patterns detected and the differences between the real data and the randomized data are at a maximum for the 12 control trials

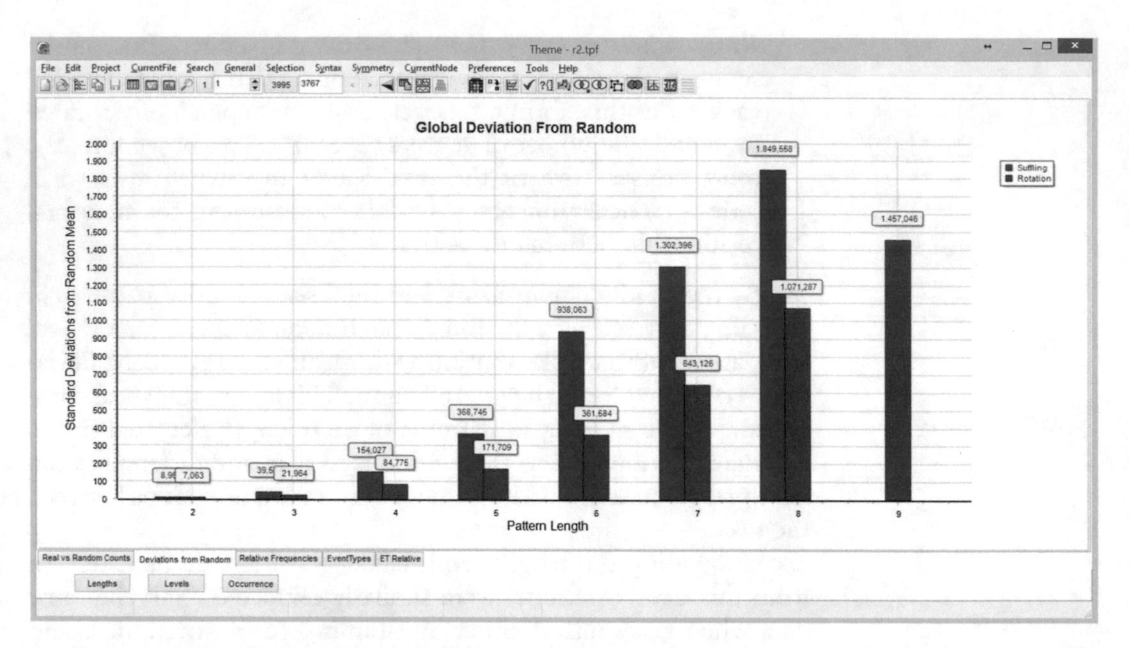

Fig. 5 For the first of the 12 control trials for animal r2 that became instable and finally had to be terminated after the 12th control trial, this chart shows the standardized differences between the numbers of patterns of each length detected in the original data versus the average over 100 repetitions of data randomization and search using the same search parameters. Here these differences are at the maximum for the 12 control trials

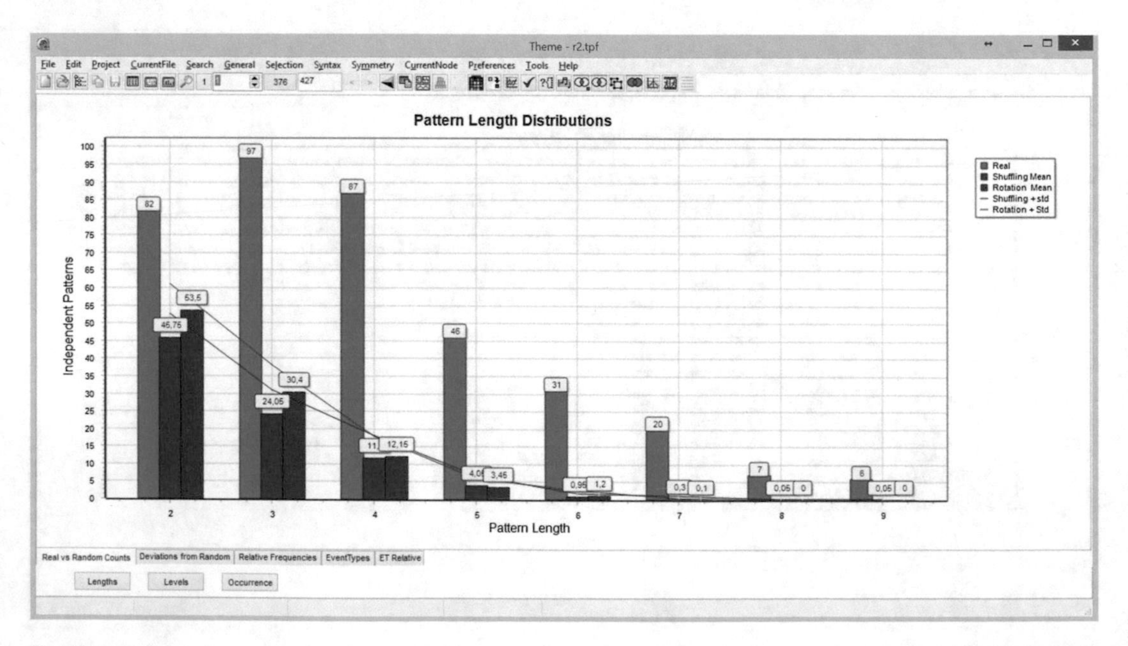

Fig. 6 This chart shows the numbers of different patterns of each length detected in the 7th of the 12 control trials of animal (r2) that became physiologically instable and finally had to be terminated after the 12th trial. It also shows the number of different patterns of each length detected on average over 100 repetitions of randomization and search for two types of randomization, shuffling and rotation. Here the number of patterns detected and the differences of detection between the real data and the randomized data are intermediate between the first and last of the 12 control trials and illustrates the gradual degradation of patterning over the 12 trials

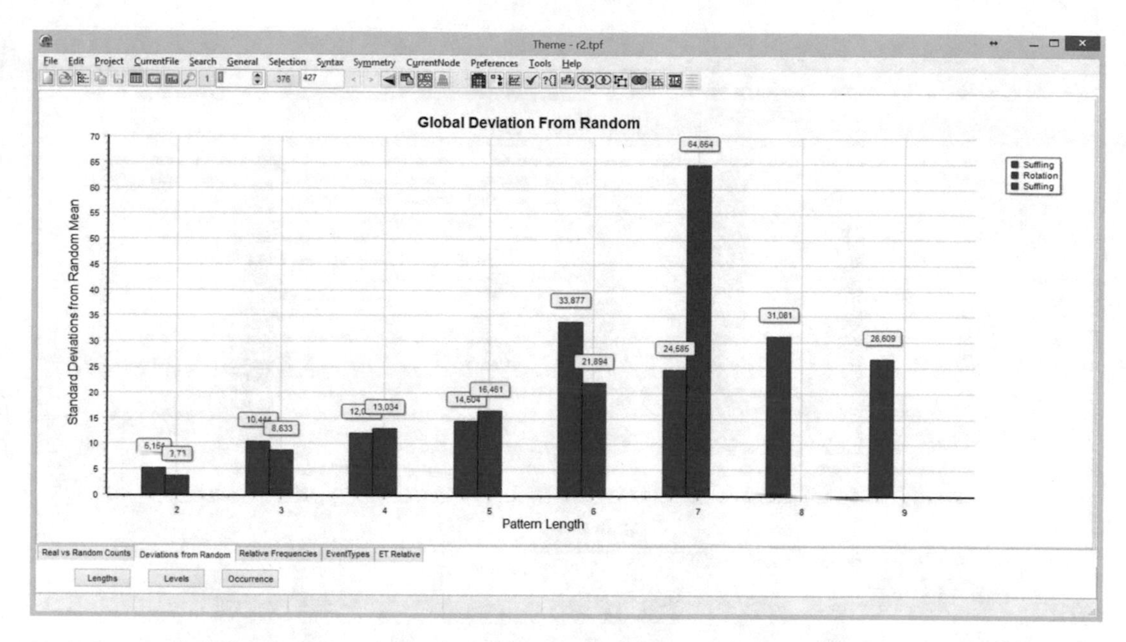

Fig. 7 For the 7th of the 12 control trials for animal (r2) which became instable and finally had to be terminated after the 12th trial, this chart shows the standardized differences between the numbers of patterns of each length detected in the original data versus the average numbers over 100 repetitions of data randomization and search with the same search parameters. Here these differences are at an intermediate level for the 12 control trials

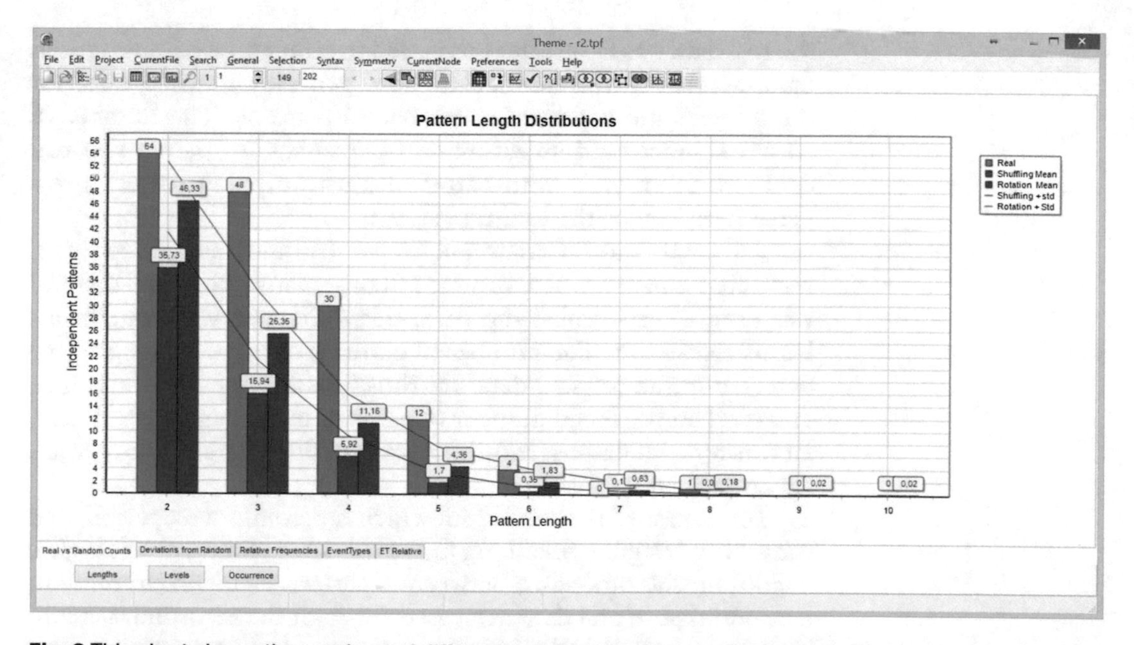

Fig. 8 This chart shows the numbers of different patterns of each length detected in the last of the 12 control trials of an animal (r2) that became physiologically instable and finally had to be terminated after this trial. It also shows the number of different patterns of each length detected on average over 100 repetitions of randomization and search for each of two types of randomization, shuffling and rotation. Here the number of patterns detected and the differences of detection between the real data and the randomized data are at the minimum for the 12 control trials

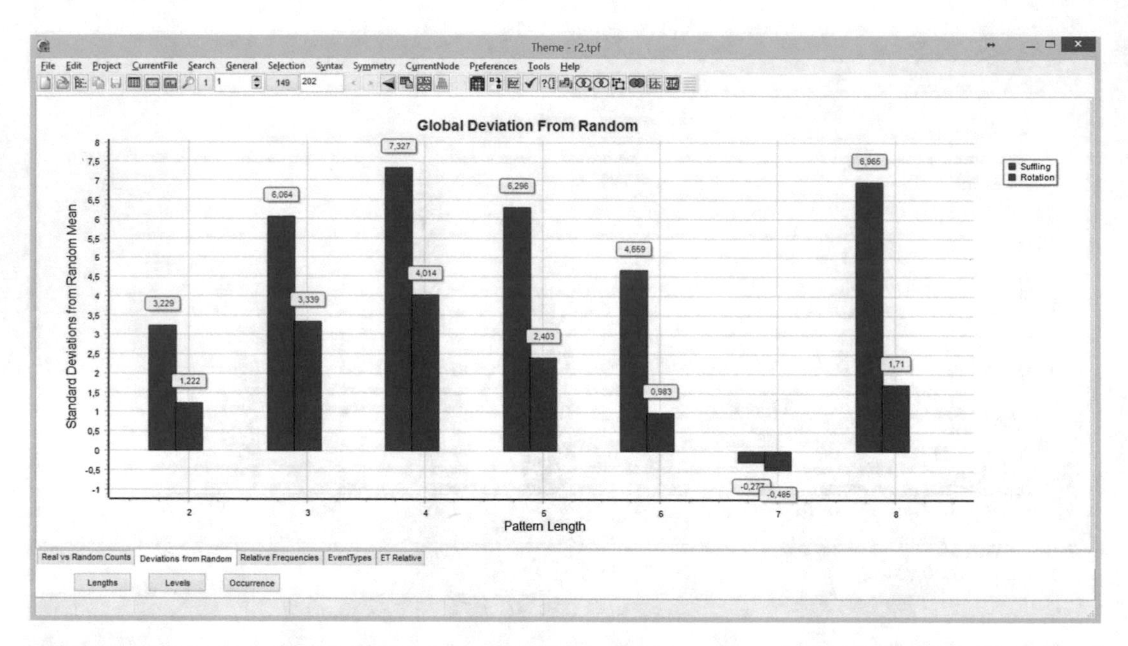

Fig. 9 For the last of the 12 control trials for an animal (r2), which became instable and finally had to be terminated after this 12th and last trial, this chart shows the standardized differences between the numbers of patterns of each length detected in the original data versus the average over 100 repetitions of data randomization and search with the same search parameters. Here these differences are at the minimum for the 12 control trials

the real data was far greater compared to either type of randomized data. Moreover, the differences were generally much greater for the longer (and a priori less probable) patterns. The number of patterns under randomization as a percentage of patterns in real data also fell rapidly with pattern length dropping to just a few percent or less for the longest patterns.

Although the difference between real and randomized data was often smaller when the more conservative rotation method was used to randomize the data, this difference was nonetheless highly significant. The number of patterns of the same length (m) detected in the original data was thus 5–1000 standard deviations greater than the mean number of patterns in the randomized data, increasing with pattern length, corresponding to significance levels far lower than 0.0001.

In the case of the subject for which recordings were terminated early when irregular breathing indicated deteriorating physiological condition, the difference between T-pattern incidence in random data and that in real data decreased through the recording session. This decrease in the distinction between random and real data continued until the termination of recordings. At that point, there was relatively little difference between t-pattern incidence in real data and that in either form of randomized data (see Fig. 10), although

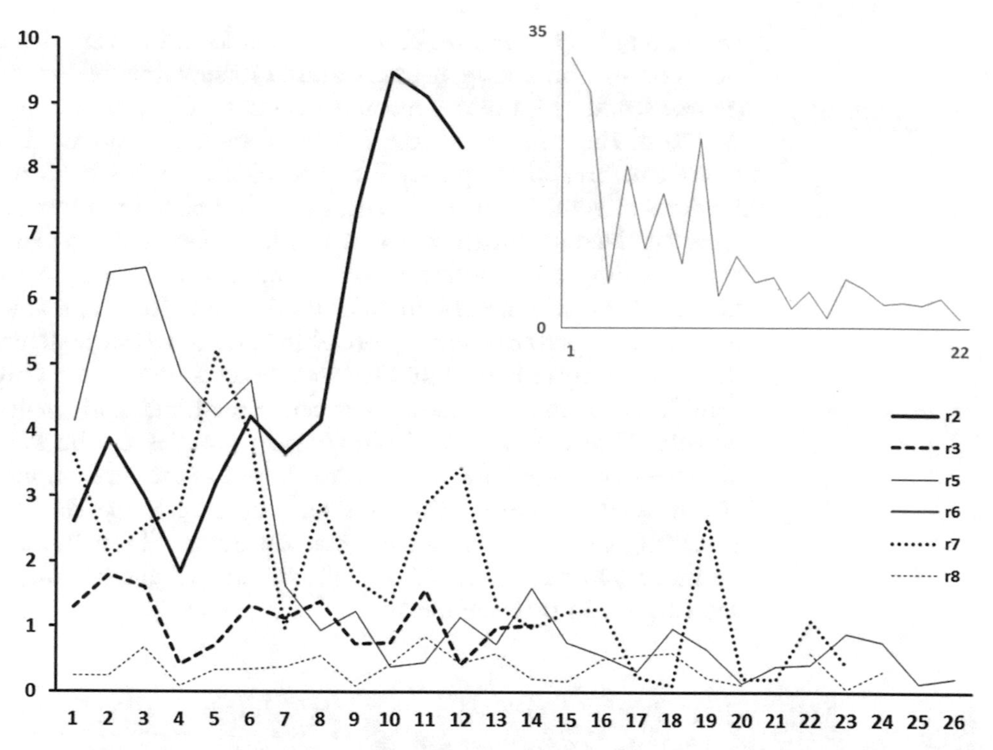

Fig. 10 Relative incidence of patterns in real and randomized data. The average incidence of patterns of any length in 20 randomizations of the data was compared to the pattern incidence in the real data. Here, pattern incidence in randomized data is expressed as a percentage of that in the real data. Data were randomized using the conservative rotation procedure described in the text. In the data from all but one of the animals, the difference in pattern detection between real and randomized data was large throughout the series of recordings, tending to increase as the recordings progressed, i.e., the ratio of patterns in randomized data to those in the real data declined across trials. However, in subject r2, the relative separation in pattern incidence between randomized and real data, whilst maintaining a level similar to that in other animals through the initial 5–6 trials, subsequently degraded (see also Figs. 4, 5, 6, 7, 8, and 9). This accompanied a deterioration in the physiological condition of the anesthetized animal, to the point when recordings were stopped. In another subject animal, r6 (*inset*), separation between randomized data and real data in the early trials was relatively poor, but improved through the series. For this animal, less time was allowed for the animal to stabilize after initial anesthesia and surgical preparation before commencing recordings. It may be that, in this case, consistent with r2, the relative incidence of patterns in randomized and real data reflects the stability of the preparation

the combined firing rates of the neurons remained relatively unchanged, and were even slightly increased, that is, from total of 3042 spikes over all 39 neurons in the first trial to 3114 over all 43 neurons in the last.

3.2 Breathing and Neuronal Firing Patterns

As would be expected, independently of neuronal firing, the breathing cycle was detected in all data sets as two-element (length = 2) t-patterns relating only the two different markers of ventilation (i.e., onset of inhalation and onset of exhalation) which regularly alternate. Numerous relationships were detected between the breathing markers and (a) the firing of individual neurons and (b) multi-neuron t-patterns. Thus approximately 3–5 % of the t-patterns detected in all data sets analyzed were related to breathing as they included either one or both of the markers of ventilation (e.g., see Fig. 2) and all ventilation events were involved in at least one neuronal firing t-pattern. When only the breathing event series were randomized in the original data and a pattern search performed using the same search parameters as before, no patterns relating breathing and spikes were detected. This was consistent over all files and subjects indicating a highly significant synchronization between breathing and neuronal activity. Figures 11 and 12 show two examples of the kind of T-patterns detected, but these were chosen as they are among the relatively few where each T-pattern occurrence covers a full breathing cycle and, moreover, contain bursts (T-bursts) in one or more neurons. (Regarding T-bursts and T-pattern diagrams see Chap. 1 in this volume.)

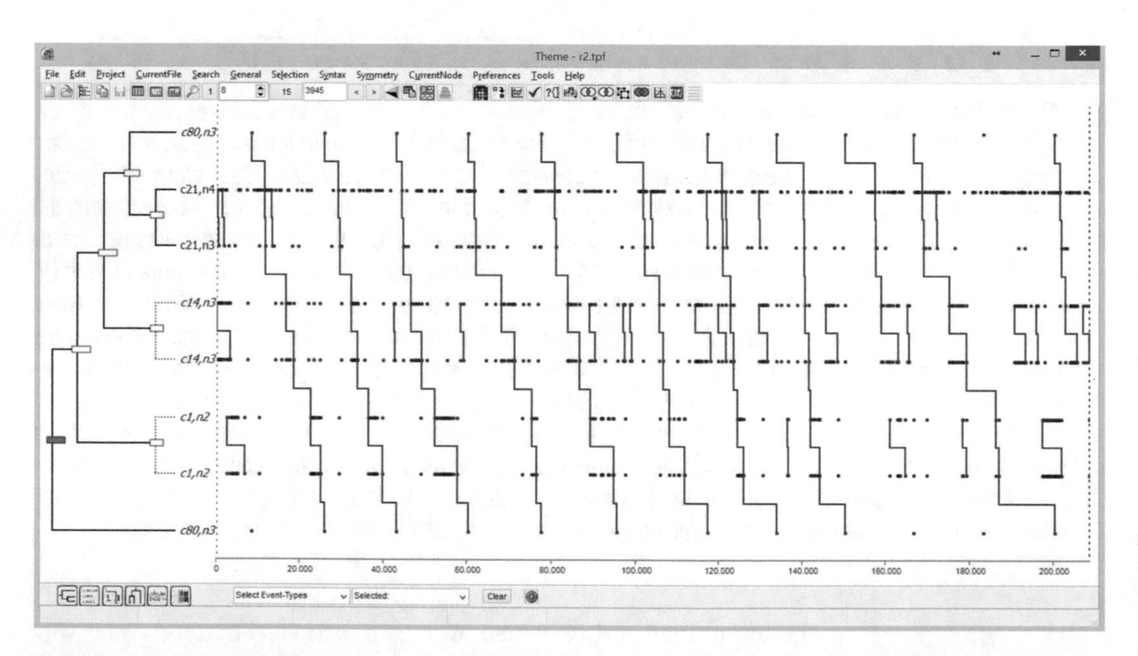

Fig. 11 This T-pattern diagram shows one pattern detected in the first of the 12 control trials for an animal (r2) that later became instable and finally had to be terminated after the 12th trial. The pattern shown is one of the 15 different patterns and with occurrences spanning a full breathing cycle, that is, from one exhale event (c80, n3) to the next, and containing bursts (T-bursts) in individual neurons. Note that this pattern includes a kind of cascade of bursts, that is, first in neuron c14, n3 and then in neuron c1, n2. About T-bursts and T-pattern diagrams see Chapter 1 in this volume (Magnusson)

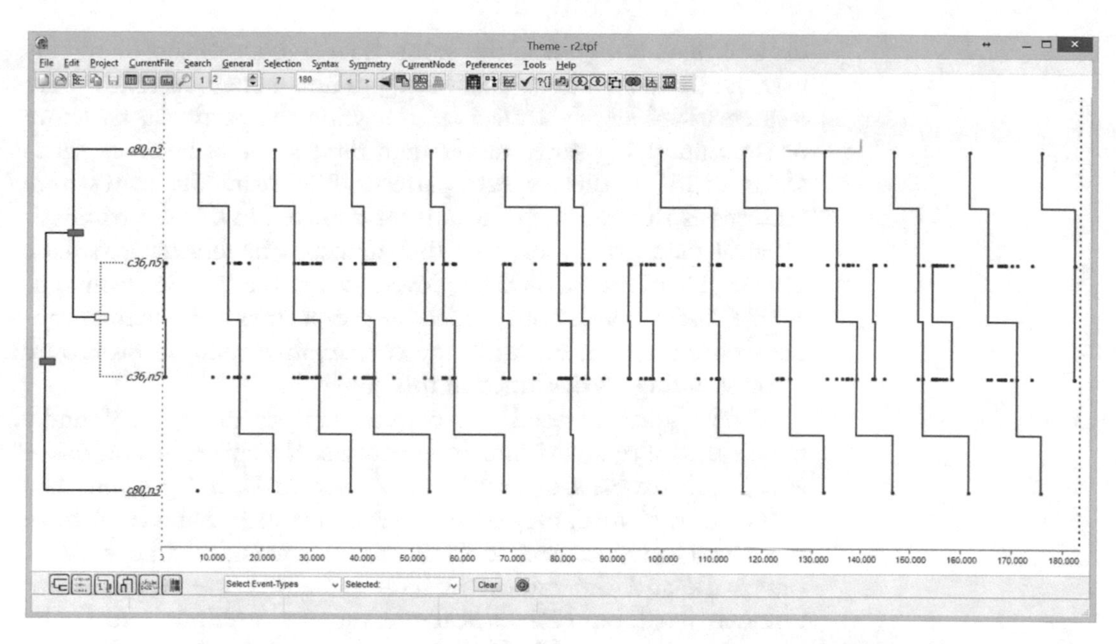

Fig. 12 This T-pattern diagram shows one pattern detected in the last of the 12 control trials in animal (r2) that had become instable and had to be terminated after this 12th control trial. At this point, all patterning had become minimal compared to the initial trial. The figure shows the only pattern containing a burst (T-burst) in one neuron with occurrences spanning a full breathing cycle, that is, from one exhale event (c80, n3) to the next. The burst connecting the two exhale events is in neuron c36, n5. About T-bursts and T-pattern diagrams see Chap. 1 in this volume (Magnusson)

4 Discussion

In the present work we show that complex sequences occur across large areas of the two-dimensional network of mitral neurons in the olfactory bulb. The occurrence of all lengths of these patterns is many times greater than that would be expected by chance. The patterns detected incorporated functional connections spanning in some cases the entire area sampled by the MEA (>2 mm^2). This is perhaps less remarkable given the presence of anatomical connections (paradoxically termed "short axons") spanning many (≤30) mitral cells across the olfactory bulb [21].

Precisely timed sequences have previously been described for neuronal firing in cortical neurons [5] and in simulated neuronal networks [22, 23]. Such "synfire" sequences in spontaneous cortical neuronal activity have been reported in intracellularly recorded postsynaptic potentials and in extracellularly recorded spikes both in an isolated tissue preparation and in vivo [24]. Synfire patterns are in some ways similar to t-patterns and may be simply described as t-patterns where all critical intervals are of the same very short length (one millisecond?), that is, $[d_1+1]$. However, critics of the synfire algorithm have noted that the incidence of sequences extracted from electrophysiological data using this procedure, may differ little (nonsignificantly) from the incidence of those similarly extracted from random data [6, 7]. Thus it may be that synfire sequences are largely artifactual, reflecting the particular structure of the underlying series rather than dependencies between those series (cf 25). In this respect t-patterns differ markedly from synfire patterns as the number of t-patterns detected in our electrophysiological data greatly exceeds that detected in randomized data, irrespective of the method employed for randomization (from 5 to 1000 standard deviations, depending respectively on whether randomization is effected using the conservative rotation procedure or by shuffling, as described in this paper).

The algorithm used in t-pattern analysis differs profoundly from the synfire algorithm. In t-patterns, the spikes of neurons A and B (or patterns of such) are connected if, more often than expected by chance, they occur in sequence such that after A there is at least one occurrence of B within particular time window (interval), and the critical interval is neither user defined nor uniquely fixed, but each critical interval relationship is detected by a special algorithm. The synfire detection algorithm does not search for critical interval relationships between events, but attempts to match the spike data to a user-defined interval that typically is very short, conforming to acknowledged synaptic physiological properties, i.e., synaptic delay in the order of a few milliseconds. The interval is thus also constant from one pair of neurons (and/or neuronal patterns) to another, whereas the intervals in t-patterns are variable and automatically detected as a function of the actual data.

The constraints imposed on the biological system in relying on a precise, inflexible, and short interval to establish sequences across multiple neurons, may be, and indeed might be expected to be, extreme. In a neuronal system, a postsynaptic neuron may receive many inputs from many afferent neurons. Some of those inputs are excitatory, others inhibitory. The activation of the postsynaptic neuron is dependent on the summative influences of those excitatory and inhibitory inputs. Moreover, the synapses themselves are subject to failure [26]. Thus, the activity of a single afferent neuron, whilst having a probabilistic influence over the susceptibility of the postsynaptic neuron to discharge, may have negligible impact on the activity of the postsynaptic neuron relative to ongoing activation by the large number of other presynaptic neurons. The statistical nature of the critical intervals used in t-pattern analysis makes this an ideal technique for the detection of sequences amongst such neuronal populations.

These analyses of spike data recorded simultaneously from multiple olfactory bulb neurons demonstrate the occurrence of repeating patterns of activity, many simple, involving perhaps two or three neurons, but many much more complex, across as many as 20 or more neurons. The incidence of these patterns in our neuronal data was very much greater than in the same data when either of two randomization procedures was used—often the separation of real and randomized data was in the order of hundreds of standard deviations, implying that these sequences are by no means a chance phenomenon. This was maintained across the series of recordings for all subject animals, with the exception of one, which for unknown reasons became physiologically unstable under anesthesia, and ultimately died before completion of the series of recordings. In the trials leading up to the point at which recordings were terminated, the separation between real and randomized data degenerated progressively until the incidence of patterns no longer distinguished them convincingly, despite there being no corresponding degradation in the combined firing rates of the underlying neurons. Again, this points to the incidence of organized spike sequences being a property of the functioning neuronal system.

Individual neurons sampled from each subject animal varied widely in their firing rates. We considered the possibility that faster neurons might have greater opportunity, by chance, to contribute to longer patterns. However, conversely to what might have been expected by chance, whilst the longest patterns involving fast neurons were longer than those involving medium firing rate neurons, the average pattern length to which faster neurons contributed was in fact shorter than for the medium neurons. This again counters the possibility that chance factors might play a part in the incidence of neuronal sequences detected by t-pattern analysis, and also suggests a dissociation in function between fast and medium neurons.

In addressing the potential physiological function of neuronal sequences in the olfactory bulb, we considered the relationship between sequences and breathing. A sizeable subset of patterns detected incorporated the onset of one or other phase in the breathing cycle, inhalation or exhalation. Thus, these patterns are timed to the animal's ventilation, and are consistent with other observations of neuronal activation in phase with breathing using in vivo optical imaging techniques [17] or electrophysiological techniques [27]. The latter study demonstrated that mitral cell membrane potential fluctuations, and therefore likelihood of discharge, occur in phase with ventilatory rhythm. Here we demonstrate that neuronal activity further involves complex sequences of discharge that are related to ventilatory activity.

These investigations demonstrate that complex sequences of spikes can be detected in a neuronal network using t-pattern analysis. Their incidence is very much greater than can be explained by chance, and this property alone can reflect the physiological condition of the animal. These patterns can be categorized according to discrete physiological functions, and hence may represent an important mechanism for coding odor information in the olfactory bulb. In our ongoing investigations we intend to expand our analyses to establish the utility of complex spike sequences detected by t-pattern analysis in the coding of sensory information in the central nervous system.

References

1. Griffith JS, Horn G (1963) Functional coupling between cells in the visual cortex of the unrestrained cat. Nature 199:893–895
2. Engel AK, Roelfsema PR, Fries P, Brecht M, Singer W (1997) Binding and response selection in the temporal domain—a new paradigm for neurobiological research. Theory Biosci 116:241–266
3. Hebb DO (1949) The organisation of behaviour. Wiley, New York, NY
4. Bliss TVP, Lomo T (1973) Long-lasting potentiation of synaptic transmission in dentate area of anesthetized rabbit following stimulation of Perforant path. J Physiol 232:331–356
5. Abeles M, Gerstein GL (1988) Detecting spatiotemporal firing patterns among simultaneously recorded single neurons. J Neurophysiol 60:909–924
6. Oram MW, Hatsopoulos NG, Richmond BJ, Donoghue JP (2001) Excess synchrony in motor cortical neurons provides redundant direction information with that from coarse temporal measures. J Neurophysiol 86:1700–1716
7. Baker S, Lemon RN (2000) Precise spatiotemporal repeating patterns in monkey primary and supplementary motor areas occur at chance level. J Neurophysiol 84:1770–1780
8. Magnusson MS (2000) Discovering hidden time patterns in behavior: T-patterns and their detection. Behav Res Methods Instrum Comput 32:93–110
9. Magnusson MS (2004) Repeated patterns in behavior and other biological phenomena. In: Kimbrough Oller D, Griebel U (eds) Evolution of communication systems: a comparative approach (Vienna series in theoretical biology). The MIT Press, Cambridge, MA
10. Magnusson MS, Burfield I, Loijens L, Grieco F, Jonsson GK, Spink A (2004) Theme: Powerful Tool for Detection and Analysis of Hidden Patterns in Behavior. Reference Manual, Version 5.0. Noldus Information Technology BV, Wageningen, The Netherlands.
11. Bolhuis JJ, MacPhail EM (2001) A critique of the neuroecology of learning & memory. Trends Cogn Sci 5:426–433
12. Kendrick KM, Levy F, Keverne EB (1992) Changes in the sensory processing of olfactory signals induced by birth in sheep. Science 256:883–886
13. Kendrick KM, Guevara-Guzman R, Zorrilla J, Hinton MR, Broad KD, Mimmack M, Ohkura S (1997) Formation of olfactory memories mediated by nitric oxide. Nature 388:670–674
14. Da Costa APC, Broad KD, Kendrick KM (1997) Olfactory memory and maternal behaviour-induced changes in c-fos and zif/268 mRNA expression in the sheep brain. Mol Brain Res 46:53–76
15. Mori K, Nagau H, Yoshihara Y (1999) The olfactory bulb: coding and processing of odor molecule information. Science 286:711–715
16. Johnson BA, Ho SL, Yihan JS, Yip S, Hingco EE, Leon M (2002) Functional mapping of the rat olfactory bulb using diverse odorants reveals modular responses to functional groups and hydrocarbon structural features. J Comp Neurol 449:180–194
17. Spors H, Grinvald A (2002) Spatio-temporal dynamics of odor representations in the mammalian olfactory bulb. Neuron 34:301–315
18. Nicol AU, Segonds-Pichon A, Magnusson MS (2015) Complex spike patterns in olfactory bulb neuronal networks. J Neurosci Methods 239:11–17
19. Price JL, Powell TPS (1969) The morphology of the granule cells of the olfactory bulb. J Cell Sci 7:91–123
20. Horton PM, Nicol AU, Kendrick KM, Feng JF (2007) Spike sorting based upon machine learning algorithms (SOMA). J Neurosci Methods 160:52–68
21. Aungst JL, Heyward PM, Puche AC, Karnup SV, Hayar A, Szabo G, Shipley MT (2003) Centre-surround inhibition among olfactory bulb glomeruli. Nature 426:623–629
22. Schrader S, Grün S, Diesmann M, Gerstein GL (2008) Detecting synfire chain activity with massively parallel spike train recording. J Neurophysiol 100:2165–2176
23. Gerstein GL, Williams ER, Diesmann M, Grün S, Trengrove C (2012) Detecting synfire chains in parallel spike data. J Neurosci Methods 206:54–64
24. Ikegaya I, Aaron G, Cossart R, Aronov D, Lampl I, Ferster D, Yuste R (2004) Synfire chains and cortical songs: temporal modules of cortical activity. Science 304:559–564
25. Gerstein GL (2004) Searching for significance in spatio-temporal firing patterns. Acta Neurobiol Exp 64:203–207
26. Zador A (1998) Impact of synaptic unreliability on the information transmitted by spiking neurons. J Neurophysiol 79:1219–1229
27. Margrie TW, Schaefer AT (2002) Theta oscillation coupled spike latencies yield computational vigour in a mammalian sensory system. J Physiol 546:363–374

Index

A

B

C

Magnus S. Magnusson et al. (eds.), *Discovering Hidden Temporal Patterns in Behavior and Interaction: T-Pattern Detection and Analysis with THEME™*, Neuromethods, vol. 111, DOI 10.1007/978-1-4939-3249-8,

N

O

P

Q

R

S

T

U

V

Printed in the United States
By Bookmasters